Aboriginal Biocultural Knowledge in South-eastern Australia

PERSPECTIVES OF EARLY COLONISTS

FRED CAHIR, IAN D. CLARK AND PHILIP A. CLARKE

Dedication

This publication is in honour of the Indigenous Elders – past, present and future – who have been and are custodians of south-eastern Australia.

'I should remark that when Tung.bor.roong spoke of Borembeep [Burrumbeep, south of Ararat] and the other localities of his own nativity he always added, "That's my country belonging to me!! That's my country belonging to me!!"' (GA Robinson Journal, 17 July 1841).

Aboriginal Biocultural Knowledge in South-eastern Australia

PERSPECTIVES OF EARLY COLONISTS

FRED CAHIR, IAN D. CLARK AND PHILIP A. CLARKE

CSIRO
PUBLISHING

A catalogue record for this book is available from the National Library of Australia.

ISBN: 9781486306114 (pbk)
ISBN: 9781486306121 (epdf)
ISBN: 9781486306138 (epub)

Published by

CSIRO Publishing
36 Gardiner Road, Clayton VIC 3168
Private Bag 10, Clayton South VIC 3169
Australia

Telephone: +61 3 9545 8400
Email: publishing.sales@csiro.au
Website: www.publish.csiro.au
Sign up to our email alerts: publish.csiro.au/earlyalert

Front cover: (top) Native encampment at Portland Bay by George French Angas, 1844. South Australian Museum Archive. (middle) Water vessels (*tarnuks*) from Smyth RB (1878) *The Aborigines of Victoria*. Victorian Government Printer, Melbourne. (bottom) Cockle shell midden on the northern Younghusband Peninsula in 1990 (photograph: Philip Clarke).
Back cover: Bark canoe from Smyth RB (1878) *The Aborigines of Victoria*. Victorian Government Printer, Melbourne.

Set in 9.5/13.5 Adobe Garamond Pro
Edited by Anne Findlay, Princes Hill, Melbourne, Victoria
Cover design by Andrew Weatherill
Typeset by Desktop Concepts Pty Ltd, Melbourne
Index by Indexicana
Printed by Ingram Lightning Source

CSIRO Publishing publishes and distributes scientific, technical and health science books, magazines and journals from Australia to a worldwide audience and conducts these activities autonomously from the research activities of the Commonwealth Scientific and Industrial Research Organisation (CSIRO). The views expressed in this publication are those of the author(s) and do not necessarily represent those of, and should not be attributed to, the publisher or CSIRO. The copyright owner shall not be liable for technical or other errors or omissions contained herein. The reader/user accepts all risks and responsibility for losses, damages, costs and other consequences resulting directly or indirectly from using this information.

CSIRO acknowledges the Traditional Owners of the lands that we live and work on across Australia and pays its respect to Elders past and present. CSIRO recognises that Aboriginal and Torres Strait Islander peoples have made and will continue to make extraordinary contributions to all aspects of Australian life including culture, economy and science. CSIRO is committed to reconciliation and demonstrating respect for Indigenous knowledge and science. The use of Western science in this publication should not be interpreted as diminishing the knowledge of plants, animals and environment from Indigenous ecological knowledge systems.

Foreword

Global warming, changed weather patterns, degradation of the environment, species extinction and unconstrained growth in human population have become the hallmarks of our time. The non-Indigenous peoples who now occupy south-eastern Australia have changed the biosphere in significant ways since first arriving in the 1830s. The introduction of sheep and cattle, as well as other animal and plant species used to acclimatise colonial settlers to their new world, changed the landscapes of south-eastern Australia forever. *Aboriginal Biocultural Knowledge in South-eastern Australia* provides an important and timely addition to contemporary Australian discussions concerning the relationship between the biosphere and the human societies that are sustained by them.

Melbourne was conceived to mirror London in both attitude and design. Although now widely acclaimed as one of the world's greatest cities, Melbourne is emblematic of settler-colonial efforts to insulate, protect and disengage themselves from the realities of the Australian biosphere on which they now live. Few non-Indigenous people in Melbourne and the other cities and towns that now occupy south-eastern Australia are aware of the rich and vibrant Aboriginal societies that preceded British colonialism and successfully occupied the country for millennia. If the non-Indigenous inhabitants of south-eastern Australia think about Aboriginal people at all today, it is usually with reference to the arid interior and the tropical north of the continent. These are places where Aboriginal cultural practices and traditions are imagined to have survived colonialism intact.

Ignorance of an Aboriginal past in south-eastern Australia is the outcome of what anthropologist WEH Stanner called the 'great Australian silence,' whereby Aboriginal peoples had by the middle decades of the 20th century become marginal, forgotten and disremembered to the 'mainstream' of contemporary Australian life and society. Since the 1980s, this situation has slowly but surely been changing, with academic researchers revisiting the Australian past in ways that attempt to be inclusive of Aboriginal peoples and their perspectives. In this way, we are gaining new and valuable insight into the ways that Aboriginal peoples experienced and interpreted the coming of British colonialism and the social, political and economic processes that led to their contemporary encapsulation in the settler-colonial state known as the Commonwealth of Australia.

Aboriginal Biocultural Knowledge in South-eastern Australia continues to build on the previous work of authors Fred Cahir, Ian D. Clark and Philip A. Clarke in documenting the ways in which Aboriginal peoples lived in south-eastern Australia, and how their knowledge systems and beliefs interacted with both the historical process of a global British colonialism as well as the intimate details of how the settlers who entered into their country came to dispossess local groups. Focused on how Aboriginal peoples in this part of the continent understood the world of nature as indivisible from human society, a unity of thought that the authors describe as Aboriginal Biocultural Knowledge, this book uses a detailed study of colonial writings in an attempt to reconstruct a past that no longer exists. The authors succeed in shedding new light upon the combination of complex beliefs and practices that directed Aboriginal peoples in south-eastern Australia to manage the environment and sustained resources in ways most beneficial to the needs of the societies in which they lived.

While the colonial ethnographers whose writings underpin this book often viewed Aboriginal explanations of their interactions with the natural world as irrational and based in primitive superstition, in revisiting these materials, the analysis provided by Cahir, Clark and Clarke emphasises the inherent rationality that underpinned Aboriginal Biocultural Knowledge. The chapters included in this book comprehensively detail all aspects of Aboriginal Biocultural Knowledge as described by settler-colonists, who as squatters, miners, amateur ethnographers and government officials documented the lives of the Aboriginal peoples they encountered across south-eastern Australia. This book provides new insights in the totemic life of Aboriginal peoples by explaining the biocultural role that terrestrial spirit beings, water spirits, space and time played in underpinning Indigenous ways of knowing plants, animals, land, sea, country. As the authors describe Aboriginal knowledge of canoe building, housing construction and the production of clothing, the most intriguing aspect of the book emerges as it becomes clear that the exchange of biocultural knowledge from Aboriginal peoples to settler-colonists was much more important to the survival and wellbeing of non-Indigenous peoples than popular national history remembers. The trade of medicines, the usefulness of bark canoes and the significant commercial trade in possum skin cloaks suggest that, during the 19th century, at least some aspects of Aboriginal Biocultural Knowledge were recognised and valued by squatters, miners, drovers and others.

This book and the details about Aboriginal biocultural understandings of south-eastern Australia that it describes builds on the work of Bill Gammage, Bruce Pascoe and many others in bringing new insight and understanding to past Aboriginal engagement with what Indigenous people today often refer to in English as Country. As noted in the subtitle, the book is based almost exclusively on settler-colonial perceptions of often bewilderingly complex Aboriginal cultural practices that were witnessed and set down in written English. Through the twin cultural lenses of British imperialism and Western science, these fragments of Aboriginal Biocultural Knowledge left to us by men including Robinson, Dawson, Howitt and women like Kirkland no doubt distorted and sometimes misrepresented the things that they witnessed. It is important to recognise that the material detailed in the book as Aboriginal Biocultural Knowledge is reliant not on the voices and cultural subjectivities of the various Aboriginal peoples of south-eastern Australia but those of the settler-colonists who encountered them.

While the voices of Aboriginal people are almost completely absent, this book and the painstaking research that underpins the detailed description and analysis of Aboriginal Biocultural Knowledge it contains will provide a valuable resource for contemporary Aboriginal peoples in south-eastern Australia as they seek to revise and strengthen their connection to Country in what is emerging as a period of sustained cultural renaissance. The Aboriginal peoples of south-eastern Australia said that the sky was held up from the earth by a series of great sticks. According to the *Wotjobaluk* of the Wimmera region of Victoria, these were placed there by the magpie and tended by an old man in the east. When the British arrived, the sticks that held the sky in place had begun to fall into a state of disrepair. To the Aboriginal peoples of south-eastern Australia, the biocultural world as they knew it was in danger of coming to an abrupt end. Cahir, Clark and Clarke are to be congratulated for their efforts, as this book does its bit, in ways that may well be significant; shoring up those great sticks and ensuring the world persists.

Barry Judd
Professor Indigenous Social Research, Charles Darwin University

Contents

Acknowledgements

Drafts of chapters were commented upon by Duane Hamacher and Christine Nicholls. For their professional advice, assistance and support, the authors would like to thank the research librarians at Federation University, and in particular Angela Thomas in her role as Client Services and eReadings and Document Delivery Officer, and Clare Gervasoni, who is the Federation University Australia Curator. Archives Collection Manager, Lea Gardam, at the South Australian Museum Archives provided permission for use of the image on the cover. Importantly, the authors would like to acknowledge and support the need for an ongoing role of Indigenous communities in maintaining the Aboriginal Biocultural Knowledge of their respective countries.

Introduction

The ancestors of Australia's Indigenous peoples first stepped on to the continent via Asia and Melanesia over 50 000 years ago. As Aboriginal people, they have not remained unchanged since their arrival but have successfully adapted to the variable Australian environment, and more recently absorbed the impact of intense British colonisation. During the period of their long custodianship of Australia Aboriginal people have seen the extinction of the megafauna and the transition between the Pleistocene and Holocene epochs. Throughout this time, they have developed unique ways of life and gained deep spiritual attachments to their country. Aboriginal people today have a strong sense of community, and have maintained their ability to draw upon their traditions when needing to respond to change. Europeans in the past dismissively judged Aboriginal cultures as 'primitive' due to their status of being one of the world's last societies with a hunter-gatherer technology based on the use of stone and wooden tools. Today, through researching Aboriginal Biocultural Knowledge and promoting its value, Australians have the opportunity to acknowledge the great depth of understanding that Indigenous people had with what the scientists regard as ecological processes. By focusing on south-eastern Australia, which is a part of the continent that has seen high levels of cultural and environmental change since the late 18th century, we can gain an appreciation of the complex relationships Aboriginal people had with a temperate landscape, while recognising the diversity of Aboriginal cultures.

Indigenous peoples and the environment

It is a popular notion that before British colonisation the Australian Aboriginal people had lived in harmony with nature, in part because as passive 'stone age' hunter-gatherers they would have had a small impact overall on the environment. As part of a broader European perception, described as the 'primitivist fallacy' (Ellen 1982, p. 73), it was assumed that tribal societies are always in harmony with their environment and are consequently a valid source of remedies for the ills of contemporary industrial capitalism (Clark 1982; Clarke 2014d; Godelier 1977). This misconception has influenced scientific thinking, with anthropologist Chris Anderson having stated that:

> Environmentalist literature frequently asserts that Aboriginal people were Australia's first conservationists. Conservationists themselves have often used a view of Aboriginal society as a model or inspiration for what they think the interaction of white Australians with the environment should be like. This view generally includes an uncritical acceptance of this notion that Aborigines were closer to nature somehow and were thus more 'natural' people. They argue that because of this we can and should learn from them how to be true conservationists (Anderson 1989, p. 219).

Anthropologist Catherine Laudine has remarked that while there are many popular ideas that champion Aboriginal environmental knowledge, many of these are based on romantic notions rather than on detailed understandings of its content (Laudine 2009). She has sought to transform popular notions of Aboriginal lore as just an integrated source of religious and scientific knowledge, arguing that Aboriginal environmental knowledge incorporates the essential mindset required for human survival. The views of environmentally harmonious Australian foragers who did not harm or alter the landscape were unsupported by palaeontological research, which has implicated human predation in the demise of the megafauna (Flannery 1994; Turney *et al.* 2008; Van Der Kaars *et al.* 2017). They were also incompatible with arguments for Indigenous people being active resource managers in pre-European Australia (Gammage 2011; Pascoe 2014; Williams and Hunn 1982).

Since the late 20th century in Australia, Indigenous people through legislation have, both nationally and on a state level, developed a voice in cultural heritage issues (Clarke PA 1994, 2003b; McCorquodale 1987) and gained greater access to land through the *Native Title Act 1993* (Hagen 2001; Sutton 2003; Weir 2009). There has also been a growing recognition of Indigenous rights to the natural resources of the inland riverine and coastal systems of Australia, which includes fishing and bird hunting (Peterson and Rigsby 1998; Smyth 1993). Across Australia, it had been noted that: 'It is apparent that there are strong developments occurring in the legislative and strategic planning arenas regarding the Indigenous peoples and local communities and their capacity to own, use, and benefit from their traditional biodiversity-related knowledge' (Langton and Ma Rhea 2005, p. 64). The generation of this kind of knowledge has seen greater Indigenous involvement in heritage interpretation (Carmichael *et al.* 2013; Hemming and Rigney 2010; Hemming *et al.* 2010; Prangnell *et al.* 2010) and has contributed to the scholarly discourse about Australian frontier history that has also benefited the tourism industry (Aicken and Ryan 2010; Buultjens *et al.* 2010). Today, the greater Indigenous involvement in caring for country has led to more engagement between environmental authorities and Indigenous communities over land management issues (Ens 2012a, 2012b; Hill *et al.* 2012).

In regional Australia, Indigenous people are increasingly becoming more involved in managing the environment. In some areas, such as in south-western Victoria, the federal Indigenous Protected Areas program offers opportunities for Indigenous communities to care for country, while protecting their traditional biodiversity-related knowledge from disappearing as part of a living oral tradition (Ens *et al.* 2012; Godden and Cowell 2016; Moritz *et al.* 2013; Ross *et al.* 2009). Indigenous communities in remote areas of northern Australia have been able to maintain closer relationships with the environment, as the onset of European colonisation was more recent than in the south and the land here was considered ill-suited for European-style agriculture, which has left large areas relatively untouched. Due to land alienation and intensive landscape transformation, this has historically not been the case for south-eastern Australia, where Indigenous communities have lost much more of their Aboriginal Biocultural Knowledge (Ens *et al.* 2015). The documentation of traditional knowledge concerning country is regarded as fundamental to enabling knowledge-holders to facilitate its proper use (Langton and Ma Rhea 2005).

Through legal processes beginning in the late 20th century, such as land rights and native title, Indigenous communities are today managing the natural resources of vast areas of Australia (Smyth *et al.* 2004), and this has created a need for trying to make the knowledge systems of modern Western Europeans and Indigenous Australians work together. It has been noted that 'at the highest levels, governments in Australia have acknowledged the lack of success of existing approaches to Indigenous service delivery and [have] committed to developing practical solutions to Indigenous disadvantage' (Smyth *et al.* 2004, p. 11). The ways that Indigenous interests and traditional knowledge are integrated into regional natural resource management (NRM) planning and the manner in which regional bodies implement their plans is critical for the engagement of Aboriginal communities (Bohensky *et al.* 2013; Davies and Holcombe 2009). In the past, scientists and bureaucracies have tended to selectively appropriate parts of Indigenous knowledge that are relatively easy to document and objectify, but have tended to ignore and dismiss that which cannot be expressed in words (Christie 2007; Muller 2012). For the two knowledge systems to work together successfully there should be trans-disciplinary research methodologies – which implies community engagement, collaboration and complex systems of knowledge production supervised from within both academic and Indigenous traditions.

There are significant economic, environmental, cultural and social benefits to be gained from cross-cultural engagement with Aboriginal knowledge systems in Australia (Daniels *et al.* 2012; Davies 2007; Ens *et al.* 2016). Such intercultural connections can be used for both translating information and transforming parties, with Indigenous knowledge being recognised as a means for establishing more ethical relationship with water (Weir 2007). There are also major economic and social benefits to be derived from greater uptake in the area of Indigenous knowledge and tourism (Butler and Hinch 2007). In parts of the Northern Territory, there is a hybrid model of sustainability combining Indigenous and Western knowledge, and this takes into account Aboriginal models for wildlife resource management and yet still achieves the stated goals of biodiversity and commercial viability (Altman and Cochrane 2003).

Internationally, it is recognised that there is an imperative for an Indigenous perspective of the environment to be incorporated into the contemporary understanding and management of natural and cultural landscapes (Lertzman 2009; Nabhan 2009). This inclusion of non-Western European points of view is not entirely based on the need to compensate Indigenous peoples for past wrongs. The same broad colonial processes that have had such a negative impact on Indigenous cultures throughout the world are also implicated in the ecological transformation of the country. Researchers in the field of ethnobotany are keenly aware of how culture change occurs when the physical environment is physically altered or people are relocated (Mohamad 2010; Wuisang and Jones 2014). An ethnobiologist has remarked that:

> We live in a world where biological and cultural diversity are being lost at dizzying rates. As ethnobiologists we know that losses of diversity in both of these realms are inextricably intertwined (Lepofsky 2009, p. 161).

Across the world, it has been noted that linguistic and biological diversity generally occur together in biodiversity hotspots (Gorenflo *et al.* 2012). The loss of any of the world's languages results in a loss of Indigenous understandings of the environment (Evans 2010; Mühlhäusler 2003; Mühlhäusler and Fill 2001).

Aboriginal Biocultural Knowledge

As a concept, past scholars have variously described Indigenous knowledge and understanding of the physical environment as 'Aboriginal Cultural Knowledge', 'Traditional Ecological Knowledge', 'Aboriginal Ecological Knowledge' and 'Indigenous Ecological Knowledge' (Ens *et al.* 2015; Pert *et al.* 2015). An early academic definition of Traditional Ecological Knowledge was as 'a cumulative body of knowledge, practice, and belief, evolving by adaptive processes and handed down through generations by cultural transmission, about the relationship of living beings (including humans) with one another and with their environment' (Berkes *et al.* 2000, p. 1252). It encompasses factual knowledge about ecological components and processes, knowledge put into practices of environmental use, and the cultural values, ethics, and philosophies that define human relationships within the natural world. For this reason, it is recommended that as scholars 'traditional knowledge as process, rather than as content, is perhaps what we should be examining' (Berkes 2009, p. 151).

Traditional Ecological Knowledge refers to the corpus of information regarding the local landscape that is possessed by native populations, and covers a wide range of subjects, not exclusively about the environment, linking the environment to spiritual, ethical and community values (Devereux 2013). It encompasses factual knowledge about ecological components and processes based upon observation over a timeframe. For applied science, Traditional Ecological Knowledge 'refers to the evolving knowledge acquired by indigenous and local peoples over hundreds or thousands of years through direct contact with the environment' (Rinkevich *et al.* 2011). Such definitions are in keeping with the notion that knowledge accumulation is the result of the process of constant augmentation of oral traditions through lived experiences.

Key elements of Traditional Ecological Knowledge are that it does not follow the empirical systems of Western science but is based in a more qualitative system and it is considered to be highly adaptive to seasonal, cyclic change within the traditional landscape and culture context, it is also oral and rural in nature (De Guchteneire *et al.* 2002). It is related to Aboriginal cultural knowledge, which can be defined as accumulated knowledge that encompasses spiritual relationships, relationships with the natural environment and the sustainable use of natural resources, and relationships between people (Andrews and Buggey 2008). All of this is reflected in a group's language, oral history narratives, social organisation, values, beliefs, and cultural laws and customs. In the literature, Traditional Ecological Knowledge has combined anthropological and ecological information in order to emphasise the relationships between the environment and culture (Parmington *et al.* 2012).

The documentation of this Indigenous realm of environmental knowledge has been performed chiefly by anthropologists, social scientists, historians and geographers, and to some extent by biophysical scientists (Pert *et al.* 2015). The diversity of names for it has come about due to the perspective of a range of researchers. The term 'ecological' emphasises how scientists strive to gain fresh insights into the environment from this knowledge, while the use of 'Aboriginal', 'Indigenous' and 'Traditional' as descriptors highlights its connections to original custodians of the country. The use of such terminology is an impediment to recognising that Indigenous-held information about the environment integrates culture and nature. The extraction of useful data from Indigenous knowledge systems by scientists is therefore fraught with ideological and methodological difficulties. When promoting the inclusiveness of Indigenous people in broad biological conservation measures, Indigenous people and their social and political environment must remain in the foreground to ensure that their knowledge does not simply become a commodity detached for broader consumption by the general public (Agrawal 2002). This is particularly so for Indigenous communities that still hold substantial knowledge about the local environment as a living form.

Ethnobotanist and Mbabaram Traditional Owner from the north Queensland rainforest region, Gerry Turpin, has been keenly aware of the problems with naming the field, and his solution has been to describe it as 'Indigenous Bio-cultural Knowledge', which he has holistically described as 'knowledge that encompasses people, language and culture and their relationship to the environment' (Turpin [cited Ens *et al*. 2015, p. 135]). As recently as 2014, ecologist Emilie Ens and her co-writers asserted that:

> … we found no other documented definition of the term "biocultural knowledge" in the literature, although we note and draw from the increasing use of the term "biocultural diversity" defined by Maffi (2001, 2007) as 'the diversity of life in all its manifestations: biological, cultural, and linguistic – which are interrelated (and possibly coevolved) within a complex socio-ecological adaptive system (Ens *et al*. 2015, p. 135).

In the current publication, the authors follow the lead of Turpin and Ens, and in doing so use Aboriginal Biocultural Knowledge as the description of the perspective being described, while recognising that scholars in many parts of the world and some parts of Australia still describe it as Traditional Ecological Knowledge.

The international recognition of Indigenous Biocultural Knowledge (IBK), or its equivalent, by conservation scientists, environmental managers and policy makers has developed only in recent times (Snow 2012). In Australia, the acknowledgment of Aboriginal Biocultural Knowledge is increasingly being used to encourage cross-cultural awareness and to solve communication problems between Indigenous people and the broad group of researchers and public servants who are involved in land management (Ens *et al*. 2015; Pert *et al*. 2015). In recognition of this development, the IBK working group within the Australian Centre for Ecological Analysis and Synthesis has designed and populated an online spatial database and website user interface (www.aibk.info) as a resource for the broad users of this knowledge. This resource highlights areas of strengths and gaps in the documentation of Aboriginal Biocultural Knowledge, as an aid to overcoming barriers to developing further engagements between knowledge-holders and non-Indigenous entities.

South-eastern Australia as a biocultural region

There are many definitions of south-eastern Australia in terms of both natural and cultural regions. The most expansive of them includes all the country south-east of an imaginary line drawn between Adelaide and Brisbane (Low Choy *et al.* 2013), while the smallest includes just the coastal high rainfall belt of southern Victoria and eastern New South Wales (Horton 1994). The description of south-eastern Australia used in the current work is the region of mainland Australia from the mouth of the Murray River in South Australia, eastwards across all of Victoria to the south-east of coastal New South Wales. It is justified on the basis that the Indigenous cultures within this region have many shared cultural attributes, and because the physical environment here is largely temperate.

The current definition of the region in the present book has some previous history, being what the late 19th century ethnographer Robert H. Mathews mapped as the territory covered by the 'nations', which are broad cultural blocs, of the 'Thurrawal' (Tharawal) in south-east New South Wales, 'Bangarang' (Bangerang) in central Victoria, the 'Kurnai' (Ganai) in eastern Victoria, the 'Booandik' (Bunganditj) in the south-east of South Australia and western Victoria, the 'Narrinyeri' (Ngarrindjeri) of the Lower Murray in South Australia (Mathews 1898, p. 343). The use of this definition is also based on the 'Area S' which was part of the bibliographical regions formerly utilised by the Australian Institute of Aboriginal and Torres Strait Islander Studies and some state collecting institutions, such as the South Australian Museum, and were loosely based on the affinities of language and material culture (Clarke PA 1989, 1994, 1995a; Craig 1969; Oates and Oates 1970).

Due in part to the demographics of modern Aboriginal Australia and the historical process of British colonisation having begun in the south, there is a discernible bias towards remote northern regions that pervades Aboriginal studies in Australia (Clark 2003), with a review of the literature on Aboriginal Biocultural Knowledge lending support to this

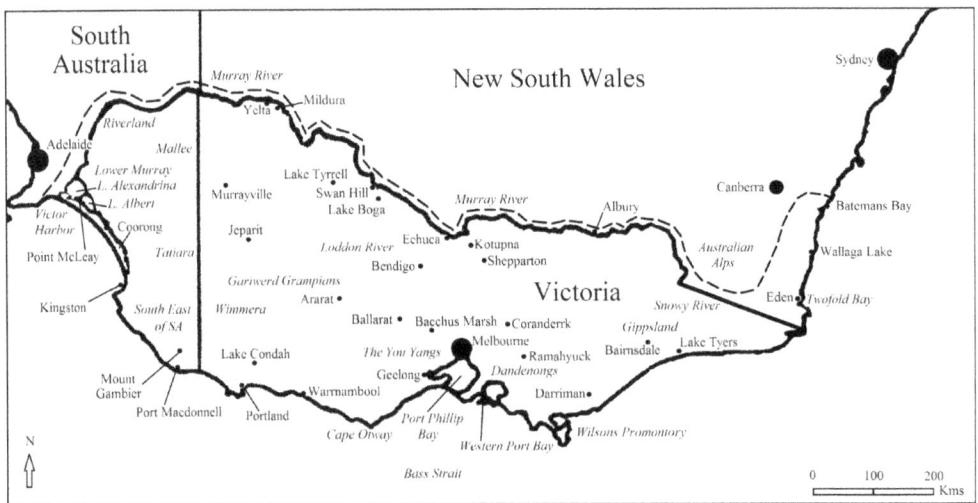

Map of south-eastern Australia. The study region is located south of the dashed line, and includes the Lower Murray, Riverland, Mallee and South East of South Australia, all of Victoria and south-east New South Wales. PA Clarke, 2017.

argument. For instance, while studies concerning northern savannah or tropical Indigenous communities which call for the inclusion of local Indigenous knowledge systems in environmental management are numerous (i.e. Altman and Cochrane 2003; Altman and Larsen 2006; Orchard *et al.* 2003; Rose 1996; Rose and Clarke 1997; Smyth 2011; Young *et al.* 1991), the concomitant to temperate Australian specific studies (Builth 1998; Porter 2004; Gott 2005; Richards and Jordan 1999; Roberts 2003) are comparatively sparse and are biased towards an archaeological perspective. Demographic studies since the late 20th century have indicated that there is an increasing urbanisation of Australia's Indigenous peoples (Barwick 1971a; Dugdale 2008; Gray and Smith 1983; Taylor 1997), making the focus of remote communities less relevant to the majority of contemporary Aboriginal people. South-eastern Australia is a major part of the region that anthropologists have sometimes referred to as 'settled' Australia (Clarke 2007b; Keen 1994).

Sources

The ethnographic base data used in the current work to describe Aboriginal Biocultural Knowledge are chiefly historical, recorded by colonists who had travelled through or lived in south-eastern Australia before and during the early phases of European colonisation. The knowledge of south-eastern Australia would have been much less without the detailed records compiled by these colonists. The historical sources quoted in this volume include phrases and descriptions of Indigenous peoples that would cause offence if used without qualification in a contemporary context. Where possible, Aboriginal sources of the Aboriginal Biocultural Knowledge are acknowledged and standard spellings of group names employed when they occur.

The recordings of colonists who became ethnographers include the works of Edward John Eyre (1815–1901), Alfred William Howitt (1830–1908), John Mathew (1849–1929), Robert Hamilton Mathews (1841–1918) and William Edward Stanbridge (1816–1894). Their work is particularly important, as it involved interviews with elderly Aboriginal people who could remember a time when community members were still living as hunter-gatherers. As sources of data, their ethnographic records were published in scholarly journals and books of the late 19th century. In the case of Howitt, Mathew and Mathews, their work was particularly focused on ceremonial life and kinship systems, and for this reason they should be considered as anthropologists. Major parts of the foundation literature for the study of Aboriginal Biocultural Knowledge in south-eastern Australia are the detailed accounts of Aboriginal life compiled by settlers and colonial administrators, such as James Dawson (1806–1900), Edward Micklethwaite Curr (1820–1889), George Henry Haydon (1822–1891), Peter Beveridge (1829–1885), Robert Brough Smyth (1830–1889) and Alfred Charles Stone (1868–1920).

Colonial scientists recording biocultural data are a diverse group: the zoologist Johann Ludwig Gerard Krefft (1830–1881); botanists such as Daniel Bunce (1813–1872), Ferdinand Jakob Heinrich von Mueller (1825–1896), John Burton Cleland (1878–1971); and naturalists such as George Bennett (1804–1893), William Blandowski (1822–1878) and George French Angas (1822–1886). Missionaries and Aboriginal Protectors who compiled important accounts of Indigenous life include George Augustus Robinson (1791–1866), William Thomas (1793–1867), Christina Smith (1809–1893), Heinrich Meyer (1813–1862), John Bulmer (1833–1913) and George Taplin (1831–1879). In addition to the above texts, there are a multitude of 19th century newspaper accounts of Aboriginal life, although many of them are from anonymous writers. These historical publications and archival documents are representative of the primary research material, and they contain little or no processing of the data.

From the early 20th century, researchers from backgrounds spread across several disciplines recorded data that today would be described as Aboriginal Biocultural Knowledge. Scholars with interests in museum collections, such as Thomas Worsnop (1821–1898), Daniel Sutherland Davidson (1900–1952), Norman Barnett Tindale (1900–1993), Frederick McCarthy (1905–1997) and Aldo Massola (1910–1975), used Indigenous perceptions of the environment as a means of interpreting Aboriginal material culture and art. Anthropologists with records pertaining to the field included Ronald Murray Berndt (1916–1990), Catherine Helen Berndt (1918–1994), and Adolphus Peter Elkin (1891–1979).

Since the late 20th century, data relevant to the study of Aboriginal Biocultural Knowledge have come from scholars from a much more diverse range of academic disciplines and fields. In the humanities, researchers with relevant work have come from linguistics (Hercus 1986; Hobson *et al.* 2010; Bell 2010; Blake 2003), historical geography (Clark 2003), anthropology (Vaarzon-Morel and Edwards 2012), archaeology (Allen and Holdaway 2009; Taçon *et al.* 2003) and material culture (Memmott 2007). In the sciences, researchers making significant contributions are from climate change studies (Berry *et al.* 2010; Green *et al.* 2010a; Green and Raygorodetsky 2010; Jackson and Morrison 2007; Jones *et al.* 2016; Low Choy *et al.* 2013, 2016; McIntyre-Tamwoy and Buhrich 2012; Petheram *et al.* 2010) and the discipline of ecology (Balme and Beck 1996; Beck and Balme 2003; Ens *et al.* 2012, 2014, 2015, 2017; Head 1989; Presland 2005). Specialists from the inter-disciplinary field of ethnobiology have also added to the growing literature of Aboriginal Biocultural Knowledge (Clarke 2007a, 2008a, 2016a, 2016b, 2016c; Clarke M 2012; Gott 1982a, 2005, 2008; Hill *et al.* 2011; Laird 2010; Yaniv and Bachrach 2005).

Australia's Indigenous peoples have a detailed knowledge of the country that has integrated what scientists would separately classify as the biological and cultural aspects of the landscape. The study of this rich body of knowledge offers many opportunities for several interlinked research fields, including wellbeing, education, cultural renewal and natural resource management. This book places equal emphasis on the sciences and the humanities, with research techniques that are multi-disciplinary and include ecology, history and anthropology. Prior to European colonisation, Aboriginal Biocultural Knowledge existed largely in oral forms held by Indigenous custodians. After British settlement, much of this knowledge was lost as Aboriginal people were forced to abandon their foraging lifestyles and to live more restricted lives on mission reserves. In some instances, a fraction of this knowledge was recorded in piecemeal fashion by Europeans who were interested in both Indigenous culture and the environment. Today, the records relevant to Aboriginal Biocultural Knowledge are held in widely dispersed records, such as artworks, private journals, linguistic dictionaries, unpublished reports, community publications, academic journal articles and books, archival photographs and films.

Aim

This book presents archival material that highlights the exposure that European colonists had with Aboriginal Biocultural Knowledge in south-eastern Australia (excluding Tasmania). It has unearthed a large corpus of printed historical data that hitherto has not been examined from the viewpoint of contributing specifically to our understanding of Aboriginal Biocultural Knowledge. The authors hope that the generation of knowledge from this research will encourage greater Aboriginal involvement in heritage interpretation and will add to the scholarly discourse about Australian frontier history, with benefits to the tourism industry, and in doing so contribute to a greater appreciation of the shared histories of Aboriginal and non-Aboriginal Australians.

This current work is a collection of histories and narratives on biocultural knowledge topics that were compiled from publicly available sources, either as part of the published literature or as archival sources in libraries and museums. The following chapters are three-pronged in their approach to the available Aboriginal Biocultural Knowledge material, and each to varying degrees provides:

1. An ethnographic summary of a category of Aboriginal Biocultural Knowledge, largely based on the records of Europeans who were in a position to produce detailed descriptions of Aboriginal relationships to country.
2. Historical evidence of the complex asymmetrical relationships between European colonisers and Aboriginal people on the frontier of European settlement in relation to the recording of Aboriginal Biocultural Knowledge, with investigations considering the extent to which colonists used this knowledge and the degree to which they understood it.
3. Consideration of the implications of the interplay between Indigenous and European knowledge systems, looking at the ways that it affected frontier relationships and changed the lifestyles/culture for both Aboriginal people and the colonists.

Recent advances in metadata websites, in particular the TROVE project at the National Library of Australia (http://trove.nla.gov.au/), have made possible the compilation of material that was previously not easily accessed. The research has focused on topics selected in order to illustrate the main Aboriginal Biocultural Knowledge themes outlined above. Among the chapters are those that have an anthropological focus through considering an Indigenous perspective of country, discussing Aboriginal beliefs in totemic ancestors and spirit beings. Others have joint ethnobiological and material culture interests by investigating the means of subsistence, such as the obtaining of food and medicine, as well as the production of artefacts and the generation of trade. There are chapters that are more ecological, through discussion of Indigenous uses of fire and investigating the perception of seasonal calendars through an understanding of time and space.

The authors of the current work have long held research interests that are in keeping with the aims of the book. Fred Cahir and Ian D Clark are historians who have focused on the complex relationships between Indigenous peoples and European settlers on the colonial frontier, and have specialised in the landscape of Victoria. Sarah McMaster is a historian who is currently researching the fire practices of Aboriginal people in Victoria. Philip A

Clarke is a social anthropologist with research interests in the ethnosciences, in particular ethnobiology and ethnoastronomy, with a broad focus across Australia.

Spelling

As Aboriginal people did not have a written language and the spellings from the original researchers were not fixed, there are spelling variations throughout the text.

Warnings

It is strongly recommended that untrained people do not experiment with any of the foods, medicines, poisons or narcotics discussed here as part of Aboriginal Biocultural Knowledge. It is stressed that this publication is not intended for use as a survival guide. Notwithstanding, this book does indicate some developing areas worthy of future research and for the application of this knowledge with the approval of custodians.

Chapter 1
Totemic life

Philip A. Clarke

The British colonists who settled in Australia from the late 18th century and the Indigenous peoples whose ancestors first arrived here over 50 000 years ago, had fundamentally different accounts of the Creation of the world and how it was first peopled. This meant that there were major differences in the ways that the two groups saw themselves in relation to the physical environment. While modern Western Europeans have seen themselves as having a separate existence from the 'natural' world, Aboriginal people consider that the social and physical aspects of their lives are closely intermeshed and therefore inseparable. In Aboriginal tradition, the actions of spiritual ancestors during the Creation gave a deep social relevance to the country, imbuing it with their power and in doing so humanised it. For this reason, in Aboriginal thought the land could never be seen as a total wilderness. Aboriginal Biocultural Knowledge reflects the merging of the cultural and natural worlds.

Introduction

The first British colonists who arrived in Australia had a model of religion based on the major religions originating from the Middle East, like Christianity and Judaism, which were structured round the worship of a single high god. They therefore found it difficult to make sense of the Indigenous forms they encountered, which were based on multiple spirit ancestors who existed alongside a wide range of other spirit beings (see Chapters 2 and 3). Europeans in the 19th century considered Aboriginal religion to be an incoherent set of 'superstitions' that was mostly focused on magic (Dawson 1881, p. 49; Smyth 1878, vol. 1, pp. xviii, lx–lxi, 30; Taplin, 1879, p. 23). There was little or no appreciation of the depth and complexity of Indigenous mythologies, with one reminiscing Victorian colonist describing Aboriginal people subjecting themselves to 'huddling up together in their *loondthals* (huts), and to the coarse, obscene, and lewd character of the stories in which are spent so many of their evenings round the camp fires' (Beveridge 1883, p. 24).

In the late 19th century, European scholars became interested in recording Indigenous customs and traditions. This was partly because they could see that these were changing and because in many areas the Aboriginal life style was no longer a lived reality (Curr 1886–87; Smyth 1878; Taplin 1879; Woods 1879). In their accounts of Aboriginal culture, they generally mistook the Aboriginal spiritual ancestors as 'high gods' rather than a class of ancestral beings (Hiatt 1996). For instance, the missionary George Taplin in the Lower Murray of South Australia remarked that Ngarrindjeri people had 'several times told me something about a bird making the world. Surely there cannot be a sort of fetishism amongst them' (1859–79, 17–18 August 1859). Similarly, in Victoria the scholar Robert Brough Smyth noted that:

> Birds and beasts are the gods of the Australians. The eagle, the crow, the mopoke, and the crane figure prominently in all their tales. The native cat is now the moon; and the kangaroo, the opossum, the emu, the crow, and many others who distinguished themselves on earth, are set in the sky and appear as bright stars (Smyth 1878, vol. 1, p. lx).

Researchers who were missionaries looked for analogous Indigenous beings in the beliefs of their convert communities upon which to graft notions of God and Satan (Clarke 1994; Kolig 1981, 1988; Swain 1991, 2000).

Aboriginal beliefs

With the development of anthropology as a discipline from the early 20th century, Europeans stopped seeing the religion of Australian hunter-gatherers as primitive, and began to acknowledge the depth and complexity of their traditions that were told in myth and song (Hiatt 1996; Strehlow 1970). It was recognised that in Aboriginal religion the totemic spiritual ancestors as a group performed heroic deeds during the Creation period, as they moulded and imparted spiritual power to the land, and then formulated the customs for all their descendants, both human and non-human, to ultimately follow (Hiatt 1975; Sutton 1988; Clarke 2003a, 2016a). In 1904, anthropologist Alfred Howitt published an account of the cultures belonging to the Aboriginal people in south-eastern Australia he had worked with and remarked that:

> A belief is common to all the tribes referred to, in the former existence of beings more or less human in appearance and attributes, while differing from the native race in other characteristics. Their existence, nature, and attributes are seen in the legendary tales which recount their actions (Howitt 1904, p. 475).

Aboriginal people believed that at the conclusion of the Creation, their ancestors took the shape of entities such as birds and other economically important animals, but many were also plants, atmospheric and cosmological phenomena or even human diseases. The totemic links that each person possessed gave them rights to country, of the type that lawyers and anthropologists would later refer to as native title (Peterson and Rigsby 1998; Sutton 1998a, 2003). The corpus of recorded Indigenous Creation narratives for south-eastern Australia is immense and therefore too large for adequate treatment in any one chapter, so for this reason only certain myth complexes are discussed below in order to demonstrate the intimacy of relationships between people and their associated totemic species.

Creation of country

Recorded accounts of the total landscape as it was believed to be during the Creation period are Utopian, concerning a golden age when food and water were easily procurable everywhere

(Clarke 2014b, 2014c, 2015a). For instance, among the Buandik (Booandik) people of the south-east of South Australia it was believed that the *bo-ong* was one of the two soul spirits within each person, and upon death it would go up to the 'cloudland' (Skyworld) 'where everything is to be found better than on the earth' (Howitt 1904, p. 434). Among these people, it was said that a fat kangaroo was just like the kangaroo of the clouds (Smith 1880). At the conclusion of the Creation period, the paths that the ancestors had made during their travels on Earth became ancestral tracks, or songlines, which connect with mythological sites where according to Aboriginal tradition particular significant events had occurred. It was Aboriginal belief that many of their spiritual ancestors migrated from the terrestrial country into the heavens, sometimes through the Underworld (Elkin 1932, 1964, 1977; Clarke 2003a). In the heavens, the ancestors remained as 'sky heroes' who maintained their influence over the total landscape (Elkin 1977, p. 32).

In Victoria, it was noted that: 'The progenitors of the existing tribes - whether birds or beasts or men - were set in the sky, and made to shine as stars if the deeds they had done were mighty, and such as to deserve commemoration' (Smyth 1878, vol. 1, p. 431). It was believed that when the Creation drew to a close, this idyllic existence was restricted to the heavens, where the spiritual ancestors and deceased human spirits who had left Earth were able to continue their foraging activities in a land of plenty. People on Earth could recognise their ancestors in the night sky. According to colonist William Stanbridge, in north-western Victoria the people of Lake Tyrrell had beliefs concerning 'Berm-berm-gle (two large stars in the fore-legs of Centaurus)', which they identified as:

> Two brothers who were noted for their courage and destructiveness, and who spear and kill Tchingal [Emu, *Dromaius novaehollandiae*]. The eastern stars of Crux are the points of the spears that have passed through him, the one at the foot through his neck, and that [one] in the arm through his rump (Stanbridge 1857, p. 139).

These brothers are the Brambimbula, who are two ancestors who were frequently recorded in the mythology of north-west Victoria and adjacent parts of the Mallee region in South Australia (Clark 2007a; Clarke 2009c; Hercus 1986; Howitt 1904; Smyth 1878).

Living people were generally barred from the Skyworld, although there were some exceptions to the rule, with healers ('medicine-men') being able to travel there to learn new songs and gain special knowledge of rituals from the spirits residing there (Elkin 1977). In the south-east of South Australia during the 19th century, a *pangal* ('doctor') was said to have spoken with the spirits of the dead by climbing a tree into the Skyworld (Smith 1880), with similar accounts coming from south-western Victoria of powerful 'doctors' and 'sorcerers' who claimed to have been regular visitors to the heavens (Dawson 1881). Stanbridge stated that in Victoria:

> There are doctors or priests of several vocations; of the rain, of rivers, and of human diseases. The office is alleged to be obtained by the individual visiting, while in a trance of two or three days duration, the world of spirits, and there

receiving the necessary initiation, but there are natives who refuse to become doctors, and disbelieve altogether the pretensions of those persons (Stanbridge 1861, p. 300).

Their crossing into the Skyworld was achieved by various means, such as by using a magic rope, climbing to the top of a large tree, walking to the top of certain high hills, or through their ritual power to pass through space itself (Howitt 1904; Elkin 1977; Clarke 1997, 2015a). In many cases, such travel was part of the initiation for healers who specialised in treating sicknesses caused by supernatural causes.

In the Lower Murray region, there was a tradition that the Supreme Male Ancestor, Ngurunderi (Nurunduri), had thrown out a line attached to his testicles to help guide his lost son from Earth towards the west and from there into the heavens (Clarke 1995b; Meyer 1846 [1879]). It was believed that after the close of the Creation period this method of finding the Skyworld remained, as it was recorded that 'after death the [human] spirit wanders in the dark for some time, until it finds a string when ... Oorundoo [Ngurunderi] pulls it up from the earth' (Angas 1847b, p. 97). Across Aboriginal Australia it was traditionally believed that strings, sometimes described as threads in a web, were regularly used as means of conducting soul spirits into the Skyworld (Elkin 1977; Howitt 1904).

Legacy of the ancestors

For Aboriginal people, their spirit ancestors could simultaneously be seen as landscape features on Earth, such as a tree or a hill, and as celestial bodies, like stars in the Skyworld. For example, the large granite boulder formation on the coast at Papajara (the Granites) north of Kingston in the south-east of South Australia represented the physical bodies of the ancestral Emus who fought with the Brolgas, all also seen in the night sky (Clarke 1997, 2016a). Across Victoria, Bunjil the Eaglehawk (Wedge-tailed Eagle, *Aquila audax*) Ancestor was associated with the Skyworld and with particular mountains on Earth (Brumm 2010), which were probably both seen as connected regions. Bunjil (Buunjill) was variously seen as Fomalhaut or Altair in the night sky (Dawson 1881; Howitt 1884, 1886), and on Earth he was said to reside in ranges at the headwaters of the Yarra River (Fison and Howitt 1880) or at places such as in Western Port Bay (Barwick 1984) and at Lal Lal Falls near Ballarat (Clark 2002).

Recorded narratives of myth explain how ancestors became animals towards the end of the Creation. At Lake Tyrrell, it was documented by Stanbridge that:

Bunya (Oppossum [sic.]), (star in the head of Crux), who is pursued by Tchingal [Emu], and who, in his fright lays his spears at the foot of the tree and runs up it for safety. For such cowardice he became an opossum (Stanbridge 1857, p. 139).

As an animal, each ancestor's behaviour continued in the Skyworld. For instance, in south-western Victoria, colonist James Dawson recorded that:

Hydra, 'Barrukill,' is a great hunter of kangaroo-rats [*Dipodomys* species]. On his right, and a little above him, are two stars – the rat, and his dog 'Karlok';

above these again are four stars, forming a log; underneath are four other stars, one of which is his light, and three form his arm. The dog chases the rat into the log; Barrukill takes it out, devours it, and disappears below the horizon (Dawson 1881, p. 101).

Such myths document the behaviour of modern animal species, with possums being secretive and Kangaroo-rats living in hollows on the ground (Menkhorst and Knight 2001) and Dingoes (*Canis lupus dingo*) used as an aid to hunting (Cahir and Clark 2013; Hamilton 1972). Similarly, in south-western Victoria there is a myth concerning how a Snake Ancestor received venom from the Tortoise (Dawson 1889, pp. 106–7). It involved the Tortoise realising that he could no longer kill people after they had mastered drinking water by scooping it up by hand and throwing into their mouths, which meant that their heads would no longer be immersed in the waterholes and streams. In this way, people were able to see dangers in the water without exposure to themselves. The Snake, who was previously without poison, received the gift of venom because it gave him better chances to bite people as they walked in long grass.

The Creation mythologies can account for the physical traits and behaviour of contemporary species. From south-western Victoria, it is recorded that an ancestor, who had stolen a fish from a group of people camped at Dunkeld on the southern edge of the Grampians, was punished by being turned into a blue heron (White-faced Heron, *Egretta novaehollandiae*), which is largely a solitary species (Dawson 1881). In the Lower Murray of South Australia, a Ngarrindjeri version of the White-faced Heron myth has him being severely attacked by other bird ancestors, resulting in him ever after flying with his legs bent up underneath as if he is hurt (Clarke 1990).

Bird ancestors feature prominently in the recordings of Aboriginal mythology. For instance, a Victorian newspaper correspondent remarked that:

> Birds and beasts figure in the tales, legends and folk lore of the Australian blacks as their gods. The eagle, the crow, the mopoke [boobook owl, *Ninox boobook*] and the crane [brolga] especially occupy prominent places. The kangaroo, opossum, emu and other Australian beasts are set in the sky and appear as bright stars (Anonymous 1888b, p. 2).

Many of the Creation ancestors were believed to have become species of bird, and due to their common ability of flight most would have been considered to have already had some access to the Skyworld (Clarke 2016a, 2016c). For instance, in south-western Victoria it was recorded that the Large Magellanic Cloud was called *kuurn kuuronn*, meaning 'male native companion' or 'gigantic crane' [Brolga, *Grus rubicunda*], while the Small Magellanic Cloud was *gnaerang kuuronn*, meaning 'female native companion' (Dawson 1881, p. 99). Related beliefs were recorded at Lake Tyrrell in north-western Victoria, with 'Kourt-chin (Magellan Clouds). – The large cloud a male, and the lesser cloud a female Native Companion' (Stanbridge 1857, p. 139). The people belonging to the clan of the 'native companion', along with the birds of this species on Earth, were both seen as the descendants of those ancestors who were present in the night sky.

Eaglehawk and Crow narratives

In south-eastern Australia, recordings of the mythology concerning the Eaglehawk and Crow/Raven are rich in ethnographic detail (Blows 1975, Clarke 2016a; Hercus 1971; Tindale 1939). Among the Ganai (Kurnai) people of Gippsland in eastern Victoria, anthropologists stated that:

> ... the Eaglehawk ('Gwannumurung') is greatly referenced. He is regarded as the type of the bold and sagacious hunter. His plumes and talons played a part in their necromancy. He figures in their tales in company with 'Ebing,' the little Owl [probably boobook owl, *Ninox boobook*] (Fison and Howitt 1880, pp. 322–3).

In contrast, the Crow Ancestor, variously known by onomatopoeic names such as Waa, Waak, Wak, Waku or Wark (Clarke 2016a), is generally portrayed as a thief (Maddock 1970). Most ethnographic sources describe the bird as a 'crow', although in virtually all cases the species involved would have been more accurately described as the Australian Raven (*Corvus coronoides*) (Clarke 2016a). Across western and central Victoria, the Eaglehawk Ancestor was recorded variously as Bunjil, Buunjill, Pund-jel and Pund-jil, being the attempts by different European recorders to write the same name (Howitt 1904; Maddock 1970).

Bunjil was a Supreme Male Ancestor and creator of both men and the physical world. The Woiwurrung (Woeworung) people of central Victoria celebrated him in their dance ceremonies, which according to anthropologist Alfred Howitt were based on mythology that:

> ... Bunjil held out his hand to the sun (*Gerer*) and warmed it, and the sun warmed the earth, which opened, and blackfellows came out and danced this corrobboree, which is called *Gayip*. At it images curiously carved in bark were exhibited. Usually *Bunjil* was spoken of as *Mami-ngata*, that is, 'Our Father,' instead of by the other name *Bunjil*. It is a striking phase in the legends about him that the human element preponderates over the animal element. In fact, I cannot see any trace of the latter in him, for he is in all cases the old blackfellow, and not the eagle-hawk, which his name denotes; while another actor may be the kangaroo, the spiny ant-eater, or the crane, and as much animal as human ... Among the Kurnai, under the influence of the initiation ceremonies, the knowledge of the being who is the equivalent of Bunjil is almost entirely restricted to the initiated men (Howitt 1904, p. 492).

Eaglehawk feathers were ritually powerful and therefore used in sorcery (see Chapter 12), and Howitt claimed that:

> The Kurnai fastened some personal object belonging to the intended victim to a spear-thrower, together with some eagle-hawk's feathers and some kangaroo or human fat. The spear-thrower was then stuck slanting in the ground before a fire, and over it the medicine-man sang his charm. This was generally called

'singing the man's name' until the stick fell, when the magic was considered to be complete. Those who used this form of evil magic were called *Bunjil-murriwun*, the latter word being the name of the spear-thrower (Howitt 1904, p. 361).

Many of the actions of Bunjil and his fellow ancestral beings during the Creation period were those of humans, rather than of animals, although the consequences of those actions were felt by human and animal descendants alike. It was recorded that:

> The Yarra blacks have their own version of the birth of the human race. Pundjel is admitted to be the first man, who made everything, including the second man, Karween, as well as two wives for the latter. But he foolishly omitted to make a wife for himself, and after a time he set envious eyes upon Karween's black treasures. Notwithstanding that Karween watched them jealously, they were both stolen one night by Pundjel. Subsequently a fight for the women took place at a corroboree convoked by the crow. Pundjel was the conqueror, and the two women became his wives, and gave birth to many children. After this the jay [Australian raven], who at that time was a man, had a great many bags full of wind, which he opened one day in a fit of anger, and such storm arose that Pundjel and nearly all his family were carried up into the heavens (Anonymous 1888b, p. 2).

On the edge of south-eastern Australia, the Wiimbaio people of the lower Darling River believed that the celestial bodies in the Skyworld were once great men who had lived on Earth. The planet Mars was Bilyara, the Eagle Ancestor, while a star was seen as Kilpara the Crow Ancestor (Bulmer [cited Howitt 1904]).

The mythologies concerning the Eaglehawk and the Crow/Raven Ancestors encapsulate the essence of the living species that they are connected with, and in doing so have therefore incorporated much Aboriginal Biocultural Knowledge. The theme of opposition between them would have been constantly reinforced through the observation of the behaviour of living birds. Ornithologists have also noted the aggression displays between the Australian Raven and Eaglehawk (Wedge-tailed Eagle) (Rowley 1973).

The antagonism between the Eaglehawk and Crow/Raven was reflected in the existence of a major division within the systems of descent among Aboriginal people in south-eastern Australia. For their moieties, the Wiimbaio people had the Eaglehawk as a totem for the Mukwara and the Crow for the Kilpara (Howitt 1904). Similarly, among the Ngarigu (Ngarigo) people on the Manero Tableland their moieties were Merung (Eaglehawk) and Yukembruk (Crow), and for the Walgalu (Wolgal) on the Upper Murray River their moieties were Malian (Eaglehawk) and Umbe (Crow) (Howitt 1904). Among these groups, it was recorded that the 'law of marriage was that of the tribes with female descent, and a man might marry a woman of any of the opposite totems; and, as was the case with the Omeo tribe, a man's proper wife was the daughter, own or tribal, of his mother's brother' (Howitt 1904, p. 197).

The Eaglehawk and Crow complex of mythology took centre place during an anthropological debate over the origin of moiety systems in Aboriginal Australia (Blows 1975). It was the theory of an amateur anthropologist writing from the perspective of Aboriginal groups in south-east Queensland, that the Eaglehawk and Crow myths were an Aboriginal account of a violent encounter between two actual groups of people (Mathew 1899). The theory, as summed up by Howitt as one of his detractors, was that the myth complex:

> ... speaks of a pristine conflict between two races of men contesting for the possession of Australia, 'the taller and more powerful and more fierce Eagle-hawk race overcoming and in places exterminating the weaker, more scantily equipped sable Crows.' This hypothesis, as I understand it, infers that the two class divisions arose from the amalgamation of two groups, the totem of one being Eagle-hawk, and of the other Crow (Howitt 1904, pp. 143–4).

It was argued that the outcome of the mythic conflict between groups, signified by the Eaglehawk and Crow, was a resolution that they would forever intermarry to keep the peace. Other interpretations of these myth narratives included that they represented a primeval struggle that exists between father and sons, which resulted in the moiety system as a compromise (Roheim 1925), and another explanation was that the apparent dichotomy between the birds as species was a convenient metaphor for the divisions between humans (Radcliffe-Brown 1958). It is the last explanation that is explored in more detail here, as it involves the Aboriginal Biocultural Knowledge of the avian species involved.

In the Lower Murray area of South Australia, Aboriginal people had patrilineal clans (Clarke 2016a; Berndt *et al.* 1993; Brown 1918), but nevertheless they had incorporated some elements of the Eaglehawk and Crow mythologies into their own accounts of the Creation. In a Yaraldi myth concerning Marangani the Crow (Australian Raven), he was camped with other bird ancestors at Rufus River, a tributary of the Murray River near the South Australian/New South Wales border (Berndt *et al.* 1993). In this account, the Crow was annoyed that he had been denied access to women, so he took the two young sons of his two nieces and hid them in a large tree that he had sung into existence. The children were eventually rescued by Tjuit (Tuit), described as the ancestor of 'a black-and-white whispering reed bird' (Berndt *et al.* 1993, p. 241). In the Murray River/Darling River version of essentially the same myth, the two nieces are young girls in the care of Kanau the Eaglehawk (Wedge-tailed Eagle), who is married to the sister of Waku the Crow (Tindale 1939). In this version, Walpu was an initiand anointed with red ochre, and it was he who was able to traverse the barrier of galls jutting out on the trunk of the large tree containing the boys to rescue them.

In most recorded Yaraldi versions of the Crow myth, Marangani was described as being generally disliked, due to his habits of calling every man he meet '*ronggi*' (brother in-law) as a tactic for gaining access to their women (Berndt *et al.* 1993). After the Creation, it was said that he and Waiyungari (Red Man) were both responsible for causing women to indulge in *kuruwolin* (ritualised sexual licence). As an old man, Marangani was always the first to appear when fish were being drawn in with nets. He would sit next to women who were cooking fish, with his enlarged penis travelling underground to enter them. Since then, women who were jabbed by something when sitting down would cry out, 'That is Old Crow again!' (Berndt *et al.* 1993, p. 240). In some Lower Murray recordings, Marangani's name is given as Wak (Waak, Wark), which is the name for the 'crow' species (Australian Raven) in the Ngaiawung language along the river immediately north of Murray Bridge (Tindale 1964).

In Lower Murray tradition, the Crow came downstream along the Murray River, spent time in the Lower Lakes, and then travelled south along the Coorong, where he was credited with creating landforms and was acknowledged as having introduced the returning boomerang (Tindale 1934). According to the associated myth as recorded by museum researcher Norman Tindale, the Owl (possibly Barn Owl, *Tyto delicatula*) and Mopoke (Boobook Owl) Ancestors challenged Waak (Wark) the Crow at a pipeclay lake (saltpan) near Salt Creek, which is a southern tributary of the Coorong, where they:

> … said they would allow him to live at Salt Creek only if he could throw five consecutive boomerangs in complete circles otherwise they would kill him. The Crow threw four of them successfully, but the fifth was caught by the wind and fell short. Thus today in the lake there are four complete islands and the fifth is a peninsula connected to the shore by a tongue of land. The men of the tribe held boomerang throwing competitions there until recent times (Tindale 1934).

The Owl and Mopoke then tried to spear Waak, but he crawled away to the south-east where his tracks remained as a series of saltpans. When the Crow eventually stopped on the edge of the Coorong, he could see that he was no longer being pursued. In order to look about from this high vantage point, he thrust his spears into the ground, where they left holes that could later be seen in the limestone cliff. Gerum Gerum was a 'sacred lake' near the Coorong where 'The Crow, an ancestral being threw boomerangs on this lake and they flew with such an erratic course that the shores of the lake which they cut out were sinuous' (Tindale c.1931–c.1991). Naramendjarang is a site south of Lake Alexandrina in the Lower Murray which was described as a 'Hill with blackwood [*Acacia melanoxylon*] trees where in the Waak or Crow story the ancestral being [Creator being] prepared boomerangs by chiselling wood of the trees which remain as a memorial' (Tindale 1987b).

The Crow being continued to travel further into the south-east of South Australia, and here he met Ngurunderi's two 'nephews', who might have been classificatory, who had also gone in this direction. According to Tindale's account recorded from Moandik (Meintangk) people, the nephews had arrived:

> Then there appeared an old Crow or *Wark*. This being had set himself up magically to cut the country up into separate living spaces. He offered the nephews each the choice of a place to live. Younger nephew took a fancy to a certain kind of shrub and found a girl who became his wife while the older one, also seeking a wife, went into the ti tree swamp and found another girl to be his wife (Tindale 1934–37, pp. 58–9).

The Crow convinced these men to keep clear of a certain emu nest, as he was planning to eat a giant egg there (Tindale 1934–37). He ended up fighting the Emu Ancestor, and in the process travelled underground. In the bay at Port MacDonnell in the south-east of South Australia, the Crow created an island, which he could magically move closer to shore. Crow eventually killed the Emu and after cooking it, he scattered the pieces along the Coorong where they turned into present-day emus. Here, the Crow showed people where to fish and taught them how to swim and to make nets. Two young girls he allowed to chase the fish into the nets were transformed into ocean seals.

As with the living species, as a thievish being Crow had problematic relationships with other bird ancestors. Tanitjari the Wandering Albatross (*Diomedea exulans*) Ancestor of the Needles prevented Marangani the Crow coming from the Coorong (Tindale c.1931–c.1991), while Throkopuri (Torakopuri) the Seagull [Silver Gull, *Chroicocephalus novaehollandiae*] Ancestor had 'challenged the Crow being' (Tindale c.1931–c.1991). Throkopuri had his camp at Tarakorinking (Tea Tree Point in the Coorong), which means 'home of the sea gull', and he was:

> … a mythical man, one of the Tanitjari or seagulls, [who] lived here, he fought the malicious Crow men who spied on his methods of fishing and stole his *ngorankuri* or hair charms. He thrashed the crows so severely about the eyes and head that the crow thereafter always walked as if lame and his eyes are white (Tindale c.1931–c.1991).

In the southern South Australian mythologies, the Crow was spoken of either as a singular entity or, as in the above account, as a group.

Emu and Brolga Ancestors

The Lower Murray mythology of the Emu and Brolga (Native Companion, *Grus rubicunda*) Ancestors is based in the country surrounding present day Kingston (Berndt *et al*. 1993; Tindale 1931–34; 1936d; 1938–56; 2005). In the south-east of South Australia, this myth was sometimes used to identify local Aboriginal groups, with the Moandik people referring to coastal clans as Porolgi (Brolga), while inland clans were Pindjali (Peindjali, Emu) (Tindale 1976). The Brolga Ancestor was the *ngaitji* (totemic species) of some Lower Murray clans based at the southern end of the Coorong (Clarke 2016a), as well as at Kingston (Berndt *et al*. 1993). It was Lower Murray tradition that if any man from Brolga country was 'knocked down and bruised by one from another tribe, and left on the ground unconscious, the Brolgas come down, lift him up, and show him the road home' (Wilson 1937, p. 49).

The main unifying theme in several recorded Lower Murray accounts is that a pair of Emus tricked a pair of Brolgas into killing all but two of their own chicks. Yaraldi man Mark Wilson said that: 'In their deep grief, they [the Brolgas] put their heads in the fire, and rubbed hot ashes on their heads and necks, and that is why Brolgas now do not have any feathers on their heads or neck' (Wilson 1937, p. 49). The Brolgas sought revenge and by pushing their long sharp beaks into the ground in the Kingston area, caused a flood that killed the Emus, which are land birds and therefore unable to fly away. It was recorded that:

Australian Raven (crow), an ancestor in many myth narratives in south-eastern Australia. PA Clarke, Laratinga Wetlands, South Australia, 2015.

> You can see to this day the young emus in the water, and the father and mother emu on the beach in the form of granite rocks [i.e. the Granites] a few miles north of Kingston … The Brolgas, knowing that the emus, if they escaped, would hunt them up to kill them, flew up into the air, and, like the pelican [*Pelecanus conspicillatus*] flew up in circles getting higher and higher, until they reached the sky, and found it a good country to live in. So they stopped there. You can now see them at night in the form of two pieces of cloud at the end of the Milky Way (Wilson 1937, p. 49).

The account of this mythology recorded from Yaraldi speakers emphasised the female gender of the main birds involved (Berndt *et al.* 1993).

For Aboriginal people, evidence of the Creation events was all round them. At the southern end of the Coorong, north of Kingston, the large boulders known as the Papajara (the Granites) were associated with the Emu Ancestors, with some of the chicks being seen as other granite outcrops at Kingston and at Boatswain Point on the eastern side of Cape Jaffa (Berndt *et al.* 1993; Tindale 1931–34; c.1931–c.1991; 1934–37). Other sites were connected too. Cadara is a hill south-west of Kingston that was called Teriteritjngola, which reportedly meant 'willie-wagtail lookout', and this was a reference to how the Willie Wagtail (*Rhipidura leucophrys*) being 'by his chatterings warned the emu people when they were about to be cut off by the rising tide during a visit to Baudin Island [Baudin Rocks]; the flooding of the land by the sea was due to the evil genius of the brolga' (Tindale c.1931–c.1991). It was asserted that this Creation event had transformed the Emus into scrub birds – no longer closely associated with the coastal wetlands (Berndt *et al.* 1993).

In the south-east of South Australia, it was a Moandik tradition that a Lower Murray myth track of Ngurunderi's two 'nephews' came from the north and briefly connected with that of the Emu and Brolga in their country. As recorded by Tindale, the men fled to Bald Hill, west of Mount Gambier, where:

> On an occasion they heard a strange noise and feared it, for they thought some devil being was close by. Thus they shifted camp to Mt. Benson. But the strange noises were only the cries of kutuk (brolga or 'bull-necked cranes'). At Mt. Benson everything was 'all right'. Then they found a great big emu sitting on two gigantic eggs. As my narrator [Alf Watson] said 'Present-day eggs are only like scraps of big ones'. They let it be and continued their kangaroo hunting (Tindale 1934–37, p. 58).

Tangani-speaking people of the Coorong knew of several songs associated with the Emu and Brolga mythology from the south-east of South Australia. One described an enraged male Emu trapped on Baudin Rocks and displaying his anger at having been tricked by his long-time foes, the Brolgas (Tindale 1941). Another two songs concerned a pair of Emu Ancestors who made their nest in the bracken near trees where the Eagle had nested (Tindale 1941).

In south-west Victoria, a structurally similar myth was recorded that also had trickery as a main theme (Dawson 1881). Here, the Brolga (Native Companion) tricked the Emu into burning her feet, wings and bill by hiding a long stick being used to get sedge roots (probably Marsh Club Rush, *Bolboschoenus medianus*) out of the cooking fire. For vengeance, when the Emu was alone she hid all but two of her chicks, and later convinced the Brolga that she had killed them for food. After the Brolga had followed the Emu's lead, the other chicks came out to expose the deception. When the Creation period ended, Emus retained their burnt colouration and Brolgas could only have two chicks at a time.

Stealing fire and creating waterholes

Across Aboriginal Australia, there is much structural similarity in the mythic explanations of how the property of making fire was transferred from the ancestors to people on Earth (Fraser 1930; Isaacs 1980; Maddock 1970). Fire and its presence in the Aboriginal Creation is discussed elsewhere in this work (see Chapter 7), with the discussion here focused on two main themes; the origin of fire in the Skyworld, and the forced seizure of fire from an ancestor who kept it to himself. An example of a narrative concerning the taking of fire in the Skyworld is as follows:

> The legend of how fire was brought on the earth has many dressings, but all agree on one cardinal point, and that is, that the fire was stolen. Each tribe has its own legend. The Lake Condah tribe, for instance, hold to the following

Emu showing its 'burnt' plumage. PA Clarke, 2009.

> myth: - A blackfellow threw a spear towards the clouds; to the spear a string
> was attached. The man climbed, up with the aid of the string and brought fire
> to the earth from the sun (Anonymous 1888b, p. 2).

Various birds were players in the corpus of Fire Creation mythology from the Lower Murray (Clarke 2001). A recorded Ramindjeri myth that is localised in the upper Hindmarsh River area near Victor Harbor concerned Kondoli (Kondole) the Whale Ancestor and the origin of fire, where at the termination of a ceremony the decorated young male performers, 'who were ornamented with tufts of feathers, became cockatoos, and the tuft of feathers being the crest' (Meyer 1846 [1879, p. 204]). There was a fight over the power to make fire, and Rilballe the Skylark (Singing Bushlark, *Mirafra javanica*) speared Kondoli in the neck, giving the whale its distinctive blowhole.

Other bird species were also noted as being involved with the release of fire into the countryside, thereby giving people access to it. A Ramindjeri version recorded from a Yaraldi-speaker chiefly involved Krilbali the Skylark and Kuroldambi the Owl (Barn Owl, *Tyto delicatula*) (Tindale 1934–37). According to a Yaraldi version, it was the Skylark and Willie Wagtail who conspired to spear Kondoli and steal his fire (Berndt *et al.* 1993). The fire released by the spearing was spread throughout the country by the Skylark flying off and dropping the sparks onto the ground, where they were transformed into flints. In another recorded Yaraldi account, which is essentially the same as above, it was Krilbali the Skylark who speared Kondoli (Tindale 1930–52). As with many myths, the Kondoli narrative was told as a song (Tindale 1931–34).

Other myths from the Lower Murray area implicate different bird ancestors in the origin of fire. For instance, Tuta the Scarlet Robin (*Petroica boodang*) was involved with the release of fire, and as a result had his chest burnt a bright red (Tindale 1930–52). The Whale Ancestor is replaced by a Shark in 'another version of the story, in the Murray basin, [when] a small species of hawk stole an ember from the shark as he was entering the sea and placed it in a grasstree (*Xanthorrhoea*) for safekeeping', from where the power of fire could be released by the use of fire-drills made from dry Grass Tree (*Xanthorrhoea* sp.) flower stems (Tindale no date). In another Lower Murray myth, the Tjelgawi (Australian Ringneck Parrot, *Barnardius zonarius*) Ancestors 'were women who attacked the Being *rat:aragi* [Scarlet Robin Ancestor]' (Tindale 1934–37, pp. 233, 235).

In the south-east of South Australia there was a recording of a myth concerning the Creation of fire, that appears structurally related to the Lower Murray narratives, which also involved birds. Here, Tatkanna (Tal Kanna), the Robin Redbreast Ancestor (possibly Scarlet Robin), stole the power to make fire by secretly placing a dry flower stick from a Grass Tree into a fire made by Mar the White Cockatoo (probably the Sulphur-crested Cockatoo, *Cacatua galerita*) (Clarke 2001; Smith 1880; Stewart c.1870–c.1883 [1977]). A fight broke out between the ancestors, causing a wildfire to spread quickly through the lank grass and brush. The burning of the country finally ended when Croom the Musk Duck (*Biziura lobata*) Ancestor clapped and shook his wings, which brought in water that settled as swamps and lakes. It was believed that after the Creation had concluded, fire could still be made by using fire-drills made from the Grass Tree flower sticks.

In south-western Victoria, there was a myth involving ngamma holes, which are typically water reserves stored in granite cavities and capped with a slab to keep it from being fouled (Magarey 1895a, 1895b). The first waterholes were said to have been created by a 'magpie lark' (Mudlark, *Grallina cyanoleuca*) and a 'gigantic crane' (probably a Brolga) during an exceptional dry season (Dawson 1881). They noticed that a Bustard (*Ardeotis australis*) was never thirsty, so flew high above to see where he had his secret water supply. Discovering that he had it hidden in a rock hole under a flat stone, they waited until he had left and flew down for a drink and a bath. They remarked:

> 'King gnakko gnal' – 'We have done him.' They flapped their wings with joy, and the water rose till it formed a lake. They then flew all over the parched country, flapping their wings and forming waterholes, which have been drinking-places ever since (Dawson 1881, p. 106).

Life among whales

The stranding of whales was a major event in Aboriginal Australia, to the extent that some coastal groups believed that among them lived people who had the power to cause strandings through sorcery or to prevent them by calling upon their totemic powers. Within the society of the Ngarrindjeri people of the Lower Murray, the whale was associated with the Latalindjera descent group, which received its name from the place, Latang, at Hindmarsh River near Victor Harbor at Encounter Bay (Berndt *et. al.* 1993; Clarke 2001). To the south of Encounter Bay, Kondilindjarung ('place of the whales') was a death place for whales, where one or two often beached themselves during the winter months (Berndt *et. al.* 1993, p. 81). It is situated on the Younghusband Peninsula on the seaside, south-east of the Murray Mouth but north-west of Noonamena on the Coorong. It was claimed that Kondilindjarung was where the people of this group and their associated beings, the whales, attempted to return before death. The Coorong ocean beach was renowned for its whale remains. Tindale recorded that:

> On the ocean beach one might be the inhabitant of another world, composed of sand, sea, spume, and giant whale bones. The scattered bones of blue whales [*Balaenoptera musculus*] lie here and there along the beach, for this is a graveyard where the great circumpolar current impinges upon the shores of Australia and casts up its burden of flotsam (Tindale 1936a, p. 18).

The Kondilindjera clan responded to a whale stranding by sending messengers with invitations to neighbouring groups to come in for a feast on whale blubber, *pailpuli*, and meat, *mami* (Clarke 2001).

Indigenous people had a close relationship with Killer Whales (Orca, *Orcinus orca*) at Twofold Bay on the south-east coast of New South Wales, over the hunting of Humpback Whales (*Megaptera novaeangliae*). Anthropologist Robert H Mathews recorded that here:

> When the natives observe a [humpback] whale, 'mururra,' near the coast, pursued by 'killers,' mananna, [killer whales] one of the old men goes and lights fires at some little distance apart along the shore, to attract the attention of the 'killers.' He then walks along from one fire to another, pretending to be lame and helpless, leaning upon a stick in each hand. This is supposed to excite the compassion of the 'killers' and induce them to chase the whale towards that part of the shore in order to give the poor old man some food. He occasionally calls out in a loud voice, ga-ai! ga-ai! ga-ai! Dyundya waggarangga yerrimaranhurdyen, meaning 'Heigh-ho! That fish upon the shore throw ye to me!' If the whale becomes helpless from the attack of the 'killers' and is washed up on the shore by the waves, some other men, who have been hidden behind scrub or rocks, make their appearance and run down and attack the animal with their weapons. A messenger is also despatched to all their friends and fellow-tribesmen in the neighbourhood, inviting them to come and participate in the feast (Mathews 1904a, pp. 252–3).

Aboriginal people cut through the layers of blubber to get to the meat for eating, and after the innards are removed 'any persons suffering from rheumatism or similar pains, go and sit within the whale's body and anoint themselves with the fat, believing that they get relief by doing so' (Mathews 1904, p. 253; see Chapter 11). Apparently, the Killer Whales ate only the tongue and lips of the Humpback Whales they had killed.

The Indigenous peoples of Twofold Bay had personal relationships with individual Killer Whales, based upon totemic relationships. It was reported that:

> The natives of Twofold Bay regard these killers as the reincarnate spirits of their own departed ancestors – and so firm is their belief that they go so far as to particularise and identify certain individual killer spirits, hence I had the greatest difficulty in obtaining the head of one of these animals as … the men had the greatest reluctance at risking all luck by killing a specimen (Brierly, 1855–51 [cited Clode 2011, p. 125]).

A European commentator observed that:

> The older race of aboriginals around Eden had strange beliefs about the Killers, holding the opinion that when they departed from this sphere of usefulness they at some later period returned as Killers [killer whales], and the one which bears the name of Cooper was so-called after an old aboriginal who had in the flesh been the king of the Kiah River tribe. Just as a small Killer, on being seen for the first time, was said to be the recently deceased child of one of the natives changed into a Killer. There is no doubt but these animals were looked upon as being supernatural, and held, consequently in great reverence. Many years ago an old whaler named Higginbotham (Flukey), in throwing the lance, by accident caused the death of a Killer. The same night the natives armed themselves with their spears, with the intention of taking his life in revenge for what they considered a great crime, and it was only owing to the intervention of some of the more powerful of the tribe that Flucky [sic. Flukey] was allowed to live (Anonymous 1912, p. 4).

This anthropomorphising of Killer Whales was said to be 'a variation of the early aboriginal legend that the dear departed "would jump up white pfeller [whitefellow, European]"' (Milne 1910, p. 3). The Indigenous perception that the first Europeans they met on the frontier of exploration being their own returned dead was commonly recorded across Australia (Clark and Cahir 2011; Clarke 2007b). Then later, when settlers had arrived, Aboriginal people incorporated the new elements they encountered, such as the worship of Jesus and Ned Kelly and presence of exotic livestock, into their own system of beliefs concerning ancestral beings and totemic animals (Kenny 2007; Maddock 1982; Swain 1991). There was significant involvement of Indigenous people across the whaling and sealing industries of Australia during the 19th century (Clarke 1994, 1998a, 2001, 2013c; Plomley and Henley 1990; Russell 2012). The full appreciation of this aspect of the European accommodation of Indigenous beliefs and traditions on the frontier requires an understanding of the totemic relationships that existed between local Aboriginal people and whales, as discussed above.

Aboriginal people were employed in the whale fishery at shore-whaling stations at Victor Harbor in South Australia (Clarke 2001), Port Fairy (Anonymous 1887a) in south-western Victoria and at Twofold Bay in south-eastern New South Wales (Anonymous 1916; Clode

2011). John (Sustie, Soostie) Wilson was one of the Ngarrindjeri men who worked at the Victor Harbor whaling station (Anonymous 1935), and he could reputedly handle a boat (Vox 1938). It is significant that the whale was not a *ngaitji* (totemic species) associated with his clan (Berndt *et al.* 1993), as if it had been he would not have been able to participate in the industry. When the explorer Thomas Mitchell reached Portland in south-western Victoria in 1836, where there was a whaling station, he noted the existence of coastal Aboriginal groups who fed on the remains of the beached whales, but 'never approach these whalers, nor had they ever shown themselves to the white people of Portland Bay but, as they have taken to eat the castaway whales, it is their custom to send up a column of smoke when a whale appears in the bay, and the fishers understand the signal' (Mitchell 1839, vol. 2, p. 243). At Eden in Twofold Bay, Europeans acknowledged that: 'It would appear that there is some kind of instinctive relationship between the aboriginal whalers and the killers [orcas], for both 'parties' arrived the same day - from opposite directions' (Anonymous 1911, p. 7).

Aboriginal people of the south-east of South Australia had a popular song about the off-shore 'whale fishery' at Rivoli Bay (Stewart c.1870–c.1883 [1977]). In the late 19th century, it was remarked that across southern and eastern Australia:

> Wherever whaling stations have been established, the natives have proved themselves to be very valuable assistants. They make the best of 'look-out' men. I have known a native sit day after day on a promontory in the keen wind or burning sun looking out to sea for a whale. They enter heartily into the sport, and make excellent 'pull-away hands' in the whale-boats (Smyth 1878, vol. 2, p. 244).

Apart from labouring jobs, Indigenous people also had higher status jobs on whaling boats. It was said in the early 19th century that: 'it was customary for the Twofold Bay Yuin to go as harpooners, or, as they put it, to go "Spearing whales"' (Howitt 1904, p. 263). It was also claimed that: 'Some of the Kurnai of the Snowy River occasionally went to Twofold Bay, to assist in whale-fishing as harpooners' (Howitt 1904, p. 353).

Discussion

In Aboriginal thought, the land could never be seen as a total wilderness. Aboriginal Biocultural Knowledge reflects the merging of the cultural and natural worlds. It is organised in such a way that it is difficult to separate the cultural beliefs in such things as spirit ancestors from the behaviours of specific animals. Such a convergence of the strands of knowledge about the past Creation and of living species in the present was not possible with a modern Western European perspective on the 'natural world'.

In spite of differing worldviews, early European settlers recognised the practical value of accessing Aboriginal knowledge and experience of the landscape. This is evident when they employed Indigenous people as trackers, guides, sealers, whale spotters and harpooners, in order to use their intimate understandings of the country and the adjacent sea (Clarke 1996, 1998a, 2008a, 2001; Reynolds 1990). European Australians have therefore appreciated the benefits of an Indigenous perspective without the need for possessing a deep understanding of it. Today, there are signs that contemporary land managers in some parts of Australia are again taking Aboriginal Biocultural Knowledge into account when formulating management plans for specific regions (Altman *et al.* 2009; Clarke 2016a, 2016c; Hill *et al.* 2012; Jones and Clarke in press; Lynch *et al.* 2015). The challenge for European Australians is how to extract useful information from knowledge systems that are based upon a perspective of the landscape that gives prominence to the past actions of the spirit ancestors, and then to apply this to contemporary situations.

Chapter 2

Terrestrial spirit beings

Philip A. Clarke

In the corpus of recorded south-eastern Australian mythology and tradition there are spirit beings who are not the 'creators' of life and country, although they were present during the Creation, but rather independent entities that influence the lives of Indigenous people. As co-residents of a landscape that they share with its human inhabitants, spirits are considered to be contributing factors that determine the nature of the local environment. Apart from the category of spirit beings there are totemic animal spirits, whose behaviour on occasion meant that they were seen as the protectors of their human kin. The study of spirit beings and totemic protector spirits offers scholars the means to delve further into the set of complex relationships between people and their country. For this reason, no account of the Aboriginal Biocultural Knowledge of south-eastern Australia would be complete without considering the perceived environmental roles of spirit beings.

Introduction

The Aboriginal cultural landscape of the past was richly imbued with meaning, not just with the topographical evidence of the former deeds of Ancestors, but by its perceived occupation by spirit beings with extraordinary powers (Clarke 1999b, 2003a, 2007b, 2016c; Howitt 1904). The records for south-eastern Australia, which will be discussed below, show that these beings are a wide-ranging group in terms of their physical form, although unified by their separateness from both living people and their ancestors. They are closely associated with particular parts of the country and often have their own sites. As spirit beings, they existed in a dimension, which can be called a spirit world, with Aboriginal people living on the periphery. Some spirit beings are associated with land, while others are closely linked to aquatic environments (see Chapter 3). Unlike the Creation ancestors, whose presence is indicated by what they left behind in terms of the order they established (see Chapter 1), the spirit beings exist contemporaneously with people. As with people, plants and animals, these beings are considered to display behaviours that are largely predictable. Aboriginal spirit beliefs contain encoded knowledge about culture and the environment.

Due in part to the success of the international Indigenous art market, a few terrestrial spirit beings in Australia have become iconic as the subject of paintings and sculptures (see Berndt *et al.* 1999; Sutton 1988). For instance, the *mimih* of western Arnhem Land are wafer-thin human-like spirit beings who are believed to live and forage along the edge of the Arnhem Land escarpment (Berndt and Berndt 1970; Taylor 1996), while the *yawk yawk* of central Arnhem are fish-tailed human-like women who are said to live in waterholes and freshwater streams (Maningrida Arts and Culture, Annandale Galleries 2007; Nicholls 2014). In coastal Queensland, the *quinkin* spirit beings are associated with the rock art of the region round Laura and are the topic of popular books (Trezise and Roughsey 1978; 1982). The *mamu* of the Western Desert are also the subject of recent Indigenous paintings, and they are generally perceived as being of humanoid form with sharp pointed teeth, and are greatly feared as shape-shifters that live underground (Eickelkamp 2004; Nicholls 2014).

An analysis of the Indigenous spirit beliefs of south-eastern Australia highlights the Aboriginal understanding of safe and dangerous places. The spirits may take the form of rocks, trees, whirlwinds, waves in the lake and various animals. Many of these spirits are said to possess certain human characteristics and, like people, they are dispersed according to a spatial plan that is determined by their complex relationships to the entire landscape. For Aboriginal people, the majority of spirit beings were either greatly feared or at least regarded as a serious nuisance. In 1847, colonist George French Angas remarked that the Aboriginal people of South Australia:

> … are in perpetual fear of malignant spirits, or bad men, who, they say, go abroad at night; and they seldom venture from the encampment after dusk, even to fetch water, without carrying a firestick in their hands, which they consider has the property of repelling these evil spirits (Angas 1847b, p. 88).

Collectively the spirit beings are a discernible class of entities that are quite separate from human ghosts, sorcerers or ancestral totemic species. In South Australia, the Lower Murray term *mooldtharp* was a recorded catchall term for 'an evil spirit', and in this context, it meant an 'evil' species of flycatcher, an earthquake or a whirlpool (Angas 1847b, pp. 96, 138). For the same area, it was said that the *muldarpe* was 'a spirit which assumes many shapes. It may come as a kangaroo, or a wombat, or a lizard' (Smith 1930, p. 349). To the Yaraldi-speaking people of Lake Alexandrina, a *melapi* or *mulapi* was considered to be a shape-changer that killed people (Berndt *et al.* 1993). From the 19th century a related term, *mull darby*, became widely used across South Australia as an Aboriginal English term for 'devil, spirit' (Foster *et al.* 2003, pp. 50–1).

Across Victoria, there are various recorded Indigenous terms for spirit beings. The missionary John Bulmer observed that among the Aboriginal people of Victoria:

> Various spirits or *mraat* existed, they had human characteristics but often took animal form. The most powerful could command the movements of the sun or moon and could exert a powerful influence on men and women. The Aborigines had no idea of worship yet they had a fear of the spirits and some of them were malignant and ever ready to injure them. Indeed some people would try to induce *brewin*, an evil spirit, to injure their enemies. The Gunai/Kurnai [Ganai] of Gippsland call a spirit being *nargan*, the Monaro; *myal* and *canoora*, the tribes of the Wimmera; *mooku garip* (Bulmer 1855–1908 [1999, pp. 48–9]).

In south-western Victoria, colonist James Dawson noted that: 'Of terrestrial spirits there are devils, wraiths, ghosts, and witches, the differences between them being somewhat indefinite' (Dawson 1881, p. 50). Here, spirit beings were represented by both genders, as it was stated that:

> There are female devils, known by the general term Gnulla gnulla gneear. Buurt kuuruuk is the name of one who takes the form of a black woman 'as tall as a gum tree.' She has for a companion the dark-coloured bandicoot [possibly the Southern Brown Bandicoot, *Isoodon obesulus*] (Dawson 1881, p. 50).

For Aboriginal people the perceived ability that spirit beings had to change form, or to shape-shift, meant that a person could never be entirely certain whether one was nearby, or even what its true identity might be.

The following sections provide detailed information on terrestrial spirit beings, focusing upon their ecological relationships to the environment. Water spirits will be discussed elsewhere (see Chapter 3). This division between land-based and water-based spirits is solely for the purposes of this book and is therefore arbitrary.

Ethnographic details

In this section, the focus is on the detailed ethnographic descriptions of the individual classes of spirit beings that were recorded in south-eastern Australia. The discussion of totemic animals as helpers to living people is included, although their intrinsic links to humans put them outside the main body of spirit beings. The role of totemic ancestors in the lives of people was discussed in Chapter 1.

Bird spirits

Birds of one kind or another featured prominently in much of the recorded south-eastern Australian mythology (Clarke 2016a, 2016c; Dawson 1881; Howitt 1904). In the pre-European worldview of Aboriginal people, many bird species were perceived as travellers between the cosmic and terrestrial landscapes, due to their ability to fly. The association of large birds of prey with carrying the spirits of the dead was a common belief across Australia (Clarke 1991, 2016c; Teichelmann and Schürmann 1840; Wilhelmi 1861). The Skyworld was perceived by Aboriginal groups of south-eastern Australia as a cosmic landscape (see Chapter 14) where many of the constellations were bird ancestors who had travelled there from Earth (Clarke 1997; Dawson 1881; Johnson 1998; Stanbridge 1857, 1861). Bird spirits remain connected to Earth, and the ethnographic descriptions of them often emphasise the appearance and behaviour of certain avian species.

In an early description of a spirit bird in the Lower Murray, colonist Richard Penney used the word, *muldaubie*, which is a generic term for bad spirit, and said:

> They believe that he appears at night when the moon is up, in the evening or just before the dawn of day, in the form of the screech-owl, although he assumes

occasionally other appearances. Those to whom he appears in dreams or who
see his form almost infallibly die (Penney 1842).

In the Lower Murray, the screeching of the 'night-owl' was believed to be a sign that
something was wrong (Smith 1930, p. 322). This is consistent with early Aboriginal beliefs
in western Victoria where it was considered an evil spirit used owls to watch over people who
had strayed from the camp during the night (Dawson 1881). The identity of the owl species
involved with these beliefs is not recorded, although more than one species may have been
involved as the call of several species, such as the Barking Owl (*Ninox connivens*) and the
Barn Owl (*Tyto alba* or *javanica*), could be described as a 'screech' (Cayley 2011).

The Southern or Bush Stone-curlew (*Burhinus grallarius*) as a spirit being was an omen
of death across southern Australia (Clarke 2016c). To the Yaraldi-speaking people of the
Lower Murray, hearing the call of this bird at night foretold the death of a close relative
(Berndt 1940b). The reason behind the Stone-curlew's call is given in the traditions of the
Creation. For instance, from Lake Boga in central northern Victoria there was a myth
recorded of a mother who lost a baby to a Wedge-tailed Eagle (*Aquila audax*) and who then
'commenced a mournful wailing, which the curlews or stone plovers ('Will') in sympathy
took up, and have continued ever since' (Stone 1911, p. 461). The call of the Southern Stone-
curlew is like 'ker-loo' (Slater 1970).

Aboriginal people in other parts of Australia had similar beliefs concerning Stone-
curlews (Berndt 1940b; Gosford 2010), as: 'in much of Aboriginal mythology, Curlew is a
taciturn and gloomy fellow. As a bird, his cries echo mournfully and are often taken as
presaging death' (Berndt and Berndt 1989, pp. 196–7). Europeans are also frightened by the
call of this bird, as it was reported in the early 20th century in a South Australian country
town that:

> … there were persistent reports of a 'ghost' which sat on a road bridge and
> prevented travellers from crossing at night – with 'huge outspread wings' and
> 'uttering a wailing cry.' The tale was investigated and the 'ghost' proved to be a
> Stone Plover [stone curlew] (Condon 1941, pp. 6–7).

In south-western Victoria, there is also another record of Europeans treating an encounter
with the Curlew as an ill omen (Anonymous 1885b), and in this area, it was commonly
known as the 'ghost bird' (B.E.C. 1944, p. 1).

In recent times, the bird spirit most commonly spoken about by Ngarrindjeri people in
the Lower Murray has been the '*mingka*-bird' (Clarke 1999b, 2014a, 2016c). Ngarrindjeri
people described the *mingka* as essentially a night-time spirit. According to one report, there
are two forms of this spirit, a northern and a southern type, both of which live in caves. The
mingka was described as a 'sphinx-like bird' that has a human head when it is carrying the
spirit of a recently deceased person (Clarke 1999b, p. 160). It is said to come to Aboriginal
homes at night, when it makes a noise like a baby crying. Some people have said that the
mingka looked like a Boobook Owl (*Ninox novaeseelandiae*), but could sound like a wild Red
Fox (*Vulpes vulpes*), while others described the call as a 'shrill whistle' (Clarke 1999b, p. 160).

Boobook owl, one of the predatory bird species that is associated with the *mingka* death spirit. PA Clarke, Adelaide Plains, South Australia, 2013.

As with many other spirit beings, its assumed presence is a commonly used threat by adults to make their children behave.

Belief in the *mingka* spirit being extended beyond the Lower Murray area. The Aboriginal name for the *mingka* (*minkar*) was said to be a Potaruwutj language term from the south-east of South Australia, and to be the equivalent of *merambi* from the Tangani language of the Coorong (Tindale 1931–34). The Potaruwutj believed that the *mingka* was a 'being, sinister, who may assume form of totem animal' and 'is an evil being, warns about death or trouble' (Tindale [cited Clarke 1999b, p. 160]). The spirit being was recorded as being able to assume the form of various *ngaitji* (totemic 'friends'), such as an eagle, dog or hawk (Tindale 1931–34). In these forms, the *mingka* carried the spirits of sinister people, connected to their owners by *nunggi* or *kortui* described as 'like a spider web'. Men could kill these beings and the sorcerer owners of the attending spirits with a 'sacred club'. Ngarrindjeri people said that the *mingka* was connected to the *kowuk* bird, which they described as a Tawny Frogmouth (*Podargus strigoides*) (Clarke 1999b, 2014a). Berndt suggested that the *mingka* in the Lower Lakes was an owl (Berndt 1940b).

Bird species commonly seen during the day may also attract the attention of Aboriginal people. In the Lower Murray today, a commonly held belief in omens concerns the *ritjaruki* (Willie Wagtail, *Rhipidura leucophrys*) (Clarke 2014a, 2016c). When this bird is observed persistently making strange erratic movements near a person's house, it is perceived as a message that someone had died. Tapping on a window was taken as a particularly bad sign. A variation of this belief is that when a *ritjaruki* is observed with a rusty colour on its beak, this means someone will die. Since its omens are invariably considered to be unwanted, the bird is often chased away. These traditions have some continuity with the past, as in the Lower Murray during the 1840s George French Angas wrote that:

An elegant species of flycatcher [probably the willie wagtail], of a black colour, which continually hovers about in search of insects, performing all manner of graceful manoeuvres in the air, is regarded by them as an evil spirit, and is called mooldtharp, or devil. Whenever they see it, they pelt it with sticks and stones, though they are afraid to touch or destroy it (Angas 1847b, p. 96).

The belief in the ability of the Willie Wagtail spirit to summon bad news appears to have been widespread across south-eastern Australia. In the experience of the current author, it is still widely believed by many Indigenous peoples. For instance, in the late 1980s it was recorded in western Victoria that an Aboriginal person said that they considered the Willie Wagtail to be a bad omen, and that she knew of two occasions when people had reportedly died soon after they had seen this bird tapping on their window (Clarke 1999b). Outside of the south-eastern Australia region, the Willie Wagtail has a similar status, such as on Yorke Peninsula where it is considered to be a 'message-carrier' (Smith 1930, p. 342). It is likely that the attention the Willie Wagtail attracts from its highly energetic movement across low and open areas, such as around homes and tracks, adds to the likelihood of this species being seen and taken as an omen.

The ability of birds to fly and to be relatively easily observed probably made them more suitable than other land-based animals as omens. The specific behaviours of certain species, such as night-time activity and the ability to fly high, appear to have been important characteristics in their designation as carriers of human souls or messages from the spirit world. Peculiar behaviour on specific occasions could also be interpreted as an indication that a spirit had taken the shape of a bird, or as an intended sign from the spirit world.

In the Riverland of South Australia, the *ngout-ngout* was seen as an individual female spirit associated with birds (Lindsay [in Education Department of South Australia 1991]). She was reportedly once a woman who was expelled from her local group for breaking custom. For revenge, the *ngout-ngout* tricked children into becoming lost, and this she achieved by using a trail of flowers to distract them. *Ngout-ngout* could turn into a bird. This story was told to children with the warning that they never wander off. Knowledge of bird behaviour was important, as all birds observed acting strangely could be seen as potential spirit beings.

Spirit men

The oral traditions of Aboriginal communities across south-eastern Australia are rich in narratives concerning humanoid creatures that lived in their own separate communities. Contemporary Aboriginal people sometimes relate them to similar beings they have heard about in stories, such as the leprechaun of Ireland and the sasquatch of North America (Clarke 2007b). The shared characteristics of these spirit men in south-eastern Australia, as discussed below, are that they are generally described as 'coloured' (not black), often of different dimensions to people, usually dangerous or at least mischievous, and generally associated with a particular type of landscape (Clarke 1999b). In Aboriginal thought, the spirit men exist on a continuum between the entities of the spirit world and living entities, such as people and animals.

Europeans among the first wave of settlers were convinced that there were human-like animals in the dense forests in remote parts of south-eastern Australia. Colonist George Haydon gave an account of a 'devil' called *bundyil-carno*, and said that: 'the cry, which much resembled the scream of a woman, was so very different to any other I had been accustomed to hear in the bush, that I have no doubt it proceeded from some animal unknown to the colonists on the plains' (Haydon 1846, p. 67). In form, the animal was said to be 'something very similar to an ouran-outang [Orangutan, *Pongo pygmaeus* and *P. abellii*]', and it was believed to live in the mountain ranges behind Westernport, south-east of Melbourne (Haydon 1846, p. 65). Haydon said that:

> An account of this animal was given me by Worrouge-tolon, a native of the Woeworong [Woiwurrung] tribe, in nearly the following words – "He is as big as a man and shaped like him in every respect, and is covered with stiff bristly hair, excepting about the face, which is like an old man's, full of wrinkles - he has long toes and fingers, and piles up stones to protect him from the wind or rain, and usually walks about with a stick, and climbs trees with great facility; the whole of his body is hard and sinewy, like wood to the touch." Worrongy also told me, that many years since, some of these creatures attacked a camp of natives in the mountains and carried away some women and children, since which period they have had a great dread of moving about there after sunset. The only person of his tribe now alive, who had killed one, he informed me, was Carbora, the great doctor, who had succeeded in striking one in the eye with his tomahawk, on no other part of its body was he able to make the least impression, (Haydon 1846, p. 66).

There is a related account of a human-like animal that preyed on people, which also emphasises its contemporaneous presence with people. The scholar Robert Brough Smyth stated that:

> There is a range with a well-marked culminating point lying to the northeast of Western Port [in Victoria], which, the Aborigines say, is inhabited by an animal resembling in form a human being, but his body is hard like stone. The mountain is called Narn, and the strange animal is named Wi-won-der-rer. Formerly this animal used to kill many blacks. So many indeed were killed by Wi-won-der-rer that at last it became necessary to consider in what way those remaining might be preserved. A council of aged and wise men was held, and much debate ensued, and many suggestions were made. Finally it was agreed that the most cunning doctor, with other learned doctors and priests, should visit Narn and ascertain the condition of Wi-won-der-rer, and, if possible, kill him and his people (of whom there were a good many). The wise men explored the mountain ranges very carefully. Armed with spears, stone hatchets, and waddies, they sought to find and slay the strange creatures with bodies like stones. And they found them at length; but their weapons, when they assaulted them, made no impression on them. It was reported, however, that these

creatures were vulnerable in the eyes and the nostrils. One doctor said he had thrust his spear into the eye of a Wi-won-der-rer, and had killed him, and another said that he had killed one by thrusting his spear into his nostril (Smyth 1878, vol. 1, p. 455).

According to Smyth, after settlement a European man was lost in those ranges, although due to the threat of the Wi-won-der-rer there were no local Aboriginal people who could be induced to enter this area to find him.

By the early 20th century, the stated presence of such human-like beings was no longer explained away as the existence of species undiscovered by science, but rather they were classified as Indigenous spirits. For instance, anthropologist Robert Mathews recorded that in south-eastern New South Wales the Dharawal (Thurrawal) people believed in Mumuga as a:

> … fabled monster … possessing great strength and residing in caves in mountainous country. He has very short arms and legs, with hair all over his body but none on his head. He cannot run very fast, but when he is pursuing a blackfellow he evacuates all the time as he runs, and the abominable smell of the ordure overcomes the individual, so that he is easily captured. If the person who is attacked has a fire stick in his hand, the stink of Mumuga has no effect upon him (Mathews 1904a, p. 345).

More recently, European scholars have once again given some credence to such Australian beliefs, equating the more human-like 'little folk' and 'little people' spirits with either a 'pygmoid race' of early Indigenous colonists (Tindale no date) or to a now extinct Hominid species like the so-called 'hobbit' of Flores and the Denisovans (see Clarke 2007b; Nicholls 2014). Therefore, Indigenous people are not alone when it comes to engaging with historical syncretism.

The existence of the spirit men demonstrates the high degree to which both pre-European and post-European landscapes were humanised. In the ethnographic records for south-eastern Australia, the spirit men were described as being human-like in many ways, both physical and in terms of their behaviour (Berndt *et al.* 1993; Smith 1930), although Aboriginal people in more recent times always described them in terms that highlight their distinctiveness from people, such as by their strange colouring (Clarke 1999b). In the Lower Murray and south-east of South Australia regions the spirit men were variously described as being red, yellow and green. Another variation is 'little white men' with spiky ears, and while some were beardless in other accounts they had excessive body hair. Many of the modern accounts of humanoid spirits concern entities that are smaller than people (Clarke 1999b). In one of the early ethnographic accounts of such beings from the Lower Murray the *raitchari* were described as 'pygmy men who sometimes act as guides to hot men, sometimes lead them astray. They used to live in the scrub' (Harvey 1939). The 'hot men' are presumably people with some form of ceremonial status.

There is the account of 'Gnutcha, which means little person', from western Victoria (Cameron-Bonney 1990, p. 19). The '*natja* men' from the Tatiara district of the south-east of

South Australia were described in the 1980s as 'red hairy men', also said to look 'like monkeys or orang-utans' (Clarke 1999b, p. 154). In popular writings, the Yowie is an ape-like monster that inhabits the forests of eastern Australia (Healy and Cropper 2006), although its name was probably derived from northern New South Wales where Yuwaalaraay word, *yuwi*, meant 'dream spirit' (Dixon *et al*. 1992, p. 111). In the earlier literature, 'yowie' was described as a 'soul' (J.D. 1931).

In many Aboriginal communities, adults would take advantage of the perceived existence of a range of spirits in order to frighten children into obedience (Clarke 1999b). In this context, there is an overlap between records of 'ghosts' and 'spirit men'. Anthropologist John Mathew, for example, stated that:

> In Gippsland the chief of the mraad, or ghosts, is Nargon. He is conceived as a hairy being who lives in the mountains. He carries naughty children in a bag. A cave at the upper end of the Toorool Arm is known as Nargon's Cave, and therefore, presumably, it was one of his haunts (Mathew 1925, p. 66).

By another account, Nargon was a group of beings, with a missionary having recalled that 'The blacks in telling me these stories were very careful to assure me that these *nargan* were not *mraat* (spirits) but real flesh & blood' (Bulmer 1855–1908 [1999, p. 50]). In 1875, anthropologist Alfred Howitt visited a cave known as the 'Den of Nargun [Nargon]' along Deadcock Creek in Gippsland, and he also knew of another cave near Lake Tyers said to be inhabited by Nargun (Clark 2007a, 2007b; Seddon 1989). There is also an unlocalised story recorded about a 'queer little red man', called 'Yara-ma-yha-who', that was reportedly told to children as a threat against misbehaving (Smith 1930, p. 343).

In the Lower Murray in the 1980s, Aboriginal children were told about the *but-but* spirit being, who had only one arm and one leg (Clarke 1999b). He was dangerous but stupid, and not associated with any particular place as he moved randomly across the country. When seen, people would lie down as if dead and place on their eyes and mouth the maggots that they kept in bags for the purpose. This fooled the *but-but* into thinking they were dead, and he would pass them by. Such analogous beliefs exist in other cultures, such as the bogeyman story of Western Europe that is also told to frighten children (Widdowson 1971).

There were parts of the landscape that Aboriginal people associated with spirit beings and therefore took extra care when passing through. It has been suggested that Monkey Creek near Darriman in eastern Victoria was named after the ape-like man known locally as a 'yowie' (Healy and Cropper 2006), although it is also possible that in this case the 'Monkey' was a koala sometimes recorded as the 'monkey-bear' (Ramson 1988, p. 400). In the Lower Murray area, the 'little red men' were said to have been seen at specific places, such as Pelican Point, Teringie and Poltalloch. Some people call these particular beings '*kintji* men' or '*kindi* men'. In the ethnographic record, they are sometimes described as 'imps' or 'dwarf beings', and it was suggested that perhaps Fairy Penguins (*Eudyptula minor*) were the source of such beliefs (Tindale [cited Clarke 1999, p. 153]). The *kintji* were associated with high places, presumably due to their connections with the Skyworld (Clarke 1999b). A small point on the mainland side of the Coorong just south of Noonamena, known by local Aboriginal people as the Kintji Cliffs, was regarded as the home of *kintji* spirits. Another site, Kindjunga Hill near the Coorong south-east of Pelican Point, was recorded as the home of *kindja* spirits who lived in its caves and amongst its rocky outcrops (Berndt *et al.* 1993). Other places related to the *kintji* were Kentjingatung and Kintjanga, which are sandy rises on southern Campbell Park Station near Lake Albert in the Lower Murray (Tindale [cited Clarke 1999b]).

In the Lower Murray and south-east of South Australia a common element to many accounts of spirit men is that only one or two of them were seen together, usually in swamps and lagoons where Aboriginal duck hunting and 'swan egging' activities occurred (Clarke 1999b). A Ngarrindjeri man reported seeing 'red men' at Pelican Point earlier in the 19th century (Clarke 2016c). He was a duck hunter who had disregarded warnings about these spirit men being there, and after shooting some ducks he started to search for them in the spot where they had fallen. It was then when he noticed that a 'red man' was picking up the ducks. The Aboriginal man grabbed his gun and fled. According to several Aboriginal sources, Lower Murray people who hunted in areas that were recognised as being inhabited by 'red men' used to leave one or two ducks behind for them. Accounts of the spirit men are often associated with hunting. For instance, in ~1932 an Aboriginal man at Lake Condah reportedly shot a rabbit in an area known as the Stony Rises, and then saw a 'small hairy man-like creature' seize the carcass and run off with it (Healy and Cropper 2006, p. 219).

In the Lower Murray, a sighting of a 'little red man' was reported in the 1950s alongside a swamp near the One Mile fringe camp situated on the outskirts of Meningie (Clarke 1999b). The spirit surprised an Aboriginal man and his young son walking along a path. The man was paralysed but the boy escaped to raise the alarm. Several men from the camp came to the aid of the father who was still on his back. As with the other spirit men of the Lower Murray, the little red men were often seen in proximity to water, but believed to dwell in higher places. A related narrative from the south-east of South Australia involved a Moandik man, Alf Watson, who was apparently duck hunting with his rifle when he saw a *kintji* man among the *winggi* (Sagent Sedge, *Lepidosperma gladiatum*) (Tindale [cited Clarke 1999b]). This spirit caused him to freeze and he fell on his back. The *kintji* man came up to Alf and sat on him, feeling his face with interest. This was because this type of spirit man was bearded, whereas Alf Watson was clean-shaven. Alf claimed afterwards that he could feel the *kintji* man's cold bottom on his chest through his flannel shirt.

In Aboriginal Australia, it was believed that dangerous humans in spirit form interacted with birds, and such beliefs have continued into modern times (Clarke 2007b). For instance, during the early 1990s an Ngarrindjeri man in the Lower Murray stated that Aboriginal 'sacred objects' were buried in a field because he had seen a pair of *multhari* (Australian Magpie, *Cracticus tibicen*) playing there (Clarke 2016c, p. 752). At this time, it was local folklore in the Aboriginal community that: 'Crows [Australian Ravens, *Corvus coronoides*] always knew when *kuratji* ['feather foots', sorcerers] were sneaking about, as they would cry, "Pick his eye out. Pick his eye out"' (Clarke 2016c, p. 753). In the Lower Murray, and in many other parts of Australia, an unexplained loose feather found in a person's home was sometimes taken as evidence of the presence of a *kuratji* being present. Across Australia it was widely believed that the 'magic slippers' of sorcerers or 'kadaitja men' were made from feathers and human hair string (Akerman 2005; Elkin 1977; Pounder 1985). A bad smell emanating from hidden material that was said to be *ngruwi* ('dead body fat') was another sign (Clarke 1994). While the community discussion in the Lower Murray region over the cause of a spate of *kuratji* sightings might lead to several conclusions, there were often suspicions voiced concerning whether the actions of certain community members involved with northern 'tribal people' had brought the 'feather foots' down upon them (Clarke 2007b).

Totemic protectors

In many parts of Australia, it was an Aboriginal tradition that certain special powers could be gained through the totemic animal, plant or object from which the clan derived their descent, and that these could be used for the benefit of their kin (Berndt and Berndt 1999; Clarke 2016c; Elkin 1977; Howitt 1904; Petri 1952 [2014]). Chief among the powers that specially trained people were believed to possess was the ability for an individual to either shape-shift or project their spirit into the form of their totem. For instance, in south-west Western Australia, anthropologist Daisy Bates reported that:

> The belief is also held in the southern districts that the natives were once animals or birds. For instance the Nagarnooks are called Wejuk (emus), and are even supposed at the present time to be able to transform themselves from men to emus at will (Bates 1906).

Everyone with a totemic familiar ('friend' or 'protector') defined through their descent system could potentially derive spiritual assistance from them, although only spiritually powerful people could take on the character of one of their spirit familiars. For instance, there is a record of a 'medicine man' in Dampier Land of northern Western Australia who had three *rai* ('alter egos') which were 'namely the wild turkey, a fish species and his own spirit-double' (Elkin 1933 [cited Petri 1952 [2014, p. 85]).

According to museum researcher Norman Tindale, in the Lower Murray of South Australia it was a tradition that a spiritually powerful man could transform himself into his *ngaitji* or totemic species, and, 'He will sometimes remark that he has been about in the ngatji [*Ngaitji*] form and you will recall seeing the strange bird' (Tindale 1938–56 [cited Clarke 2016c, p. 757]). The wife or female relative of a man who as a spirit was travelling as his *ngaitji* along a *ngildi*, a 'flying spider web', would watch over his inert body and as a sign:

> She will see perhaps an owl on a branch of a tree nearby. Then the bird disappears over the hill and she will know the spirit is returning and remark 'He will come soon'. Then the man stirs; he breaks the *ngildi* and sitting up describes his journey and the events he has witnessed (Tindale 1938–56 [cited Clarke 2016c, p. 757]).

From the Lower Murray in the early 20th century, there was an ethnographic record of a man projecting his spirit as a bird. This account involved a group of men who were at Marunggung near Wellington collecting wood grubs for fishing bait when a large black eagle landed in a tree nearby (Berndt *et al.* 1993). Later, a man named Manuel Karpany claimed to have been that eagle, and he was believed since he could list the people who were present. Karpany was warned that he might have been shot, but he responded by saying that: 'they could not have shot him; he himself had remained back in the camp and the eagle was his other self, his ngatji [*ngaitji*], a projection of his real self' (Berndt *et al.* 1993, p. 249).

It was part of the 20th century Aboriginal folklore that Joe Lock and his brother Chalker, who were Aboriginal men originally from the Mid North but living at places in the Lower Murray and south-east of South Australia, were able to project their spirits into the form of their totem, which was a species of grey 'eaglehawk' not found in the region (Berndt *et al.* 1993; Clarke 1999b). Joe Lock stayed at the Blackford Aboriginal Reserve in the south-east of South Australia, and he reportedly left on occasion by turning into an eagle that flew away. To do this, Joe would lie down on a blanket out in the open, and then enter a trance. No one was allowed to move him, as in bird form he may not have been able to relocate his body. Once when soaring high as an eagle, Joe had seen several people in a horse and buggy travelling towards Blackford, and later, when back in his human body, he was able to accurately describe the travellers before they had arrived.

Chalker was said to have been involved in a similar transformation event, which reportedly took place when a party was out gathering fruit on the Coorong Hummocks and a woman saw a large bird that resembled a Cape Barren Goose (*Cereopsis novaehollandiae*) (Berndt *et al.* 1993). Chalker later claimed that he had been that bird and rather than looking for potential victims to attack, he was just out there protecting people from other 'flying sorcerers'. It was claimed that: 'doctors such as Chalker wore the feathers like a skin and were able to turn themselves into birds, even birds which were not their own *ngatji* [*ngaitji*]' (Berndt *et al.* 1993, p. 250). There are other recorded examples of transformations, and these have become part of contemporary Aboriginal community folklore. For instance, Ngarrindjeri woman Rhonda Agius told a story in a schoolbook about Old Jolok who could turn himself into her *ngaitji*, the *nuri* (Pelican, *Pelecanus conspicillatus*) (DETE 2001).

The living animals associated with each *ngaitji* could also act as messengers on their own (Berndt *et al.* 1993; Clarke 2016c). The Yaraldi-speaking people of the Lower Murray had a belief concerning the special powers of the *kiling-kildi* ('black magpie', Grey Currawong, *Strepera versicolor*).

> If Jarildekald [Yaraldi] men are hunting singly and hear a kelinkeli [*kiling-kildi*] bird (*Strepera*) calling they at once return to their companions for the cry of this bird is a warning of the presence of visible or invisible spirit enemies. The miwi (mind) of a man (believed to be in his belly) cannot resist this warning, so strong is their fear of the sound (Tindale 1930–52 [cited Clarke 2016c, p. 757]).

Tangani-speaking clans of the Coorong believed that: 'Kilindi – bird was [a] messenger going ahead and warning people of strange tribes, before he came a bird' (Tindale 1934–37 [cited Clarke 2016c, p. 757]). They described the *kelindi* (Grey Currawong) as a 'crow with white tail; makes lot noise' and a *ngaitji* species, although the associated clan was not recorded (Tindale 1931–34 [cited Clarke 2016c, p. 757]). An ornithologist described the Grey Currawong as having a loud metallic 'clink' or ringing call (Slater 1974, p. 282).

There is a mythological basis to the Grey Currawong being an omen species. A myth based in the southern Fleurieu Peninsula area referred to a pair of hills, one of which is Mount Hayfield and represents Wati-eri the Grey Currawong ancestor, and the other is Mount Robinson, which is a monument to his companion Lepuldalingul the Marsupial Possum ancestor (Berndt *et al.* 1993). During the Creation period, these ancestors were involved in the rescuing of a boy stolen by a *melapi* (*muldapi*) spirit being and taken up a large gum tree. A song was recorded and is based upon what appears to be the same mythology, giving the name of 'the two hills' as Watirangenggl (Watarangalang, Wataraberingg) and stated that Watiari (Watiriorn) and Lepidawi (Lepuldawi) were 'ancestral men of the forest' identified as the 'Swallow' and 'Ring-tailed Mouse' ancestors respectively (Tindale c.1931–c.1991 [cited Clarke 2016c, p. 758]; 1941, pp. 242–3). These ancestors used their calls to lure a 'sandhill-man' deep into the forests of southern Fleurieu Peninsula, where he became lost. Because of the type of vegetation in the country involved in the record above, it is likely that Watiari actually referred to *watiyeri*, which in the Ramindjeri dialect of the Lower Murray is the name for the Grey Currawong ('black magpie', *Strepera versicolor*), which seems more probable than to a species of swallow as recorded here (Clarke 2016c).

Aboriginal beliefs in spirit beings and totemic spirits had an impact on hunting practices. Riverland Aboriginal man, Barney Lindsay, is recorded telling a story about the female spirit Witj-witj (Wich Wich) in the Murray River area that alerts game animals that hunters are about by imitating the call of the Masked Lapwing (Spur-winged Plover, *Vanellus miles*) (Education Department of South Australia 1991). Such knowledge that is based on spirit lore assists the hunter. In the Lower Murray, the *wate-eri-on* (*kiling-kildi*, 'black magpie', Grey Currawong) was treated as a spirit, and 'If a hunter killed one, the other *wate-eri-on* would have their revenge and make it very difficult for him to catch other birds', of any species (Berndt *et al.* 1993, p. 124). It was said that: 'This bird warned a kangaroo when it was being tracked down by a hunter, by calling out *kiling-kildi*, which was also the name of this bird on the Lower Murray' (Berndt *et al.* 1993, p. 234).

In the narrative of the Tjirbruki myth of the Lower Murray that extends to the Adelaide Plains, the Kelindi (Kelendi, Grey Currawong) ancestor was: 'a great messenger who travelled around the country singing songs and telling people of the coming meetings for initiation of their young men' (Tindale 1987a, p. 9). Tjirbruki killed a *kelindi* bird and used its feathers and flesh to transform himself into a bird, the Glossy Ibis (*Plegadis falcinellus*). The Grey Currawong was described as an omen for living people, such as, 'Once when Karloan and Frank Blackmoor [Blackmore] were at Waitpinga [Beach] even in relatively recent times and were prospecting for metals they desisted and returned home abandoning their pick, as soon as they heard a jay [grey currawong] screaming at them' (Tindale 1938–56 [cited Clarke 2016c, p. 753]). The Currawong is significant to other Aboriginal groups in south-eastern Australia, such as the 'Wa-noo-rong or Yarra tribe' of Victoria who believed that at the conclusion of the Creation period: '*Ballen-ballen* (the Jay [Currawong]), who at that time was a man, had a great many bags full of wind, and being angry, he one day opened the bags, and made such a great wind that BUND-JEL [Supreme Creator] and nearly all his family were carried up into the heavens' (Smyth 1878, vol. 1, p. 427).

Pelicans are totemic species and therefore 'friends' to their clan members. PA Clarke, Coorong, South Australia, 2016.

For the Lower Murray, there are records of several stories told during the 1980s of observed bird *ngaitji* behaviour having been used to notify family members of a death. For example, an elderly woman living on the Coorong claimed that her mother's *ngaitji*, which was a Swallow (probably *Hirundo neoxena*), would fly up close to them to 'sob like a child' if someone close was dying, and that this had happened when her cousin's baby had died (Clarke 2016c, p. 754). Another account of a *ngaitji* portent involved an incident during the 1950s when an Aboriginal hunter noticed the unusual behaviour of a Pelican, *nuri* (Clarke 2016c). He had observed this bird flying overhead near Narrung, and he raised his gun to shoot it. Although it was the man's *ngaitji* and its flesh was considered bad eating, the feathers would be used for ornament making. The Pelican acted strangely, so was not shot. It flew several times over the man's head and then out over Point McLeay and across the lake towards Adelaide. This was interpreted as a message that the hunter's brother, who was known to be in an Adelaide hospital, had just died. The narrative concludes with the later surprise of the Point McLeay Mission superintendent – who had come down to the house of the deceased with the sad news received via phone – at being told the family already knew. In earlier times, totemic species and objects would have been considered to have the power to warn people of impending danger or even death.

That birds (with the exception of Emus) have the ability to fly grants them a characteristic that appears to make them more acceptable to act as messengers, than is the case for most other land-based animals. While the specific behaviour of certain species, such as the Willie Wagtail (a spirit being in its own right), appears to be a major factor in their designation as message carriers, for other birds like the Australian Magpie and Pelican who are totemic species, peculiar behaviour on single occasions may be interpreted as a sign of impending disaster.

Spirit beings and Aboriginal Biocultural Knowledge

The Aboriginal cultural landscape of the past was imbued with meaning, not just with the topographical reminders of the actions of Creation ancestors, but by the perceived occupation of spirits. Across south-eastern Australia there are beliefs in certain spirits, often seen as birds, which serve as portents and therefore provide a means by which Aboriginal people could prepare themselves for the calamities ahead of them. Fundamental to many of these beliefs is the Aboriginal notion of a person's soul surviving death and being able to be summoned and taken away by spirit carriers to the Skyworld.

Aboriginal people believed that many of the spirit beings are able to take on the form of birds, because they can fly and therefore have privileged access to parts of the landscape that living humans would find hard to reach. The characteristic behaviours and appearances of particular avian species add to the likelihood of their assigned identity being that of a spirit. For instance, nocturnally active owls have large forward-facing eyes that are reminiscent of those of humans, while the mournful call of the Stone-curlew makes this bird a spirit. Aboriginal Biocultural Knowledge comprises detailed observations, such as species' behaviour and ecology, and these are cross-referenced with beliefs concerning spirit beings and ancestors in accordance with the Indigenous worldview.

The beliefs in spirit beings contain encoded knowledge about the landscape, particularly about dangerous places and the movements of human spirits after death. Concerning these 'monstrous beings' an anthropologist has stated that:

> All of these figures materialise fear; bringing it to the surface. At the psychological level, the stories about these entities are a means of coping with terror (Nicholls 2014).

In the 20th century Aboriginal culture in 'settled' Australia has often been described in sociological rather than cultural terms, although there were nevertheless elements of the pre-European worldview that persisted here (Barwick 1971b; Clarke 1994). In many Indigenous communities of south-eastern Australia this situation continues today. Beliefs and traditions concerning spirits, albeit in greatly modified form, continue to serve as a means to explain the rural landscape related to how they were used in pre-European times (Clarke 1999b, 2007b). Much of the Creation folklore formerly known by Aboriginal people has been lost or significantly altered since European settlement in the early 19th century, but traditions concerning spirit beings have remained because they are part of a body of knowledge that is not only still relevant but is still being augmented. The Indigenous practice of seeking information from signs in the environment has persisted.

As with other parts of Aboriginal Australia, most of the spirit beings in south-eastern Australia are fearful beings, although Aboriginal people still nonetheless derived much enjoyment in the belief of them as stories to tell round the campfire. The knowledge of spirits is important as signifiers that local Aboriginal people employ to express their regional identity and for when arguing for a greater role in environmental management (Clarke 2016c). In the case of the Ngarrindjeri people, the close relationship they have with their country is supported by such beliefs as those concerning their *ngaitji* or totemic ancestors and that of the *mingka* bird, which they believe targets them alone when foraging for souls.

Chapter 3
Water spirit beings

Philip A. Clarke

South-eastern Australia is in the heart land of a large region where from the early nineteenth century Aboriginal people gave British settlers accounts of fabulous amphibious beings that dwelt in the relatively quiet inland waters and which were extremely dangerous to all people. Generally called bunyips, they have become an important part of Australian folklore. So poignant were these Indigenous descriptions, that European scholars initially believed that large prehistoric beasts still remained undiscovered in Australia, and that eventually the bunyip would be found to exist.

Introduction

Throughout south-eastern Australia there are many accounts in the popular literature of Aboriginal water spirits that are generally called 'bunyips' (Barrett 1946; Beatty 1969; Cameron-Bonney 1990; Gunn 1847; Holden and Holden 2001; Kerr 2004; Massola 1957; Mulvaney 1994; Ramson 1988; Smith 1996; Wignell 1983). Some have been described as being animal-like, and others as predominantly aquatic humanoid creatures. As with the terrestrial spirits (see Chapter 2), Indigenous accounts of them reflect a territoriality. The consistent theme of most Aboriginal bunyip descriptions is the threat they pose to children who have strayed too close to the water's edge. Most reports have generally reinforced it as a symbol of the dangerous nature of the inland waters.

Bunyips

The Australian English term, bunyip (or sometimes 'bunyup'), is believed to have been derived from the Indigenous word, *banib*, from the Wemba Wemba language of western Victoria (Ramson 1988; Dixon *et al.* 1992). In the Gippsland area, bunyips were known as *tanutbun* (Locke 1866; Smyth 1878, vol. 2). Along the lower reaches of Murray River, the Ngarrindjeri people still refer to the spirit being as the *mulyawonk*, which in the past was recorded in many different ways – *moolgewanke, moolgiewankie, mul-ja-wonke, muldewangk, mulgewanke, mulgewongk* and *multyewanki* (Taplin 1874 [1879]; Rankine [cited Isaacs 1980]; Cameron-Bonney 1990; Education Department of South Australia 1991; Berndt *et al.* 1993; Clarke 1999b). While many of the earlier descriptions of bunyips were from Victorian and South Australian waterways, these spirits were also described in other parts of Australia. For instance, at Hunter River in eastern New South Wales Aboriginal people referred to equivalent water spirits as the *wawee*, or variously as *wauwai, whowie* and *wowee* (Smith

1930; Mulvaney 1994; Holden and Holden 2001). At the Murrumbidgee River in southern central New South Wales, the bunyip was referred to *katenpai*, sometimes written as *kajanprati*, *kayan-prati*, *kiampratee*, *kine pratia* and *kinepraty* (Gunn 1847; Anonymous 1847c; 1857; Barrett 1946; Holden and Holden 2001).

In south-eastern Australia, there are bunyips mentioned in the corpus of Creation mythology. For instance, Bunjil the Wedge-tailed Eagle Ancestor was attacked by a bunyip (Mathew 1925). Bunjil's mother-in-law had orchestrated this on the Little Wimmera River, as revenge for him dropping her. This occurred when he, his two sons and mother-in-law, had all jumped from the edge of a high cliff in the Grampians. The bunyip hid in a pile of vegetation, which looked like the nest of a Kangaroo-rat (probably the Eastern Bettong, *Bettongia gaimardi*), which the mother-in-law had dragged from the river. When attacked, Bunjil was ripped into many pieces, although other bird ancestors using rainbows as nets were eventually able to gather up the fragments for spreading out on a possum skin rug so that he could be put back together again. It was said that on the Little Wimmera a 'big tree, where the river cannot be bottomed, still marks the spot' (Mathew 1925, p. 66). To memorialise this event, the figures of Bunjil and his two dogs were reportedly painted in a cave near the Grampians. This cave appears to be that which is still known today as 'Bunjils shelter' (Wettenhall 1999, p. 55). As with other ancestors, Bunjil (Buunjill) went up into the Skyworld at the conclusion of the Creation period, where he was seen in the night sky as the bright star Fomalhaut (Dawson 1881).

Other ancestors also had violent encounters with the bunyip. The Booroung (Boorong) people at Lake Tyrrell in north-western Victoria had a tradition concerning 'Totyarguil (Aquilla), the son of Neilloan [Lyra the Mallee Fowl], and who, while bathing was killed by the Bunyips, his remains were afterwards rescued by his uncle Collenbitchick ['Double Star in the head of Capricornus', Ant species]' (Stanbridge 1857, p. 139). In a Lower Murray myth, the bunyip was more of a nuisance to the ancestors, with the creator Ngurunderi said to have had his nets torn by the *mulyawonk* that lived near Wellington in the dangerous waters where the Murray River exits into Lake Alexandrina (Tindale and Pretty 1980).

Bunyips shared some of the same powers possessed by the ancestors. In western Victoria, it is recorded that the bunyip was also believed to have access to the Skyworld, as colonist James Dawson reported that:

> The coal sack of the ancient mariners – that dark space in the milky way near the constellation of the Southern Cross – is called 'torong,' a fabulous animal, said to live in waterholes and lakes, known by the name of bunyip, and so like a horse that the natives on first seeing a horse took it for a bunyip, and would not venture near it (Dawson 1881, p. 99).

It was also stated elsewhere that:

> The [Aboriginal] doctors alone, says the Rev. Mr. Hageuauer [sic. Hagenauer, Moravian missionary at Ramahyuck], are able to point out where the Bun-yip has his dwelling. Sometimes they indicate a deep waterhole as the place of his

abode, and sometimes a swamp surrounded by scrub and reeds (Smyth 1878, vol. 1, p. 437).

From Indigenous accounts, the bunyips also possessed powers of sorcery, which added greatly to the fear of them. In the Lower Murray of South Australia, the missionary George Taplin recorded that:

> The blacks say that the Moolgewanke [*mulgyewonk*] has power to bewitch men and women and that he causes disease by the booming noise which he makes. I am now convinced that the noise does come from the lake. They say that [Sub-protector of Aborigines] Mr Mason shot at one over on Pomont [Pomanda] and it made him ill afterwards by its power. They say he is very much like a pungari (seal [Australian Fur Seal, *Arctocephalus pusillus*]) but has a face with a menake (beard) like a man (Taplin Journals, 2–3 July 1860).

Ngarrindjeri people believed that hearing the *mulyawonk* (*mulgewanke*) would cause rheumatism (Taplin 1874 [1879]).

The form of a starfish is one theme in the earliest reports of the bunyip. Colonist George French Angas reported that Aboriginal people at Moorunde near Swan Reach in the Riverland of South Australia during the 1840s:

> … believed in the existence of a water spirit, which is much dreaded by them. They say it inhabits the Murray; but although they affirm that its appearance is of frequent occurrence, they have some difficulty in describing it. Its most usual form, however, is said to be that of an enormous star-fish (Angas 1847b, pp. 97–8).

Similarly, it was reported in the Lower Murray that the:

> Natives of the Narrinyeri [Ngarrindjeri] tribe on the Coorong used to shrink at the mention of an animal which is alleged to have haunted the shallow shores of Lake Alexandrina. The Moolgewanke, they would tell the enquirer, was a 'big-um pfeller star fish.' Any search for additional information brought the inevitable reiteration. 'Moolgewanke! Big-um pfeller him. Live longa lake. No good black feller.' (Anonymous 1924, p. 1)

By some accounts from Victoria the bunyip was an amalgam of several known creatures, with the Emu (*Dromaius novaehollandiae*) prominent among them. In 1847, lieutenant governor Charles Joseph La Trobe reported that in the Port Phillip district it was said to be: 'a fearsome beast, as big as a bullock, with an emu's head and neck, a horse's mane and tail, and seal's flippers, which laid turtle's eggs in a platypus's nest, and ate blackfellows when it tired of a crayfish diet' (CJ La Trobe [cited Norman 1951, p. 6]). The Emu-like features are a major theme in the early Victorian accounts of the bunyip. It was recorded that:

> The Western Port blacks call the Bun-yip *Toor-roo-dun*, and a picture of the animal, made by *Kurruk* many years ago, under the direction of a learned

doctor, is that of a creature resembling the emu … On the Western Port plains there is a basin of water - never dry, even in the hottest summers - which is called *Toor-roo-dun*, because the Bun-yip lives in that water. *Toor-roo-dun* inhabits the deep waters, and the thick mud beneath the deep waters, and in this habit resembles the eel. The natives never bathe in the waters of this basin. A long time ago some of the people bathed in the lake, and they were all drowned, and eaten by *Toor-roo-dun* (Smyth 1878, vol. 1, p. 436).

In 1847, another account of the bunyip from Port Phillip Bay in Victoria shares more characteristics with the Platypus (*Ornithorhynchus anatinus*). A colonist reported that:

The statements of the aborigines relative to the Bunyip is, that it is of the size of a bullock, with a head and neck like an emu's, and a mane and tail like a horse's. In their rude drawings of it they give it two tusks, or front teeth, curved downwards; and feet like those of a seal; they say that it is oviparous and burrows, commencing its burrow under water, and working upwards until it is above the water level, where in a chamber, accessible only through the water, it deposits its eggs, which are as large as a bucket, enclosed in membranous skin like a turtle's, and not in a hard shell (Gunn 1847, p. 2).

An account of a bunyip attacking a human hunter was recorded from 'old chief, Morpor' in south-western Victoria (Dawson 1881, p. 108). In this narrative, there were two brothers out in a swamp collecting Black Swan (*Cygnus atratus*) eggs and they found a great many, which they took and started to cook on the bank of the lagoon. The shorter of the brothers wanted more eggs and ignored the pleas from his taller brother not to go back into the swamp alone. After finding a nest and gathering its eggs, the shorter brother was returning back to camp when he saw the waterfowl ahead of him starting to hurry away. With his feet stuck in the mud, he was then hit by a large wave that carried him back to the swan nest and into the mouth of a large bunyip. After the water had become calm, the taller brother ventured into the swamp on a sheet of bark with a fire on it and managed to retrieve from the bunyip what was left of his brother for cremation.

In 1850, a colonist visiting Challicum Station in western Victoria was taken to three large waterholes, where according to local Aboriginal tradition a bunyip had sometime previously been speared and dragged dead on to the grass, where its outline was marked out and the turf within the outline removed (Wathen 1855). The colonist drew the figure, which had the appearance of a large Emu, and remarked that in this area people believed that bunyips still existed in deep waterholes of creeks. For some time during the mid-1850s, Aboriginal people would annually visit the site in order to clear it and retrace the outline of the figure (Clark 2010; Massola 1957, 1969). In what appears to be a related account from 1855, a naturalist on a field trip at Seymour in central Victoria reported that his three Aboriginal assistants refused to stay at his camp near a lagoon because:

… an animal somewhat resembling an emu, but with much longer legs, and of so formidable a character as to threaten them with danger, usually makes its appearance during the commencement of the warmer season, and prowls about

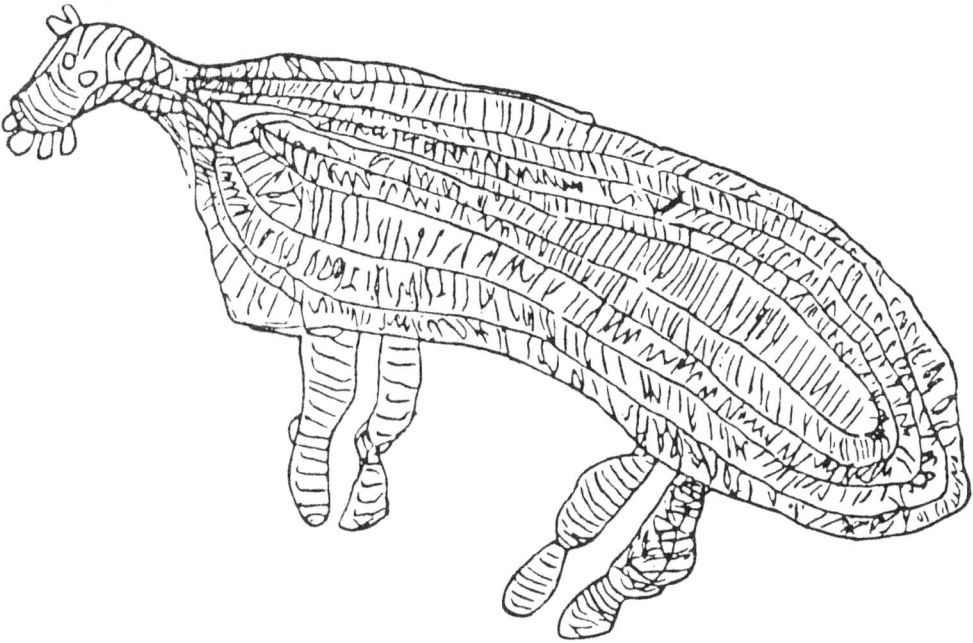

A 'Bun-yip', drawn by an Aboriginal person of the Murray River in Victoria, 1848. R Brough Smyth, 1878, Fig. 245.

the lagoons; but whether this statement contained any degree of truth I will not now venture to say (Blandowski 1855, p. 73).

From the mid-19th century, most reports from the inland reaches of the Lower Murray feature a more humanoid-type of spirit than that recorded above. Here, Ngarrindjeri people call the bunyip the *mulgyewonk*, and as a class of water spirit they have been greatly feared. It was said that: 'He is represented as a curious being, half man, half fish, and instead of hair, a matted crop of reeds' (Taplin 1874 [1879, p. 62]). A Yaraldi-speaking informant, Mark Wilson (no date), said that the *mulgyewonk* would lie submerged in the shallow waters near the edge of the lake waiting for human victims. He said that its long trailing hair in the water looked like waterweed. The smell of fish and duck grease, especially when children are washing their hands in the lake after a meal, was said to attract the *mulgyewonk* (Berndt *et al.* 1993).

In 1862, the missionary at Point McLeay was told that in the past a man, who had rubbed himself over with oil, had descended by a rope to the bottom of the lake to rescue a child captured by a *mulgyewonk* (Taplin Journals, 16–17 September 1862). The man managed to drag the child out from among a group of sleeping *mulgyewonk*, and then get back safely. A consistent theme in the folklore recorded in this area by later ethnographers was that a boy was taken by a *mulgyewonk* to live underwater (Tindale 1930–52; Berndt 1940a; H. Rankine [in Isaacs 1980]; Berndt *et al.* 1993; Wilson no date). In a narrative published in a teaching manual (Education Department of South Australia 1991), a man is fishing when attacked by a *mulgyewonk*.

All reports of the *mulgyewonk* have reinforced it as a symbol of the dangerous nature of the waters in the Lower Lakes and Murray River region. Children are associated with both the general story and many of the sightings. In 1870, the son of the missionary at Point McLeay also claimed to have seen the *mulgyewonk* in the lake (Linn 1988). It is likely that mounds of vegetation and earth trapped in the lake have resulted in some *mulgyewonk* 'sightings' (Clarke 1994, 1999b). Before the building of the barrages, when seasonal fluctuations in the flow of the river into the lakes were much greater, large amounts of material were to be found floating there. For instance, George French Angas remarked:

> Floating islands, covered with reeds, are frequently to be seen on this [Murray] river. These masses of earth, originally detached from the banks by floods or otherwise, are frequently drifted from side to side, and not a few find their way to the lake (Angas 1847b, p. 54).

In 1862, Ngarrindjeri people reported to the Point McLeay missionary that they had actually seen the *mulgyewonk* and that one of these spirit beings had died and was rotting in the water close to the shore of Lake Alexandrina near Rankine's ferry (Taplin 1859–79). It is interesting to note that Aboriginal people at Point McLeay attributed a booming sound emanating from across the nearby lake to the *mulgyewonk* breaking up gumtrees, which eventually floated downstream (Taplin 1859–79). Elsewhere, some of these *mulgyewonk* noises were probably caused by the sudden expulsion of mud under shifting sand in the Coorong (Clarke 1999b).

The water origin of the booming noise appears certain. In 1860, the Point McLeay missionary recorded that he was convinced that the booming sound of the *mulgyewonk* originated from within the lake, and during one evening he heard it 12 times in 10 minutes (Taplin 1859–79). It is possible that the advent of frequent mechanical noises in the Lower Murray, such as gun blasts, quarry activity etc., has hidden this phenomenon. Also, it is likely that the barrages constructed at the Murray Mouth and Coorong have altered the conditions that produced the booming effect (see Tindale 1936c). Rather than claiming that the cultural experience of the *mulgyewonk* resulted from a misinterpretation of the environment, the earlier flow conditions in the Lower Lakes and Murray River would probably have promoted more 'sightings' than now possible. The memory of the circumstances of each incident became incorporated into the Aboriginal Biocultural Knowledge of local groups.

Recorded within the corpus of Aboriginal Biocultural Knowledge were the places, such as particular waterholes in the riverbeds, where the bunyips might be encountered. For instance, missionary John Bulmer in Victoria noted that:

> Along the Murray I found they had an idea of a monster called *natchea*, this would be a water sprite as *natchea* means a spirit. It only occupied certain deep parts of the river, but in these places the people would not swim, though they would go within a short distance of the place and stay in the water for some time but they would carefully avoid *natchea* hole (Bulmer 1855–1908 [1999, pp. 50–1]).

Similarly, it was reported in a late 19th century newspaper that:

… it is tolerably well authenticated among Bacchus Marsh residents that the word Merrimu was applied in the early days by the aborigines of this district to a whirlpool or deep hole in the bed of the Werribee, in which they supposed the bunyip to dwell (Anonymous 1883a).

A place known to Ngarrindjeri people as Maragon, which is on the eastern banks of the Murray near the opening of the river into Lake Alexandrina, was the site of many of the oral history accounts of *mulgyewonk* encounters during the last hundred years (Clarke 1999b, 2014a). Bubbles often seen in the water here were considered to be proof of their existence. Several Aboriginal sources stated that there are large rock holes under the cliffs at Maragon beneath the water level where these spirits lived. Presumably these features are exposed during low river levels at drought time. Understandably, for Aboriginal people swimming was not allowed at this spot. According to Yaraldi-speaker Mark Wilson (no date), both Maragon and the cliffs of Pomanda (Pomond), ~8 km to the south, were the underwater homes for colonies of the *mulgyewonk*. According to Wilson, local Aboriginal people passing in canoes at night apparently avoided Pomanda altogether, said to be the *mulgyewonk* 'headquarters'. Another possible bunyip site along the banks of the Murray River is at Mypolonga, about 15 kilometres upstream from Murray Bridge in South Australia. Yaraldi-speaking man Albert Karloan considered that this place name was possibly derived from the Ngarrindjeri term, *mulgyewonk* (Berndt 1940a).

In 1952, Nganguruku man Joe Mason from the Riverland was recorded as stating that the river spirit, which he called the *muldjewangk,* still lived along the Murray River at Ranginj (Devon Downs) near Mannum in South Australia and he claimed that: 'I reckon it weighs 150 pounds [68 kg] in weight. It makes ripples on the water when it swims' (Tindale 1930–52 [cited Clarke 1999b, p. 159]). Apart from the Murray River, the *mulgyewonk* was also considered to have existed in the waters of Lake Albert and Lake Alexandrina. According to oral history from a local European settler family, another site associated with this water spirit is at the arm of Lake Albert, and for this reason it was claimed that Aboriginal groups avoided the area (Clarke 1999b).

Although the *mulgyewonk* is still sometimes reportedly seen in the Lower Murray, several Aboriginal sources of this century have said it was either extinct or at least very scarce. Mark Wilson (no date) thought that the arrival of paddle steamers and other boats on the river caused their destruction. This opinion was reinforced by Ngarrindjeri man Henry Rankine from Raukkan (Point McLeay), who gave an account of a violent encounter between a riverboat captain and a *mulgyewonk* (Rankine 1991). While conducting fieldwork with Indigenous people in south-eastern Australia, the current author has often been told accounts of bunyip sightings in particular quiet reaches of large rivers. The telling of such narratives contributes to Aboriginal identity building by strengthening connections to the local landscape. There is much historical syncretism in the more recent accounts, with some Aboriginal people having expressed opinions that it was a 'prehistoric remnant', and 'like the Loch Ness monster' (Clarke 1999b, 2007b). The folklore associated with the bunyip is rich and still being added to by both Indigenous and non-Indigenous people.

There are many officially gazetted places in south-eastern Australia that appear to relate to this spirit being. It was remarked in the newspaper that:

> There is a place now called Toor-roo-dun [= bunyip] on the northern shore of Western Port Bay. It is situate on Stawell's Creek, which discharges part of the overflow of the Koo-wee-rup Swamp into an inlet of the sea. The great swamp (Koo-wee-rup) has an area of 120 square miles [311 km²]; it receives the waters of the Bun-yip River and the Kardinia, Toomuo, and Ararat Creeks, and its overflow is conveyed to the sea by numerous creeks and channels. It is a place where one might expect to find the seal in such a situation as to give rise to the wild stories told by the natives (Smyth 1878, vol. 1, p. 436).

In the region surrounding Melbourne there were landscape features that settlers called 'Bunyip Swamp' (Koo-wee-rup) and 'Bunyip River' (Anonymous 1888c), and these were probably so named from local Indigenous traditions.

There was a European perception that true bunyips were restricted to the temperate parts of the Murray–Darling Basin. It was claimed that in western New South Wales there were:

> … haunts of evil, where the Aborigine dared not venture. Sometimes malevolence lurked in the very waterholes. The bunyip frequented several along the Murray but never ventured up the Darling – it had terrors of its own (Hardy 1969, p. 16).

The wide distribution of the recorded 'bunyip' beliefs undoubtedly adds to the variation of its reported appearance.

The European discovery of bunyips

In this section, we consider how Europeans appropriated Aboriginal Biocultural Knowledge and then applied it to their own understandings of the newly colonised land. The first European who claimed to have encountered a bunyip was the escaped convict, William Buckley, who from 1803 to 1835 lived among the Wathaurong people on the western side of Port Phillip Bay in Victoria. In his reminiscences, Buckley claimed that his adoptive band once moved to a lake they called Moodewarri, and:

> In this lake, as well as in most of the others inland, and in the deep water rivers, is a very extraordinary amphibious animal, which the natives call Bunyip, of which I could never see any part, except the back, which appeared to be covered with feathers of a dusky grey colour. It seemed to be about the size of a full grown calf, and sometimes larger; the creatures only appear when the weather is very calm and the water smooth. I could never learn from any of the natives that they had seen either the head or tail, so that I could not form a correct idea of their size; or what they were like (Buckley [Morgan 1852, p. 48]).

Buckley later came across the bunyip at a lake that local people called Jerringot, which was part of a chain of water bodies that flowed into the Barwon River (Barwin), and:

> Here the Bunyip - the extraordinary animal I have already mentioned - were often seen by the natives, who had a great dread of them, believing them to have some supernatural power over human beings, so as to occasion death, sickness, disease, and such like misfortunes. They have also a superstitious notion, that the great abundance of eels in some of the lagoons where these animals resort, are ordered for the Bunyip's provision; and they therefore seldom remain long in such neighbourhoods, after having seen the creature (Buckley [Morgan 1852, pp. 108–9]).

Buckley considered the bunyips to be extremely dangerous and he claimed:

> They told me a story of a woman having been killed by one of them, stating that it happened in this way. A particular family one day was surprised at the great quantity of eels they caught; for as fast as the husband could carry them back to their hut, the woman pulled them out of the lagoon. This, they said, was a cunning manoeuvre of a Bunyip, to lull her into security - so that in her husband's absence he might seize her for food. However this was, after the husband had stayed away some time, he returned, but his wife was gone, and she was never seen after. So great is the dread the natives have of these creatures, that on discovering one they throw themselves flat on their faces, muttering some gibberish, or flee away from the borders of the lake or river, as if pursued by a wild beast (Buckley [Morgan 1852, p. 109]).

In spite of the fears of his fellow band members, Buckley tried to kill a bunyip. He stated that:

> When alone, I several times attempted to spear a Bunyip; but had the natives seen me do so it would have caused great displeasure. And again, had I succeeded in killing, or even wounding one, my own life would probably have paid the forfeit - they considering the animal, as I have already said, something supernatural (Buckley [Morgan 1852, p. 109]).

After settlement at Port Phillip Bay, Europeans in the Geelong district made inquiries among local Aboriginal people concerning the creature from which an unusually large leg bone discovered on the shore of Lake Timboon may have originated (Anonymous 1845; Fred S. 1893). It was reported that among those who were shown the bone:

> One declared that he knew where the whole of the bones of one animal were to be found; another stated that his mother was killed by one of them at the Barwon Lakes within a few miles of Geelong, and that another woman was killed on the very spot where the punt crosses the Barwon at South Geelong. The most direct evidence of all was that of Mumbororan, who showed several deep wounds in his breast made by the claws of the animal … They say that the

reason why no white man has ever yet seen it, is because it is amphibious, and does not come on land except on extremely hot days, when it basks on the bank; but on the slightest noise or whisper they roll gently over into the water, scarcely creating a ripple (Anonymous 1845).

This 1845 account from the Geelong district came with a detailed description of the appearance and behaviour of the spirit being:

The Bunyip, then, is represented as uniting the characteristics of a bird and an alligator. It has a head resembling an emu with a long bill at the extremity of which is a transverse projection on each side with serrated edges like the bone of the stingray. Its body and legs partake of the nature of the alligator. The hind legs are remarkably thick and strong, and the fore legs are much longer, but still of great strength. The extremities are furnished with long claws, but the blacks say its usual method of killing its prey is by hugging it to death. When in the water it swims like a frog, and when on shore it walks on its hind legs with its head erect, in which position it measures twelve or thirteen feet [3.7 or 4 metres] in height. Its breast is said to be covered with different coloured feathers; but the probability is that the blacks have not had a sufficiently near view to ascertain whether its appearance might not arise from hair or scales. They describe it as laying eggs of double the size of the emu's egg, of pale blue colour; these eggs they frequently meet with, but as they are 'no good for eating,' the black boys set them up for a mark, and throw stones at them (Anonymous 1845).

A newspaper correspondent who considered the above account later queried, 'Does this fearsome beast still lurk in lonely waterholes, coming up to bask in the sun on hot days, but sinking again at the slightest sound ... Turn up a work on palaeontology, and you will find queerer beast than the bunyip. Besides, queer beasts have been found not only dead but living' (Fred S. 1893, p. 33).

In 1847, it was reported in the newspaper at Port Phillip that physical evidence of the bunyip had come forth again:

From a respectable Settler in the bush, we have found a skull of an unknown animal, which corresponds exactly with the strange animal lately discovered at Geelong and named the Bunyip. There can be no doubt that the skull does not belong to any species of animal already known in Australia, and it is a matter of some interest, the owner is to give it to Dr. Johnson to be examined (Anonymous 1847a).

In the same year, another newspaper report claimed that the bunyip or Kine Pratie of the lower reaches of the Murrumbidgee River to the north had 'a head and neck like an Emu, with a long and flowing mane – feeding on crayfish (with which the river abounds) and occasionally on a stray blackfellow; that it inhabits the darkest and deepest parts of the river, and in some of the lakes and lagoons that longest retain water' (Anonymous 1847b).

Europeans speculated that the inland stranding of sea mammals explained the alleged bunyip sightings. For instance, geographer George Windsor Earl (1846, pp. 248–9) thought that the 'certain monster amphibia' (probably the bunyip) known by New South Wales and South Australian Aborigines was probably the dugong, which he had personally encountered in the north. In 1858, the arrival of a 'sea-lioness' (Australian Sea Lion, *Neophoca cinerea*) at Western Port in Victoria was reported in the newspaper: 'Some persons assert that this creature is the much-talked of bunyip of the aborigines' (Anonymous 1858, p. 4). In 1862, the zoologist Gerard Krefft gave an account of his mammal-collecting trip along the Murray River. In this he related that an animal that he shot one night near Mildura had been identified by his Aboriginal companions at the time as a 'bunyip' splashing about in the river, although next morning it turned out to be an 'aged billy-goat' (Krefft 1862, p. 7).

Among the European observers, the sea mammal theme in the reporting of the bunyip continued into the second half of the 19th century. In 1865, it was stated that:

> There is scarcely a doubt of the existence of the animals as mysteriously known to the natives as a bunyip, and as the following facts are perfectly authentic, we may safely pronounce it to be of the seal tribe. As we have fresh water turtles, and fresh water mussels, there is no reason why we should not have fresh water seals, and we quite believe they exist – although evidently very scarce – in the Murray, its lagoons, and tributaries (Anonymous 1865a).

In 1865, a correspondent from another newspaper in north-eastern Victoria wrote:

> A great story comes from the Ovens about the discovery of a veritable bunyip, near that now notable place Peechelba, which turns out, however, to be a freshwater seal … It is undeniable that something very like a seal has frequently been seen in our rivers and deeper lagoons and we could quote many instances even in the neighbourhood of Echuca (Anonymous 1865b).

The same writer believed that the ancestors of Aboriginal people in south-eastern Australia had brought with them their traditions concerning the 'alligator' (Saltwater Crocodile, *Crocodylus porosus*) and attached them to the seal. It was claimed that:

> There can be no question that the blacks have populated the country from the northern coasts, and we believe that the actual bunyip or moolgiewankie, is a tradition of the alligator. The description given by the blacks tallies with that of this reptile. The feet, they say, are turned back, the jaws enormous, the teeth formidable, and the skin impervious to a spear (Anonymous 1865b).

Among the European settlers, the 'alligator' explanation of bunyips gained traction. In 1867, a European claimed to have seen a 'freshwater crocodile' among the reeds in the Wongungarra River of Gippsland, and it was suggested that this was also the identity of the bunyip (Anonymous 1867b). Given the tropical restricted range of the true crocodile, a large specimen of a Lace Monitor (*Varanus varius*) may have been a better identification. Many 'bunyip' sightings would have proved to be something else if fully investigated. For example, in 1867 it was reported in an Ararat newspaper that a European had thought he was watching a bunyip moving among ducks in a swamp during the night, only to find that it was a light framework of twigs and rushes under which three Aboriginal men armed with nooses attached to a stick were catching waterfowl (Anonymous 1867b).

A newspaper report in 1870 mentioned the 'seal' theory of the bunyip's identity, but also mentioned the associated folklore may have come about through a variety of causes, such as when:

> … the old blackfellows deliberately frightened the youngsters with the yarn, to prevent them from stealing the eggs of wild fowl whose haunts were usually identical with those of the fabled bunyip. A still more fanciful explanation was that the bunyip was a personification of the cramp which seizes swimmers, and apparently draws them down to the bottom at particular parts of a river where the water is unusually cold. These places were known as bunyip holes (Oedipus 1870).

This report also suggested that a still unknown species of otter (Mustelidae) may have been behind some bunyip accounts. Also, on one occasion, it was recalled that dogs had emerged from a swamp at Lake Corangamite with wounds, with an enormous pig being discovered after a party had flushed it out. The author cited above wisely stated that: 'each bunyip story must stand on its own merits'.

During the 1870s, at least three specimens of 'bunyips' were allegedly procured by Europeans. For instance, in 1871 a 'real bunyip' was shot by a farmer, and it turned out to be an Australasian Bittern (*Botaurus poiciloptilus*) (Anonymous 1871a, p. 3). In the same year, it was announced in a newspaper that:

> The mystery that has attended the existence of the bunyip will now perhaps be elucidated: - A Stratford correspondent of the *Gippsland Mercury* states that a few days ago the Rev. Mr Hagenaeur [Moravian missionary at Ramahyuck] and some blackfellows caught a veritable bunyip near Lake Tyers. The animal is of a light drab colour, with white spots. It is about 8ft. [2.43 m] long, and about 34in. [86 cm] in circumference near the flippers. It resembles a seal, but without doubt it is the animal which the blacks have often expressed terror of, and whose roaring has been the cause of so much mystery. The Rev. Mr Hagenaeur will send the creature to Melbourne for exhibition (Anonymous 1871b, p. 2).

This account sounds like a young leopard seal, which are occasionally stranded along the south-eastern coastline of Australia (Strahan 1983).

In 1877, a regional newspaper stated that Mr Moon of Oxley had in his care an unusual-looking bird that could have been the source of the bunyip belief, and it was described as:

> … a biped, distantly resembling an eaglehawk [Wedge-tailed Eagle, *Aquila audax*], with the legs of a wading bird, and no sign of a tail … which makes the same trumpet-like noise that many of us have heard at night in the bush without

being able to account for it – something like the low of a bull – and which was generally attributed to the bunyip (Anonymous 1877, p. 2).

This description is possibly of an Australasian Bittern. In 1878, a large number of bunyips were apparently heard in the marshes near Mitiamo in north central Victoria, with a 'booming noise' that was 'said to resemble the bellowing of a bull', being heard irregularly through the night, with those of a softer tone thought to be from females (Anonymous 1878, p. 2).

The seal theory for the origin of bunyips remained a strong theme in the late 19th century, although some scholars had doubts. It was remarked that:

Whether the seal which the blacks have named the Bun-yip is the eared seal (*Arctocephalus lobatus* [Australian Sea-lion, *Neophoca cinerea*]) or the large spotted sea-leopard (*Stenorhynchus leptonyx* [Leopard seal, *Hydrurga leptonyx*]), or some other animal unknown as yet to naturalists, is doubtful. That the blacks in former times ate the seals which frequented the coast is certain, and it is probable, therefore, that some other creature was the cause of the terror which afflicted them at nights when they heard growlings and bellowings on the margins of the swamps. Seals proceed inland often for a considerable distance; many during certain seasons may have frequented the samphire-bound inlets of Western Port, and by their bellowings at night frightened the natives; but there is reason to believe that the seals known to them and to the whites were not the same as *Toor-roo-dun* [= bunyip] (Smyth 1878, vol. 1, pp. 437–8).

In spite of the writer's doubts, he also reported that in central Victoria:

Major Couchman, the Chief Mining Surveyor in the Mining Department, says that he and Mr. Lavender saw an animal resembling a water-dog swimming in

A 'Bun-yip', *Toor-roo-dun*, drawn by Kurruk, an Aboriginal person of Western Port, Victoria, in the 19th century. R Brough Smyth, 1878, Fig. 244.

the reservoir at Malmsbury. It was large, and of a very dark color. He watched the animal for some time, when it dived and disappeared. He saw it again when it was nearer, and then knew that it was not a dog. Its head resembled that of a seal. Both Mr. Lavender and he watched it for some time, and its form and the period during which it remained under water after it had dived satisfied them that it was not any animal known to them. Are there fresh-water seals in Victoria, and is the Bun-yip a fresh-water seal? (Smyth 1878, vol. 1, pp. 438–9).

In 1882, the Australian Water-rat (*Hydromys chrysogaster*), which has an otter-like appearance, was probably the species that one European observer encountered and thought related to the bunyip. He described it thus:

On Sunday last while walking along the bank of the River Loddon I heard a loud splash, and on looking towards the spot where the sound came from, I saw an animal floating on the surface of the pool, which very much resembled an otter, and was about the same size and color. It seemed to have short fore-legs, a round head, and of a dark brown, or nearly black color, and from what I saw of it would weigh from fifteen to twenty pounds [7 to 9 kgs]. Will it not be a species of the Aboriginal's bunyip, or are there any others inhabiting the rivers of this country? (Anonymous 1882, p. 3).

In northern parts of Australia, it was believed that 'sea cows' or Dugongs (*Dugong dugon*) might be other possible 'bunyips' (Anonymous 1886, p. 4). In 1888, it was reported that it had been thought that the remains of a bunyip had been found in the Riverina of New South Wales, but it had 'eventually turned out to be the mangled remains of a horse; to which a singular appearance had been given by partial immersion in water, river weeds and sand' (Anonymous 1888a, p. 2).

By the late 19th century, at least some writers were treating the bunyip just as an Indigenous component of the Australian folklore. It was claimed that:

The Bunyip is the Australia sea-serpent, only it differs from that much-disputed fact or fiction in that it does not inhabit the ocean, but makes its home in lagoons and still, deep water-holes. For rivers and running creeks, it appears to have an aversion. No black fellow will object to bathe in a river because of the Bunyip, but he will shake his woolly head mysteriously over many an innocent-looking water hole, and decline to dive for water-lily roots, or some such delicacy dear to the aboriginal stomach on the plea that 'Debil-debil [devil] sit down there.' Debil-debil and Bunyip are synonymous terms with the black fellow while he is on the bank of a lagoon ... The Bunyip is said to be an amphibious animal, and is variously described – sometimes as a gigantic snake; sometimes as a species of rhinoceros, with a smooth pulpy skin and a head like that of a calf, sometimes as a huge pig, its body yellow, crossed with black stripes (A.S.T. 1888, p. 6).

Another scholar noted that: 'The size varied, according to the mind of the terrified narrator, from that of a large Newfoundland dog to that of a large bull' (Worsnop 1897, p. 167).

Particular aspects of the bunyip attracted specific explanations, with incidents when only the call was heard attracting different explanations to those of just sightings. It was noted that:

> A correspondent, an old settler and one well acquainted with the natural history of the colony, tells me it is his belief that in most cases the noise that frightens the natives is caused by the movements in the water of the musk-duck [*Biziura lobata*]. When on the banks of the River Wannon [in western Victoria], I approached a dense growth of reeds, and one of these birds that had been hidden in the reeds made a dash into the water, and the noise and its appearance, I thought at the time, would create alarm in the dusk of evening; but it is scarcely credible that so many strange tales should arise from this source. The natives are good naturalists, and are probably better acquainted with the habits of this duck than we are (Smyth 1878, vol. 1, pp. 439–40).

A frequent bird-related claim was that the bunyip's call came from the Australasian Bittern, 'being popularly known as the bunyip bird, its oft repeated note 'boom, boom,' being attributed by the aborigines to the mythical bunyip' (Milligan 1894, p. 3). This was later reiterated by other writers (J.D. 1931). Barrett (1946, p. 112) said that: 'When a booming call breaks the silence of the lonely swamp, it is the voice of the 'Bunyip-bird'.' The bird, which is also known as the boomer and bull-bird, is illusive, as:

> It is partly nocturnal in habits, and, keeping as it does to the depths of reedy swamps, is seldom seen during the day. The call consists of three or four deep 'booms', with a distinct interval between each, suggesting the bellowing of a bull, suggested as the origin of the fabulous bunyip, said to dwell in swamps and other such places (Cayley 2011, p. 646).

European scholars theorised that when seals were more common along the coast, before the activities of European sealers, they would more often stray into the Murray River system and become stranded in the swamps and billabongs when water levels dropped (Le Souëf c.1890). The seals tended to hide in the rushes and in hollow logs during the day, with the eerie booming call of the Australasian Bittern taken as its voice. It was suggested that it was the combination of the two that gave rise to the bunyip beliefs.

The changes resulting from European settlement brought many new elements to the regional ecology of south-eastern Australia, probably resulting in further alleged bunyip sightings. In 1898 in the Koondrook district of north central Victoria there was a debate in the newspaper about the 'alleged dimensions of the wild pig that recently played up with Mr. Reaper ... [and that] stood 4ft [1.2 m] high at the shoulders; therefore, it must have been little short of a ton [1016 kgs] in weight' (Anonymous 1898). Others were said to have come to the conclusion: 'that the monster was not a wild pig at all, but the historical Barham bunyip of whose depredations there exist many legends among the local aborigines'.

Bunyips remained newsworthy items well into the 20th century. In 1924, a correspondent to an Adelaide newspaper remarked that:

> Whether there is sufficient justification to say that the bunyip is a memory of a prehistoric age is a matter for argument. Perhaps this monster is of a similar mythical origin as the dragons and unicorns of ancient Europe. Or perhaps it has a more homely origin. It may have been invented by ancient lubras [Aboriginal women] as a 'big-um pfeller bogey' to stop their too venturesome offspring from straying from the paternal 'wurly [brush shelter].' It is one of those prehistoric riddles which even the best scientific brains cannot explain. As Professor Walter Howchin [of the University of Adelaide] naively says, 'Nobody knows, for nobody was there to see.' (Anonymous 1924)

For Europeans, the bunyips were a category of Aboriginal spirit beings. It was stated by the same source as above that:

> North, East, South, and West the Australian blacks have their bunyips. In the Northern Territory it takes the shape of a reptile, in Victoria it is a huge bird, and in Western Australia it is akin to a sea serpent. Its name and form maybe are different, but its character never alters. It is invariably of fearsome aspect, gigantic size, and possessed of a distinct partiality for blackfellows as an article of food (Anonymous 1924).

In 1934, a Western Australian newspaper reported the sensational news that: 'It is claimed by a Scotch zoologist that the bunyip is identical with the strange monster of Loch Ness', having migrated there from Australia (Anonymous 1934b). The same article gave a response from the director of the National Museum in Melbourne, DJ Mahoney, which was that: 'It is possible, of course, that the myth of the bunyip had its origin in fact … [although] Seals and sea-lions have sometimes been seen far up the Murray, and a stranded specimen may have given rise to the aboriginal myth.' This claim resonated with the records of the explorer Captain John Douglas Forbes who was exploring the Namoi River area of eastern New South Wales in 1832 when his Aboriginal guide Liverpool drew him a 'Wawee'. This was thought to be: 'a species of walrus, but without feet, having only short finny legs' (Forbes 1832 [cited Mulvaney 1994, p. 37]).

In 1934, another newspaper correspondent 'brought forward an old theory and adduced fresh support therefore that seals have in many cases penetrated far inland along our rivers and billabongs, and that these errant marine carnivores may have given rise to the bunyip myth of the aborigines' (Tellurian 1934, p. 42). He believed that a seal recently seen at Mildura in northern Victoria was the same that was seen some months earlier on the Murray River at Loxton in the Riverland of South Australia. Carnivorous mammals and fishes from the ocean were probably seasonal visitors to certain waterways, as it is known that before the completion of barrage construction in 1940, dolphins and sharks were occasionally seen in the Lower Lakes and even further upriver as far as Tailem Bend and Murray Bridge (Anonymous 1902; 1914b; 1927a; 1927b). In the early 20th century Ngarrindjeri people killed a small whale in the Coorong (Ely 1980).

By the 20th century the chances of discovering another large animal species unknown to science in the temperate parts of Australia must have appeared remote, so scholars looked to the ancient past for answers. There was speculation that the large prints of animals featured in Aboriginal rock engravings were of animals that had become extinct many thousands of years before the arrival of Europeans (see Hale and Tindale 1929; Mountford 1929). While the anatomical evidence was compelling, there were also attempts to find supporting evidence from the traditions of Indigenous people whose ancestors would have lived in a landscape alongside such creatures. In a paper concerning the possibility that engravings of giant extinct birds, such as *Genyornis newtoni* and *Pachyornis queenslandiae*, existed at Pimba in the deserts of northern South Australia, museum researcher Norman B Tindale cited evidence from Aboriginal mythology. After analysing the physical evidence, he claimed that:

> The discovery does not stand alone. In Western Victoria a Tjapwurong tradition exists about a giant bird called [*mihirung paringmal*] much larger than the ordinary emu, which Tjapwurong called [*paringmal*] (Tindale 1951, p. 381).

The author cited Dawson (1881) as his source for the giant bird myth, and palaeontologists have followed him by naming fossil species, such as the *Wonambi* genus of giant snakes and the *Quinkana* genus of crocodiles, after Indigenous spirit beings (Flannery 1994).

The acknowledgement by scientists of the Indigenous custodianship of the Australian biota is commendable, but nevertheless the link between long extinct species and Aboriginal Creation ancestors is tenuous. In the narratives of the Creation period all animals were bigger and more powerful before events took place that caused the ancestors to travel to the Skyworld, leaving smaller versions of them on Earth. The meanings within Aboriginal mythologies are complex, and not easily equated with an Indigenous form of history (see Sutton 1994). In New Zealand, it was found that Maori knowledge of the Moa (*Dinornis maximus*) had disappeared only a few generations after the bird's extinction (Cocker and Tipling 2013). This would suggest that since Indigenous Biocultural Knowledge is dynamic, any new data from Indigenous experiences of the environment will quickly replace redundant information.

From the early 20th century, the media's interest in solving the bunyip mystery was more pragmatic. In 1941, a newspaper correspondent remarked that the bittern: 'is commonly known as the bunyip bird on account of its curious booming call; but there are other birds, such as the frogmouth and several owls, whose calls at night might have suggested the name bunyip to the aboriginals' (Oriolus 1941, p. 12). The theme of bunyips being a manifestation of the dangerous waterways in south-eastern Australia continued. In 1950, a former settler suggested that the 'bunyip beliefs' at one waterhole in the mid-north of South Australia may have been a reference to dangerous conditions, which included cold water under the surface, and possibly undercurrents and a whirlpool (Osborne 1950).

In many ways, bunyips in Australia today are the sum total of all the animals and environmental dangers attributed to them, and much more as well. As a key iconic entity of Australian folklore, bunyips will continue to capture the imagination of European Australians and remain an element in the oral traditions and cultural heritage for Indigenous peoples. For Aboriginal authors such as Eileen Morgan:

> … restoring the bunyip to its fearful place in their lore has been part of a broader process of self-realisation as well as a statement of ethnic pride. In this way the bunyip remains a real-life creature, part of an ongoing tradition of storytelling, still making its presence felt (Holden and Holden 2001, p. 204).

Discussion

In south-eastern Australia, the water spirit beings were considered to exhibit particular spatial behaviours and with their own territoriality. During the 19th century, the descriptions that Europeans received of spirit beings, that were generally classed as bunyips, were so detailed from Indigenous sources that it was widely believed that they must have existed, either then or as extinct species from the past. The ability of Aboriginal trackers was legendary, so European settlers took very seriously the Indigenous assertions that dangerous beasts existed at certain places. By poorly understanding an Indigenous perspective of the environment, Europeans selectively appropriated information from the large body of Aboriginal Biocultural Knowledge and then made scholarly deductions about potentially undiscovered species that were 'new' to science. For non-Indigenous people, the risk of specific information imported from Aboriginal Biocultural Knowledge being rendered invalid is high when the worldview of the knowledge-holders has not been taken into consideration.

Many of the water spirit narratives recorded from south-eastern Australia either relate directly to children, or concern the actions of beings that could be used as threats by adults to help control their behaviour. This aspect of spirit traditions has been recorded elsewhere in Aboriginal Australia (see Eickelkamp 2004; Musharbash 2016). Anthropologist and educationalist Christine Nicholls remarked that:

> Importantly, in Aboriginal Australia, these [spirit] figures and their attendant narratives provide a valuable source of knowledge about the hazards of specific places and environments. Most important of all is their social function in terms of engendering fear and caution in young children, commensurate with the very real environmental perils that they inevitably encounter (Nicholls 2014).

In the case of bunyips, the attendant mythology as discussed here contains many warnings about youngsters straying too close to the edge of waterholes and rivers. The meaningfulness of such warnings would not be diminished by the circumstances of Aboriginal people interpreting the sighting of a seal or platypus as being a bunyip.

In south-eastern Australia, the knowledge of the purported presence of bunyips in certain waterholes is still being used by knowledge-holders to demonstrate their connection to country. The building of weirs and locks on the major rivers of south-eastern Australia has undoubtedly decreased the chances of marine animals, such as seals and sharks, penetrating the inland waters and posing a danger, although the risks of drowning have remained. For the local Indigenous community, their Aboriginal Biocultural Knowledge concerning the location of safe and dangerous places in their foraging territories is important, even when expressed in terms of their beliefs associated with spirit beings. By engaging in the discussion of the activities of spirits as fellow occupants of their land, Aboriginal people highlight their identity with respect to their country.

Chapter 4

Plant food

Philip A. Clarke

The temperate region of Australia in the south-eastern corner of the continent is dominated by the Murray River Basin, with its numerous water courses, lakes, lagoons and swamps. With ample food available, Indigenous population levels here were high in comparison to other parts of Australia. Foraging band members used their Aboriginal Biocultural Knowledge to position themselves seasonally within their foraging territories so as to maximise their chances of obtaining highly favoured foods while reducing the amount of labour required to produce it. The techniques they used to gather food were sophisticated and appropriate for maintaining their food sources in the longer term.

Introduction

Aboriginal foragers in south-eastern Australia lived in a region that was rich in resources. In such regions, the limiting factors for human population growth would have been the ability of people to cope with severe climate events, such as droughts and flooding, along with a variety of social factors, including warfare and sorcery (Clarke 2003a). It is no accident that the British colonists chose the temperate zone of Australia to settle first, as this landscape has climates most like that of north-western Europe. The British made the land they were colonising even more European-like by land clearing and the introduction of many of their familiar plants and animals (Crosby 2004). European perceptions of the physical environment of Australia, that they first encountered as a 'wilderness', differed greatly from the Aboriginal views of it as the creation of the ancestors, with the evidence of their feats all around them. For foraging bands, the means for procuring food was part of their Aboriginal Biocultural Knowledge, which was a dynamic body of experiences and understandings that were specific to their territory, passed down through the generations.

The historical records of Indigenous languages contain entries for words that help document the cultural significance of specific food sources. For instance, in 1839 botanist explorer Daniel Bunce was travelling with local Aboriginal people through Gippsland in eastern Victoria when his guides gathered for him 'Some long tuberous roots, of a composite plant [yam-daisy], … of which we partook. These plants produced a bunch of tubers like the fingers on the hand, from whence they were called *myrnong-myrnongatha*, being the native word for "hand"' (Bunce 1859, p. 71). The Yam Daisy is a strong candidate for the many plants in the temperate zone that historical language records often simply catalogue as an 'edible root'. The tuber is recognised as one of the major food sources for Victorian Aboriginal people, who generally called it *myrnong* or by a variation of this word (Frankel 1982; Gott 1983; Gott and Conran 1991; Low 1989; Zola and Gott 1992). Similarly, in the language spoken at Lake Boga in northern central Victoria the Common Nardoo (water fern plant, *Marsilea* species) was recorded in a missionary's Aboriginal word list as *dullum dullum*, while the edible sporocarps were said to be *jerinyuk* (Stone 1911). Here, a nardoo mill roller was known as *boolpa jerinyuk laar*, and was used with a nardoo millstone called *jerinyuk cotthup*. True seeds, such as from Wattles (*Acacia* species), were milled on small grindstones termed *woodtheuk*, and even smaller grindstones known as *yeretheuk* (Stone 1911). There were other types of grinding stone, too, that were used to grind hatch heads and for sharpening mussel shell knives.

Indigenous population

A strong indication of the abundance of food in a region is the number of people who were able to live there, although it is difficult to calculate Aboriginal population densities for south-eastern Australia from data extracted from the historical record alone, as Indigenous people moved frequently across their country according to season (Clarke 2003a, 2009b) and because they generally avoided contact with the first European explorers (Clarke 2008a; Donaldson and Donaldson 1985). In 1845, colonist Edward Eyre, who was an explorer and Aboriginal Protector on the Murray River in South Australia, remarked:

> … there is scarcely any point connected with the subject of the Aborigines of New Holland, upon which it is more difficult to found an opinion, even approximating to the truth, than that of the aggregate population of the continent, or the average number of persons to be found in any given space (Eyre 1845, vol. 2, p. 368).

The rapid decline of Aboriginal people due to the spread of diseases across the British settlement frontier has also confounded estimates of the pre-European human population size (Beveridge 1883; Butlin 1993; Campbell 2002; Clarke 2003a).

There have been attempts to estimate population levels before the negative impact of British colonisation. Colonists in the south-east of South Australia believed that the local Bunganditj (Booandik) people were not numerous before European settlement, considering that their density was 'never equalling one to a square mile [2.6 km²]' (Stewart c.1870–c.1883 [1977, p. 62]). Here, it was conservatively estimated that the average population density immediately before European settlement to be 5 square miles [13 km²] or less to one person (Davidson 1938). Areas close to the coast, along with its adjacent wetlands, probably supported a much larger population, as it estimated that for south-western Victoria the inland-based groups had a density of 2.5–3.3 km² per person (0.3–0.4 people per km²), while at the coast this was 1.4–2.5 km² per person (0.4–0.7 people per km²) (Lourandos 1977, 1997). These are low population densities when compared to that of horticultural communities in nearby parts of Asia and Melanesia, but they are high for Aboriginal Australia (Clarke 2003a).

On a day-to-day basis, Aboriginal people in south-eastern Australia would have lived in bands of variable size, depending on factors including the seasons (Clark 1990a; Stanner

1965; Sutton 2003). Each band would have been formed around one or more senior people, and included children, in-laws and more distant relatives. Through their totemic system (see Chapter 1), individuals would have had close cultural connections to one or more clan territories, which anthropologists have often called 'estates'. Throughout the year, bands would have utilised a variety of links within their group in order to forage across various clan territories. In pre-European times, it is likely that people were multilingual, which makes defining Aboriginal groups according to language alone difficult.

Food diversity and distribution

Aboriginal food sources were highly seasonal and widely dispersed, which meant that the number of species used by Aboriginal people across Australia is immense in comparison to the foods brought out by the first Europeans (Brand Miller *et al.* 1993; Cherikoff and Isaacs 1989; Cribb and Cribb 1987; Isaacs 1987; Low 1991; Maiden 1889). In some areas, the proportion of animal food in the overall diet with plant food as a baseline varied through the year from being a major element to a minor one, with meat sources including insects, crustaceans, shellfish, fish, reptiles, birds and mammals (Lawrence 1968; see Chapter 5). The contribution of plant foods to the overall diet would have been high (Clarke 1985a, 1986b, 1988, 1998b; Gott 1982a, 1983, 2008). Nutritionists have stated that in Aboriginal Australia:

> If plants provided 20-40% of the energy in the diet (the most likely range), then plants would have contributed 22-44g protein, 18-36g fat, 101-202g carbohydrate, 40-80g fibre and 90-180mg vitamin C in a 12 500kJ (3000kcal) diet. Since all the carbohydrate came from plant foods, the traditional Australian Aboriginal diet would have been relatively low in carbohydrate (especially starch) but high in dietary fibre in comparison with current recommendations (Brand-Miller and Holt 1998, p. 5).

Records made by early European observers rarely recognised this high diversity of food. This is in part due to their lack of knowledge about the local plants they encountered, which is shown by their use of a mixture of European and Indigenous names in their records (Clarke 2008a). For instance, a British settler observed that for the Bunganditj in the south-east of South Australia: 'Their food consisted chiefly of kangaroo, fish, emu, opossum, fine roots [probably Yam Daisy, *Microseris lanceolata*], candart-seed [Wattle, *Acacia* species], 'meenatt,' [Bulrush, *Typha* species] and honeysuckle [*Banksia* species]' (Smith 1880, p. xi). When available, animal food would have generally been highly desirable over more reliable plant food sources, such as roots, that were either less palatable or required more time to collect and prepare.

In listing the specific food sources that Aboriginal people in south-eastern Australia utilised, it is often not clear the extent to which any of them could be classified as staple food sources in terms of continual consumption. Aboriginal groups living here had such a wide variety of food seasonally available to them (see Clarke 2015b, 2015c; Gott 1983, 2008, no date; Gott and Conran 1991; Zola and Gott 1992) that it is unlikely that they had any true staples. A botanist remarked that 'No sources of plant food were ignored, although some were resorted to only when more preferred foods were scarce' (Gott 2008, p. 221).

At times of the year when food was more difficult to collect, such as during the cold and wet season, Aboriginal people in the coastal and riverine areas appear to have chiefly relied on the roots and tubers that women mainly collected (Clarke 1985a, 1986b, 1988; Gott 1982a, 1983, 2008). This food category was particularly important when needing to feed an increased number of people, such as during the holding of ceremonies in Gippsland when Aboriginal women gathered large amounts of *dura* (Bulrush, Cumbungi, *Typha* species) rhizomes for baking in ashes (Howitt 1904). An early account from along the Murray River suggests that Bulrush rhizomes were harvested from late summer and lived on for several months, being cooked in an earth oven and carried about as provisions (Krefft 1862–65). Across south-eastern Australia, it has been determined that for Aboriginal foragers increasing rainfall produced a greater reliance upon roots and a declining use of seeds (Gott 2008). As shown in the detailed descriptions below, when particular types of highly desirable fruit, gums and greens were in season, often during the warmer months, considerable effort was taken in their collection (Clarke 1985b, 1986a; Gott 1982b). When an abundance of food was available for a relatively short period, a surplus was gathered to preserve for later use and for trading (Clarke 2003a, 2007a).

Through their movement patterns across the landscape, Aboriginal people were able to position themselves in order to take maximum advantage of the availability of seasonal resources that are restricted to particular habitats (Clarke 2003a, 2009b; Thomson 1939; Tindale 1938a, 1974). Seasonal indicators, including the flowering of certain plants and the movements of star constellations, were used as a guide (Clarke 2014c, 2015a). Aboriginal people would have generally spread themselves thinly across the landscape in order to avoid hardship caused by the overuse of resources. In favourable seasons, after rains had made travel across large areas possible, big camps were formed at key places for holding ceremonies, resolving disputes and trading (Clarke 2003a).

Not all parts of the country were regularly visited in south-eastern Australia. Some areas, like the interior of Mallee-covered deserts and some mountain areas, appear to have been only sporadically visited when the season was benign, while flat country near wetlands was heavily utilised (Builth *et al.* 2008; Clarke 2005a, 2009c; Tindale 1981). A colonist in Victoria remarked that:

> The dense forests of South-Western Gippsland and Cape Otway were not often entered, if at all; and the blacks who fished on the shores at the mouth of the Parker had probably no communication with their near neighbours, the natives of the Gellibrand; and it is almost certain that the Cape Otway blacks never travelled through the forest to Colac (Smyth 1878, vol. 1, p. xxxiv).

In the Mallee areas of the western parts of south-eastern Australia there were times when surface water was scarce. It was recorded that in 'The neighbourhood of the Salt Lake, Tyrril, the country more particularly alluded to, is so devoid of surface permanent water, that it is inhabited by the natives only in winter, except when they have recourse to the water in the roots of the Mallee [such as Red Mallee, *Eucalyptus oleosa*]' (Stanbridge 1861, p. 304).

There is evidence to suggest that the use of resource-rich zones, such as at Lake Condah in south-western Victoria, encouraged local Aboriginal bands to build eel traps and more substantial shelters constructed from stone that enabled them to have a quasi-sedentary lifestyle (Builth 1998; Coutts *et al.* 1978). At times of physical stress, such as during a prolonged summer drought or during a particularly cold wet spell, Aboriginal people would go outside of their normal practices to gain food, resorting to gathering the emergency sources normally left untouched that are known in Aboriginal English as 'hard time foods' (Clarke 2007a). Their Aboriginal Biocultural Knowledge guided them on the location and preparation of such marginal food sources that were crucial for their survival.

Gender specialisation and food classification

In Aboriginal Australia, the men and women had different roles in the foraging for food and its preparation (Berndt 1981; Clarke 1986b, 2003a, 2013a; Dawson 1881). Women as foragers generally focused on collecting static foods such as plants, shellfish, ground insects, lizards and small burrowing animals. Colonist Carl Wilhelmi noted that in northern central Victoria:

> The principal vegetable used for food by the Murray natives, near Swan Hill, is the root of *Typha Shuttleworthii* [probably *T. orientalis*], native name, *gortong*, or the common bulrushes. With them it is a rule that all vegetable food is prepared by the women, whilst animal food can only be dressed by the men (Wilhelmi 1857, p. 6).

Similarly, settler William E Stanbridge remarked that in the south-east of South Australia:

> At Mount Gambier the females collect large quantities of the roots of the fern [probably rhizomes of Bracken Fern, *Pteridium esculentum*], which are eaten when baked, as well as the pretty green and gold frogs [possibly the Growling Grass Frog, *Litoria raniformis*], and a very fleshy mushroom, which is red on the upper and green on the other side [possibly *Boletellus* species]; these are brought home strung on rushes. Our mushroom [*Agaricus campestris*] is very rarely used. In spring they gather cakes of wattle (mimosa [Golden Wattle, *Acacia pycnantha*]) gum, and use it dissolved in water (Stanbridge 1861, p. 291).

Here it was also observed that: 'The men eked out a precarious living by hunting; the women dug roots and gathered small fruit in season' (Stewart c.1870–c.1883 [1977, p. 62]).

As with most Aboriginal peoples, the chief foraging tool for the women of south-eastern Australia was the digging-stick or 'yam stick', and this was made from dense timber (Angas 1847a; Clarke 2012; Mathews 1903; Smith 1880; Stanbridge 1861). Aboriginal women and the children in their care would eat the food they had gathered at day camps and then later bring back a supply for the rest of the band in the main camp (Clarke 2003a). The men fished and hunted more mobile animals, such as wallabies, kangaroos, koalas, possums, Emus and other birds. When it was necessary to ascend trees, stone hatchets were used to cut notches in the trunk as toe grips as well as climbing bands made from straps from the inner bark of the Stringybark (Messmate, *Eucalyptus obliqua*) tree (Bulmer 1855–1908 [1999]; Clarke 2012).

Plant foods

Through the detailed analysis of historical records and comparisons with records across temperate parts of Australia, many of the edible plants of south-eastern Australia have been documented in terms of the methods of collection and preparation (Clarke 2015c; Gott no date; Gott and Conran 1991; Zola and Gott 1992). Aboriginal people here utilised a wide variety of plant parts: fruits, seeds, nuts, sporocarps (seed-like growths from ferns), foliage, stems, galls, gums, roots, rhizomes, tubers and fungi (Brand Miller *et al.* 1993; Cherikoff and Isaacs 1989; Cribb and Cribb 1987; Isaacs 1987; Low 1991; Maiden 1889). The following account is somewhat artificially divided into the parts of plants utilised by Aboriginal foragers.

Seeds

Aboriginal people in southern Australia did not heavily utilise ephemeral sources of edible seed, such as from grasses, although it is known that some Wattle (*Acacia* species) seeds were used when still green (Clarke 2005b). The small millstones that are sometimes found in the Lower Murray Valley were once used for the dry pounding and crushing of coarse seed from wattles and other shrubs (Tindale 1977).

Wattle seeds, while still green like peas, were probably the main source of seed in the diet of Aboriginal people in south-eastern Australia (Clarke 1986b; Gott and Conran 1991; Zola and Gott 1992; Smyth 1878; Tindale 1977, 1981). In the lower south-east of South Australia, a colonist recorded that: 'Kundurt. (Bushy Wattle) [the] Blacks make bread out of the seed' (Davidson 1898 [2001, p. 329]). The Coastal Wattle (Boobialla, *Acacia longifolia* subsp. *sophorae*) was a prolific food source in southern South Australia and probably in many parts of south-western Victoria (Bonney 1987, 2004; Clarke 1985b, 2015c; Campbell *et al.* 1946; McCourt and Mincham 1987). In the 1980s, Lower Murray people described this species as having edible 'beans', as they mainly collected it when the pods were still green (Clarke 2014a, pp. 34–5). In south-eastern Australia, the seeds from other acacias, such as the Coastal Umbrella Bush (*Acacia cupularis*) and the Red-stemmed Wattle (*A. leiophylla*) were probably also collected for eating (Bonney 1987; Clarke 2015c).

During the summer months, Moandik people at Kingston in the south-east of South Australia made cakes by crushing Coastal Wattle seed and mixing it in with mashed fruit, and then adding to it the nectar shaken from the Native Honeysuckle (*Banksia* species) cones for cooking in the ashes (Clarke 2015c). The wild fruit that was used included the Coastal Ballart (Doll's Eyes, *Exocarpos syrticola*), White Currant (*Leucopogon parviflorus*), Muntry (Monterry, *Kunzea pomifera*) and Pigface (*Carpobrotus rossii*), which all grow along the coast. Naturally formed limestone hollows, often with some modification, were used to mash the seed and fruit together (Anonymous 1934a). A surplus of sweet cakes from the Coorong area of South Australia were used to trade for stone for toolmaking (Tindale [cited Clarke 2015c]).

Flour made by grinding the seed of ephemeral sources would have been more often used for food in the drier parts of south-eastern Australia, with the main sources being Lovegrass (*Eragrostis* species), Panic Grass (*Panicum* species) and Common Purslane (*Portulaca oleracea*) (Gott 2008). In some more arid riverine areas on the northern fringe of south-eastern

Australia, the sporocarps or seed-like structures of the Nardoo Water Fern (*Marsilea* species) were eaten after extensive preparation to remove toxins (Smyth 1878; Bulmer 1887; Worsnop 1897; Zola and Gott 1992). Since only the dry sporocarps are suitable to collect and prepare, the removal of the hard-outer covering required grinding on hard stone slabs (Clarke 2013b). Hard seed as part of the diet was probably a hard time food category that was only used when more easily obtained sources were not available. The species mentioned above were akin to those more heavily relied upon by Aboriginal people in Central Australia (Latz 1995).

Fruits

As sweet food, fruits would have been highly desired by foragers, and therefore extra effort taken in its collection. Most fruit harvests were highly seasonal. For instance, missionary John Bulmer recorded that:

> In Gippsland the kangaroo apple [*Solanum vescum*] was enjoyed when plentiful, but they did not grow every season. They are a sort of tomato and are fairly good to eat when properly ripe (Bulmer 1855–1908 [1999, p. 66]).

A botanist has remarked that this species: 'can be managed as a fireweed, abundant after fire and disappearing a few years only to re-appear with the next fire from the soil seed store' (Gott 2008, p. 219). With the growth of such desirable plant species being so heavily dependent upon fire, it is apparent that in Aboriginal Australia foragers managed their food sources in the short and medium time with the judicious use of vegetation burning as a land management tool (Burrows 2003; Cahir *et al*. 2016; Hallam 1975; Jackson 1968; Jones 1969; Latz 1995; Nicholson 1981; Rose 1995; see Chapter 7).

The Dillon Bush (Nitre-bush, *Nitraria billardierei*) has a common name which is thought to have been derived from *dilanj*, a term for this plant in the Wemba Wemba language of western Victoria (Clarke 2008a; Dixon *et al*. 1992; Ramson 1988; Zola and Gott 1992). This plant appears to have a close association with humans, which predates British colonisation. It was noted during the late 19th century at Swan Hill in northern Victoria that large Dillon Bushes were often found growing in the vicinity of Aboriginal camps, probably because of their seeds lodging there after being eaten (Clarke 2008a; Beveridge 1883; Gott and Conran 1991). Aboriginal foragers ate the saline Dillon fruits, which look like European grapes, raw including the stones (Eyre 1845; Low 1989; von Mueller [cited Smyth 1878]; Zola and Gott 1992). The Dillon Bushes prefer growing in disturbed situations, a characteristic that has become more apparent in recent times by this species spreading along stock routes in overgrazed areas in the semi-arid and temperate zones (Low 1989).

Fruit of the Muntry (Native Apple, *Kunzea pomifera*) was often recorded as an Aboriginal food in south-eastern Australia during late summer and early autumn (Smith 1880; Tindale 1981; Gott 1982b; Clarke 1985b; Zola and Gott 1992). The Australian English name of this plant was possibly a borrowing, as *mandharri*, from a Lower Murray Aboriginal language (Clarke 2008a, 2014a; Dixon *et al*. 1992; Ramson 1988). In the south-east of South Australia, these fruits were dried before pounding into large cakes for trade, such as in exchange for stone tools (Clarke 1985b, 2003a; Tindale 1974, 1981). During the summer time, large

Muntry (*mandharri, nurp, nurt*) fruit, one of the reasons for Aboriginal bands to move towards the coast during the summer. PA Clarke, South East of South Australia, 1987.

numbers of Aboriginal people who camped along the coast collected the Muntry (Clarke 1985b; Gott 1982b; Tindale 1974). An explorer passing through to the South East of South Australia in 1844 remarked that on the Coorong 'When the monterrie [Muntry] is ripe, the natives disperse themselves over the sand-hills in search of them, returning in the evening with their neatly-made grass bags filled with them' (Angas 1877, p. 4).

In the Glenelg River area on the southern end of the border between South Australia and Victoria, large groups of Aboriginal people from a wide-ranging area came to the coast to collect Muntry fruit, which was known locally as *nurp* (G.S. Lang, cited Smyth 1878, vol. 1, p. 219). Colonist James Dawson in south-western Victoria claimed that the *nurt* (Muntry), which is a 'red-cheeked cherry without a pip', grows on the hummocks near the Glenelg River mouth and that:

> It is very much sought after, and, when ripe, is gathered in great quantities by the natives, who come from long distances to feast on it, and reside in the locality while it lasts. In collecting the berries they pull up the plants, which run along the surface of the sand in great lengths, and carry them on their backs to their camps to pick off the fruit at their leisure (Dawson 1881, p. 22).

A 19th century newspaper correspondent commented upon the location of what (from the Aboriginal name) appears to be the Muntry and stated that:

> 'Swan Lake,' or the Lake of Swans, in Victoria, is situated 25 miles south west of Portland, and is a desolate region on its sea-girt margin, consisting of white sand hummocks, which are monotonous and displeasing to the eye at first view. In summer these hummocks are covered with sand grapes which people call 'Napps,' a species of wild berry which makes very excellent jam, and could be collected there in great abundance (Grassie 1878, p. 4).

There were many other fruits used as Aboriginal food, including Coastal Ballart (*Exocarpos syrticola*), Dodder (Devil's Guts, *Cassytha* species), Inland Wild Cherry (Ballart, *Exocarpos cupressiformis*), Mistletoe (Snotty Gobbles, *Amyema* species), Native Cranberry (*Astroloma humifusum*), Pigface (*Carpobrotus rossii*), White Currant (*Leucopogon parviflorus*), White Elderberry (*Sambucus gaudichaudiana*) and Wild Juniper (*Myoporum insulare*) (Bonney 1994, 2004; Clarke 2015c; Gott and Conran 1991; Zola and Gott 1992). Some of these foods are best eaten when they are over-ripe, such as the Apple Berry (Wild Date, *Billardiera cymosa*), which Aboriginal people have described as having a 'jelly bean'-like fruit that is best eaten after it has fallen to the ground (Clarke 2015c, p. 261). The fruit of the Quandong (*Santalum acuminatum*) was another food source, which is found on low-growing trees across the semi-arid southern regions of Australia (Bulmer 1855–1908 [1999]; Eyre 1845). This species, along with the Muntry, has the distinction today of being one of a few Indigenous foods that is the basis of the Australian 'bush tucker' industry (Ryder and Latham 2005; M Clarke 2012).

In Aboriginal Australia, the definitions of foods and medicines are blurred. For instance, Aboriginal people in south-eastern Australia ate 'sheoak apples', which are the whole female cones from various Sheoaks (*Allocasuarina* and *Casuarina* species), as a tonic (Clarke 1985b, 1987, 2014a; Gott and Conran 1991). Before the cones have hardened they are red and can be either consumed raw or cooked. Ngarrindjeri living on the Point McLeay Mission used to gather 'sheoak apples' from the remnants of scrub on nearby limestone hills and boil them in the pot before eating them sprinkled with sugar as 'blood medicine' (Clarke 1987, p. 5).

Exudates

The nectar of several species of heavy-flowering plants provided Aboriginal people with the basis for making sweet beverages. The flower cones of the Native Honeysuckle (*Banksia* species) were soaked in water held in a wooden or bone container, producing a sweet drink (Angas 1847b; Clarke 1986a; Zola and Gott 1992). Cones were left soaking all day for drinking in the evening. The flower stems of Grass Trees (*Xanthorrhoea* species) were similarly used. Flowers from other plants that produce copious amounts of nectar, such as species of Gums (*Eucalyptus* species) and Bottlebrushes (*Melaleuca* and *Callistemon* species), were taken for sucking while walking or for making drinks later back in camp.

Edible gum, particularly from the Golden Wattle (*Acacia pycnantha*), was a significant food source during the summer that was softened by soaking in water and then taken as a tonic, making it another example of a food that could also be categorised as a medicine

Golden Wattle gum, used as a food, medicine and for making artefacts. PA Clarke, Mount Barker, South Australia, 1986.

(Clarke 1986a, 1987; Gott 1985; MacPherson 1925; Zola and Gott 1992). The Golden Wattle is likely to have been the species that James Dawson in south-western Victoria referred to when he wrote:

> The gum of the acacia, or common wattle tree, is largely consumed as food, as well as for [artefact] cement; and each man has an exclusive right to a certain number of trees for the use of himself and family. As soon as the summer heat is over, notches are cut in the bark to allow the gum to exude. It is then gathered in large lumps, and stored for use (Dawson 1881, p. 21).

A colonist living along the Murray River described edible gum from several types of 'Mimosa' (*Acacia* species) (Eyre 1845, p. 273).

Greens

The foliage of ephemeral annual plants was a highly seasonal food source that is commonly found near watercourses, particularly after fires (Clarke 1986a, 2015c). Along the Coorong, greens like Peppercresses (*Lepidium* species) and other crucifers became available in greater quantity from June, followed by Blady Grass (*Imperata cylindrica*) (Tindale 1981). Aboriginal people ate Australian Thistles (such as *Sonchus hydrophilus*), the young leaves and roots of which were stripped off and discarded, with only the supple stem eaten (Clarke 1986a, 1987). Sick people, in particular, used this food as a tonic. In times of food scarcity, Aboriginal people ate the green tops of the Australian Stinging Nettle (*Urtica incisa*) after baking it between heated stones (Angas 1847b).

The fleshy leaf bases of the Sagent (Coastal Sword Sedge, *Lepidosperma gladiatum*), which grows as tussocks in sandy coastal areas, were eaten after baking (Bonney 2004; Campbell *et al.* 1946; Clarke 1986a; McCourt and Mincham 1987). The fresh sweet leaf bases of the Sagent were chewed during the summer months as a thirst quencher (Clarke 2015c). The leaf bases of Grass Trees (*Xanthorrhoea* species) were chewed when fresh, and these have a sweet nutty flavour (Bonney 2004; Clarke 1986a). Sheoak needles and cones were also said to be sucked to quench the thirst (McCourt and Mincham 1987).

The savoury juice squeezed from Pigface leaves was used to flavour cooked kangaroo and emu meat (Clarke 1986a; Gott and Conran 1991; Zola and Gott 1992). In Victoria, John Bulmer recorded that:

> As they require salt they used the pig-face plant which contained a good percentage of saline matter. This would be eaten mostly when eating fish and often it was also eaten alone. The fruit of the plant was very sweet and formed an article of diet and I daresay had a laxative effect. There were two kinds. The small leaf variety was called *nakalu* [possibly Round-leaved Pigface, *Disphyma crassifolium*] and a form with a larger leaf called *karnbie* [probably Native Pigface, *Carpobrotus rossii*] (Bulmer 1855–1908 [1999, p. 66]).

Another source said, 'I saw the natives between the Grampians and Victoria Ranges eat the fleshy leaf of this plant [Pigface] with kangaroo flesh, as a substitute for salt' (Wilhelmi 1857, p. 6).

Underground plant structures

Underground parts of plants referred to here are variously described in the literature as roots, tubers, bulbs, corms, yams and rhizomes (Pate and Dixon 1982). From the historical record, it appears that Aboriginal people living in temperate Australia relied heavily upon the rhizomes of the Bulrush (Cumbungi or Broad Flag-reed, *Typha* species) that women pulled from the swamps and lagoons (Smyth 1878; Gott 1983; Clarke 1988; Zola and Gott 1992). It was observed that: 'The roots of a flag [bulrush] which is common in Gippsland were roasted producing good food – tasting very much like the arrow-root plant [*Maranta arundinacea*]' (Bulmer 1855–1908 [1999, p. 66]). Carl Wilhelmi visited the Swan Hill area of northern central Victoria and recorded the cooking method for Bulrush rhizomes in earth ovens. He said that:

> The women dig a hole in which they keep a fire lit for some time to heat the surrounding earth as much as possible. This being done the fire is taken out, the bundle of roots are placed inside covered over with earth, and a strong fire

is then lit on the little eminence thus made. After about an hour has thus elapsed they are taken out and distributed to those present. Of the chewed roots which are gathered in they manufacture their nets, in which their knick knacks and provisions are carried. The women can twirl and twist the fibres of these roots on their legs to any length, and they then have the solidarity and appearance of flax [Native Flax, *Linum marginale*]. Thus prepared the root possesses no taste, but containing a deal of starch (Wilhelmi 1857, p. 6).

The Yam Daisy (Native Dandelion, *Microseris lanceolata*) was also among the foods most depended upon by Aboriginal people living in dry sclerophyll woodlands and grasslands of temperate south-eastern Australia (Cahir 2012; Clarke 1988; Gott 1983, 2008; Smyth 1878, vol.1; Zola and Gott 1992). A colonist in Victoria remarked that the '*Mirr-n'yong* [Yam Daisy], a kind of white radish bearing a yellow flower, is dug up and eaten by the children and adults in all places where it grows' (Smyth 1878, vol. 1, p. 49). Although once common, the Yam Daisy has disappeared from many regions through being trampled out by European grazing stock (Clarke 2014a; Gott 2008).

Some sedge species were also important food sources found in abundance on river flats. For instance, the walnut-sized corms of the Marsh Club Rush (*Bolboschoenus medianus*) were prepared by being roasted and pounded between stones into a thin cake (Eyre 1845; Gott 1982a; von Mueller [cited Smyth 1878, vol. 1]; Zola and Gott 1992). As with Bulrushes, Club Rushes grow in the vicinity of the wetlands where Aboriginal people also obtained many of their aquatic animal foods, such as shellfish, fish and wildfowl. Such close proximity of food sources in the wetlands made these areas of great importance for foraging.

In the historical records, there are other root species mentioned as being used for food, although identifying them is generally difficult due to poor descriptions. For instance, John Bulmer provided a record of:

> ... a yam like root and the roots of wild convolvulus [probably Blushing Bindweed, *Convolvulus erubescens*] which were roasted and eaten ... One root or bulb they eat on the Murray was something like a chestnut. It was very hard and required a lot of jaw force to get at its kernel [possibly Sea Club-rush, *Bolboschoenus caldwellii*]. This kind of food may account for the wearing away of their teeth till they were level with gums in old people (Bulmer 1855–1908 [1999, p. 66]).

Ethnographic records from the Lower Murray (Berndt *et al.* 1993) and Victoria (Gott and Conran 1991; Macdonald 1917) indicate that the Native Convolvulus (Bindweed, *Convolvulus* species) was another edible root utilised in south-eastern Australia.

It has been suggested that many varieties of orchids (Orchidaceae) and the Vanilla Lily (*Arthropodium* species) were important food sources that were widespread, with some even found in alpine areas of New South Wales and Victoria (Clarke 1988, 2015c; Gott 2008; Gott and Conran 1991; von Mueller [cited Smyth 1878, vol. 1; Zola and Gott 1992). In the case of the Vanilla Lily, the tubers are known to persist for more than a year, and are large and numerous. Importantly, Vanilla Lily tubers would have been available to Aboriginal groups who moved to the high country of south-eastern New South Wales in summer for the annual feast of Bogong Moths (*Agrotis infusa*).

The available historic literature on Aboriginal plant foods in south-eastern Australia is dominated by plants simply described as 'wild' versions of known European food sources. For instance, the 'wild onion' (Milkmaid, *Burchardia umbellata*) is known to have edible tubers (Bonney 2004; Clarke 1988; Gott and Conran 1991; Zola and Gott 1992), and this might also be the identity of 'Marrine. (bulb) Like an onion used as food', that was recorded by a colonist in the south-east of South Australia (Davidson 1898 [2001, p. 328]). Any references in this region to the 'wild carrot' may be the Australian Carrot (*Daucus glochidiatus*), as Aboriginal people have described them as like the true Carrot (*Daucus carota*) in having a long root and a similar top (Clarke 2015c). A South Australian 'native carrot' was identified as *Pelargonium dissectum* (Anonymous 1887b, p. 3), which is an obsolete name that probably now refers to the Austral Cranesbill (*Geranium solanderi*) or to another *Geranium* or *Pelargonium* species (Clarke 1988; Gott and Conran 1991; Maiden 1889). There is an historical reference to a 'species of geranium' that had 'an acrid flavour until roasted' that was eaten in parts of Victoria and southern South Australia (Stanbridge 1861, p. 291). Fringe Lilies (*Thysanotus* species) are also known to have edible tubers (Bonney 2004; Clarke 1988; Gott and Conran 1991).

Tubers were generally a food source that had a longer season of availability in contrast to other food categories, such as gums, fruit and seed. For instance, the Water Ribbon (*Triglochin procera*) is commonly found in the wetlands of south-eastern Australia and it has numerous edible soft tubers that grow from a rhizome and are able to survive an extended drought (Gott 2008). For this reason, the roots and rhizomes of plants such as Bulrush and Water Ribbon formed the basis of the Aboriginal diet when other foods were scarce or due to the season were not as easily gathered. Indigenous people appear to have possessed detailed knowledge of the life cycle of underground plant foods, as from elsewhere in Aboriginal Australia there are records of Aboriginal women replanting the undersized underground parts of yams with the intention of helping them to grow for gathering next season (Clarke 2003a; Irvine 1970).

The cooked root bark of Mallee trees (such as the Congoo Mallee, *Eucalyptus dumosa*) was an emergency or 'hard time' food that was eaten either by itself or with other plants (Maiden 1889; Clarke 1988; Zola and Gott 1992). Mallee roots were also a source of drinking water in the semi-arid parts of the Murray Mallee (see Chapter 6), to the extent that the Ngarkat people were only forced towards the Murray River in periods of extreme drought (Clarke 1988; Gara 1985; Magarey 1895a, 1895b; Noble and Kimber 1997; Tindale 1974). A species of subterranean fungi was sometimes described in the early records of Aboriginal food as 'native truffle' and 'blackfellows bread' (Australian Truffle, *Laccocephalum mylittae*), and surface fungi were also widely eaten (Clarke 1985a; 1986a; 1988).

The underground parts of certain plants had the added advantage of being a source of fibre, which was collected from chewed scraps and used to make string for knotting into nets, bags, various ornaments and for binding spearheads onto their shafts (Beveridge 1883; Clarke 2012). Examples of such fibre sources are the Flood Mallow (*Malva weinmanniana*) and the Bulrush (Clarke 2012, 2015c). These particular foods were of further benefit producing the material necessary for making the tools for activities such as fishing, collecting and hunting.

Making wild foods 'useful'

Since the British first colonised Australia, wild plants as food sources have had bad press, partly because many of them are widely dispersed in the landscape and are labour intensive to prepare (Clarke 2007a; Santich 2011). For instance, in 1868 scientist John Crawfurd was writing a paper titled the 'Food of the Natives of Australia' and commenting on the list of wild foods provided by botanist Anthelme Thozet when he remarked that 'none of the native fruits of Australia are comparable in flavour to a crab-apple, and the rest are but very poor food' (Crawfurd 1868, p. 113). He went further and stated that:

> There is probably not one of the 'roots, tubers, bulbs, and fruits' [quoting Thozet] that would ever be touched by those who could get a supply of wholesome grain or cultivated food-roots, and perhaps none either that could by any process of cultivation be brought up to the standard of those which have already been modified for human use (Crawfurd 1868, p. 116).

In the 19th century botanists and plant collectors were among those who often tried introducing exotic plant and animal species to Australia as a means, in their eyes, for improving the country's wild resources (Clarke 2008a; Rolls 1984). In the case of knowing how to find and prepare food sources that were otherwise toxic, the transfer of the necessary Aboriginal Biocultural Knowledge between local Aboriginal people and Europeans was generally poor in most frontier regions (Clarke 2013b). Due to failed attempts of the trans-Australian explorers Robert O'Hara Burke and William John Wills to use Nardoo properly in order to sustain themselves while awaiting rescue, Aboriginal food sources like the Nardoo Aquatic Fern in the arid zone were said to be 'no fit food for white men' (Barratt [cited Bonyhady 1991, p. 141]). In many areas, the poor understanding by Europeans of local Indigenous languages was a major barrier to their discovering how to use Australian plants (Clarke 2008a).

For much of the 20th century, the view of a poor Australian flora continued. In 1941, botanist and amateur anthropologist, John B Cleland, remarked that: 'Though European settlement has now been in existence in Australia for over 150 years, not a single species of plant, with one or two trifling exceptions, has been taken into cultivation as a food plant' (Cleland 1939, p. 4). This lack of appreciation, for both the flora and the Indigenous uses and perceptions of it, was not seriously challenged until the late 20th century when more detailed studies of ethnobotany were published (Clarke 2003b).

Introducing 'new' foods

Baron Ferdinand Jacob Heinrich von Mueller was a prominent botanist based in Victoria, who became a driving force in Australian botany during the 19th century (Pearn 1990, 2001; Ross 1996). As a botanist-explorer, he relied on Indigenous food sources to help sustain him. Von Mueller was interested in the possible horticultural uses of Australian plants and he communicated the results of his food discoveries to European scholars. He produced a detailed list of Aboriginal foods from south-eastern Australia (Smyth 1878, vol. 1). He also had a keen interest in the introduction of various exotic organisms into Australia, particularly those plants and animals he considered to be of potential benefit for agriculture (Clarke 2008a; McPhee 1996; von Mueller 1858, 1876, 1888; Webb 2003). He understood the value of Aboriginal Biocultural Knowledge, having once advised pharmacologists to employ a knowledgeable Aboriginal person from Central Australia in order to work out the medicinal dosage for the narcotic known as Pituri (*Duboisia hopwoodii*) (von Mueller [cited Bancroft 1877–80]).

As the result of a field trip in March 1855 to the Lake Wellington area of Gippsland in Victoria, von Mueller wrote a letter to botanist William Hooker at Kew Gardens in England, which said:

> Here on the coast and in other parts of Gipps land I observed a Solanum, called by the aborigines Gunyang [Kangaroo Apple], which promises to become an additional fruit shrub of our gardens. I have not yet obtained the perfect ripe fruit, which is said to be of excellent taste and of which the natives are passionately fond (F. von Mueller correspondence to W.J. Hooker, March 1855 [cited von Mueller 1825–96 [1998–2006, vol. 1, pp. 201–2]).

This plant has been identified as *Solanum vescum* and along with similar-looking plants, such as the *mookitch* or *mayakitch* (*S. laciniatum*) from south-western Victoria and the more widespread species but toxic species (*S. aviculare*), it is also often referred to today as a Kangaroo Apple (Gott and Conran 1991; Zola and Gott 1992).

Von Mueller was impressed enough with the *gunyang* fruit's potential to quickly publish an account of it in the prestigious *Hooker's Journal of Botany and Kew Garden Miscellany*, based in London. In this paper, he claimed:

> The number of fruits indigenous to this Colony [Victoria] is so limited, that any addition to them cannot fail to attract a far more general attention than even the most important discoveries in the medicinal properties of our plants, or in their geographical distribution or affinity likely would secure. With this view I selected from a series of new plants, which were obtained during my last journey through the eastern parts of this Colony, the 'Gunyang,' for an early publication. That the natives apply a special name to this production of our Flora warrants its usefulness in their nomadic life; and as, in fact, the Gipps' Land tribes collect this fruit eagerly, and as probably cultivation will improve it so much as to render the plant acceptable for our gardens (von Mueller 1856, p. 336).

The *gunyang* was later mentioned in von Mueller's list of Victorian Aboriginal plant foods (Von Mueller [cited Smyth 1878, vol. 1]).

The plant proved to have a fairly wide distribution, and the name *gunyang* from the Ganay language of Gippsland was widely adopted for this and other closely related species in different parts of Australia (Clarke 2008a; Dixon *et al.* 1992; Ramson 1988). Von Mueller's open-mindedness was out of step with European colonists to adopt Australian food species used by Aboriginal Australians, so his wish for the *gunyang* to become a widespread garden fruit has not yet happened.

Bush food market

The need to understand and acknowledge Australian Biocultural Knowledge is apparent when considering the desire of broader Australia in recent times to develop distinctive local foods to help support a cuisine that acknowledges the long history of Aboriginal use of wild sources (Bannerman 2006; Jones and Clarke in press; Link 2012). European Australians have been interested in the Muntry (Monterry, Wild Apple, *Kunzea pomifera*) since the time of the early British settlers, when it was brought back to England for growing in gardens (Elliot *et al.* 1993). In more recent times, the Australian horticultural development of 'bush foods' has focused on the Muntry as the main fruit species from temperate Australia with a commercial potential (Ahmed and Johnson 2000; Hele 2001; Jones and Clarke in press; Ryder and Latham 2005; Stynes 1997). Research behind the marketing of the Muntry has produced some extraordinary facts to help attract consumers. For instance, it is reported that the fruit contains up to four times more antioxidants than commercial Blueberries (*Vaccinium* species), it boosts the immune system and is able to prevent cancer, and it produces natural waxes that are good for skin nourishment (Levin 2010; Netzel *et al.* 2007; Schultz *et al.* 2009; Tan *et al.* 2011). In select restaurants, raw fruit is used fresh in desserts and fruit salads, and at home when cooked it is excellent in pies, jams, chutneys and in sweet or savoury sauces (A Fielke pers. comm.).

During the initial stages of the commercial development of the Muntry, local Indigenous communities with the knowledge of how, when and where to collect it were heavily involved, although this level of involvement is changing (Low Choy *et al.* 2016; Jones and Clarke in press). Over the past decade, commercial growers' groups have established the optimal horticultural conditions for the Muntry, finding for instance that excessive soil moisture in waterlogged soil produces less flavour in the fruit. While the wild Muntry is a groundcover in exposed areas, like the tops of sand dunes and on banks, the commercial growers use a trellis for the plants in order to provide improved access for harvesting and for better plant management (Hele 2001).

Muntry fruit appears set to become more widely commercialised, but to begin with mainly in niche markets, particularly 'bushfood' restaurants and novelty food shops (Berkinshaw 1999; Cherikoff and Isaacs 1989; Cherikoff and Johnson 2000; Graham and Hart 1997; Hegarty *et al.* 2001; Hele 2003; Herbert 1999; King 1997, 1998; Robins 1996; Roff 1983; RIRDC 1998). Venues for selling wild Australian spices, sauces and jams are chiefly hotels, railway and airport cafes, gift stores, museum/art gallery shops and other tourism outlets. In such places, there is already a strong market for Rosella Bush relish, Bunya-bunya toffee and Quandong preserves. The highlighting of Indigenous cultural aspects of the food is generally part of the marketing strategy, both for supermarket products and for specialist 'bush tucker' restaurants (Craw 2012). There is the need to protect the rights of the Traditional Owners of the Aboriginal Biocultural Knowledge that is accessed when developing methods of food production and producing information for marketing (Janke 2009; Jones and Clarke in press).

Aboriginal communities are involved in the collecting of wild foods for restaurants and food outlets (Archer and Beale 2004), although the bush has only a limited capacity for the harvesting of its bushfoods on a commercial scale. For this reason, the successful adoption of most 'new' foods to the market will probably lead to some form of cultivation to boost supply. Developments such as this allow horticulturalists to develop superior plant strains in response to the market, as has happened with Macadamia Nut (Bauple Nut, *Macadamia integrifolia*) from the central east coast of Australia, although in Hawaii, not Australia (Cribb and Cribb 1982; Low 1988; Orchard and Wilson 1999).

Discussion

Aboriginal people in south-eastern Australia had a holistic view of their landscape, which resonates to some degree with how contemporary academic ecologists model the physical environment (Ens *et al.* 2012, 2014, 2017; Pert *et. al.* 2015). Deep understandings of the distribution and seasonality of plant foods were important for Aboriginal foragers, particularly during hard seasons and severe climate events. With many food sources the knowledge of how to procure and prepare them was also important, particularly since some sources, such as Nardoo, are highly toxic in the raw state (Clarke 2007a, 2013b). The unpredictable lives of foragers were controlled by the practice of collecting a surplus of some plant foods, such as fruit and gum, for later use and for trade (Clarke 2003a; Tibbett 2004). Aboriginal communities had rules that gave clans ownership over key plant resources, for example, access to Wattle trees that were sources of edible gum.

In recent years, historians and writers have argued that Indigenous people actively managed their food production from the Australian environment through their burning and harvesting regimes, to the extent that they should essentially be seen as 'farmers' (Gammage 2011; Pascoe 2014). A basic assumption of such an argument is that hunting/gathering modes of subsistence are less developed than those classified as agricultural, and therefore the description of Indigenous people in Australia as hunter-gatherers fails to recognise their deep links to country. Other researchers have refuted this, stating that all human societies are of the present, and therefore do not exist within an evolutionary hierarchy (Clarke 2003a; David 2002; Hiscock 2014). Scientific evidence from palaeoenvironmental research indicates that human firing in parts of Australia did result in some local changes in the landscape, but the extent to which this was a deliberate action or an unintended consequence of long-term human activities is difficult to determine (Butler *et al.* 2014; Williams *et al.* 2015). Ecologists have remarked that 'the impact of any such anthropogenic forcing may have been entirely overshadowed by the effects of natural climate change and variability, as well as the generally low nutrient status of Australian soils' (Bird *et al.* 2013, p. 439).

The greater use of the foods that formerly sustained Aboriginal people for many thousands of years will potentially help produce a greater appreciation of Indigenous culture in Australia. European Australians, however, have historically been conservative when it comes to adopting different food sources. The challenge for future development of the bushfood industry is how to keep Indigenous interests as an integral part of it (Jones and Clarke in press).

Chapter 5
Animal food

Philip A. Clarke

The Indigenous peoples of south-eastern Australia used their Aboriginal Biocultural Knowledge to place themselves within their foraging territories according to seasonal food availability. Plant foods would have been the mainstay, although animal sources rated high as favoured foods and therefore considerable effort was expended in obtaining it. The division of animal food occurred according to the rules of their society. The techniques used to forage food were sophisticated and allowed for the recovery of the food sources.

Introduction

Aboriginal foragers in south-eastern Australia lived in a region that was rich in resources, which meant that their diet changed according to season (see Chapter 15) and their own food preferences. Victoria is a large and environmentally complex part of the south-eastern Australian region, leading a colonial scholar to remark that:

> Victoria, like other parts of Australia, presents diverse physical features; in one area the larger animals are numerous, in others rare. In some parts the natives had to depend for their means of subsistence mainly on fish; in other parts mainly on the kangaroo [such as the Eastern Grey Kangaroo, *Macropus giganteus*]; in well-timbered tracts opossums [such as Ringtail Possum, *Pseudocheirus peregrinus*; Brushtail Possum, *Trichosurus vulpecula*] were numerous, and on the plains they caught the emu [*Dromaius novaehollandiae*], the turkey [Australian Bustard, *Ardeotis australis*], and the native companion [Brolga, *Grus rubicunda*]. In and on the margins of the forests they took the bear [Koala, *Phascolarctos cinereus*], and in the volcanic tracts wombats [Common Wombat, *Vombatus ursinus*] multiplied. Many of these animals, the larger weighing as much as 150 lbs. [68 kgs], were not very difficult to capture; and the black [Aboriginal man], with his family, lived in comfort as long as the flesh of these was procurable (Smyth 1878, vol. 1, p. xxxiv).

Animal food division

Each day, much of the vegetable food brought back to camp would have been widely available for all band members (see Chapter 4), but the fish and game meat the men obtained was butchered and then pieces of it distributed according to established rules, depending on the species and often its age and gender (Clarke 2003a). Such practices effectively resulted in sharing the success of the hunt across the whole band. Anthropologist Alfred Howitt recorded that the protocols concerning the division of food at Bairnsdale in the Gippsland was such that when:

> A man catches seven live eels [Short-finned Eel, *Anguilla australis*]; they are distributed thus. (It is supposed that his family group consists only of these named).
>
> 1. Front half, himself; hind half, his wife.
> 2. Front half, his wife's brother; hind half, his sister.
> 3. Front half, his elder sons; hind half, his younger sons.
> 4. Front half, his elder daughter; hind half, his younger daughter.
> 5. Front half, his brother's sons; hind half, brother's daughters.
> 6. One whole eel to his married daughter's husband.
> 7. One whole eel to his married daughter.
>
> A second instance: A native stork [possibly White-faced Heron, *Egretta novaehollandiae*] is caught. Right side to father and mother, who keep right leg and give right arm away. Left arm and leg to wife's father and mother … Left foot is given to brother. Head, backbone and liver belong to self and wife, who first eat liver, then head. Ears go to wife, who gives right ear to sister. Husband has neck. Both eat the back (Howitt 1877 [cited Stern *et. al.* 1930, p. 262]).

Across south-eastern Australia, people generally avoided eating the totemic species they were associated with, if it was an animal (Clarke 2015b; Howitt 1904; Pepper and de Araugo 1985; Smith 1880). This was because it was 'their own flesh' (Stewart, c.1870–c.1883 [1977, p. 63]), although people outside the totemic group were allowed to do so. In the Lower Murray, the *ngaitji* (*ngaitye*, totemic familiar) meant 'friend', and the 'ngaitye, or totem, may be killed and eaten by those who possess it, but they are always careful to destroy the remains, such as bones, feathers, &c., lest an enemy should obtain them, and use them for purposes of sorcery' (Taplin 1879, p. 35). Consumption of certain animal-based foods had specific perceived risks for sections of the community. For instance, it was recorded in south-western Victoria that: 'Boys are not allowed to eat any female quadruped' (Dawson 1881, p. 52). Similarly, here, 'Women are not permitted to eat the flesh or eggs of the gigantic crane [Brolga], or of the emu, till they are old and greyheaded' (Dawson 1881, p. 53).

Ngarrindjeri people of the Lower Murray in South Australia believed that parts of the Emu's flesh contained sacred power, and therefore the butchering of them was highly ritualised. Alfred Howitt explained that:

> … when an emu is killed, it is first plucked, then partly roasted, and the skin taken off. The oldest men of the clan, accompanied by the young men and boys, then carry it to a retired spot away from camp, all women and children being warned not to come near them. One of the old men undertakes the

dissection of the bird, and squats near it, with the rest standing round. He first
cuts a slice off the front of one of the legs, and another piece off the back of the
leg or thigh; the carcass is turned over, and similar pieces cut off the other leg.
The piece off the front of the legs is called *Ngemperumi*; that off the back of the
leg or thigh, *Pundarauk* (Howitt 1904, p. 763).

Here, when butchering an Emu back at the main camp, it is placed on the ground and:

The bird is then opened and a morsel of fat taken from the inside and laid with
the sacred or *Narumbe* portions already cut off on some grass. The general
cutting up of the whole body is then commenced, and whenever the operator is
about to break a bone, he calls the attention of the bystanders, who, when a bone
snaps, leap and shout and run about, returning in a few minutes only to go
through the same performance when another bone is broken. When the carcass
has been cut up into convenient pieces for distribution, it is carried by all to the
camp, and may then be eaten by men, women, and children, but the men must
first blacken their faces and sides with charcoal (Howitt 1904, p. 763).

Apart from the species and gender of the game, the manner in which animals were
caught was also a consideration when deciding how the meat was to be prepared for
consumption. For instance, in south-eastern New South Wales there was the practice of
hunters ritually addressing the footprints of a kangaroo they were tracking in order to make
it easier prey, but it was said that: 'Before cooking such an animal, the man and his
companions dance round the body for the purpose of exorcising the magic which it has
absorbed from his incantations' (Mathews 1904a, p. 255).

Invertebrates

Across temperate Australia, insects and shellfish were a common Aboriginal food source
(Smyth 1878; Tindale 1966). In terms of their nutritional benefits, Witchetty Grubs have
high protein and fat contents, lerps are rich in carbohydrates, while shellfish contains much
protein and is rich in minerals (Miller *et. al.* 1993; Peterson 1978; Latz 1995; McDonald
2003). For Aboriginal groups that could not rely on abundant sources of vertebrate meat, the
contribution of invertebrates to their diet would have been relatively high.

Lerps

The leaves and stems of some eucalypts were sources of edible manna. Although this appears
to be a plant food source, it is produced by insects. Across Australia, eucalypts are hosts to
minute sapsuckers called sugar lerp (from the family Psyllidae) or lac scale (family Kerriidae),
that appear during summer on leaves, often tended by ants, and produce a sweet white flaky
substance described as manna that are the edible small protective shelters (Clarke 2007a).
These can be scraped off simply by running the leaf through the teeth. It has been claimed
that Aboriginal people could collect 18 to 23 kg of sugar lerp in a single day when in season
(Beveridge 1883). Lerp manna from the Peppermint Box (*Eucalyptus odorata*) was infused in

Sugar lerp produced by Psyllid insects during the warmer months of the year. PA Clarke, Adelaide Plains, South Australia, 1986.

water and the resulting solution used as a drink (Taplin 1874 [1879]). The Australian English name of lerp is derived from the Wemba Wemba word, *lerep*, from western Victoria (Ramson 1988; Dixon *et. al.* 1992). Here, it was recorded that: 'The natives of the Wimmera prepare a luscious drink from the laap [lerp], a sweet exudation from the leaf of the mallee (Eucalyptus dumosa)' (Westgarth 1848, p. 73).

Grubs

In contemporary Australian English, grubs eaten by Aboriginal foragers are generally referred to as either 'bardi grubs' or 'witchetty grubs', regardless of species (Dixon *et. al.* 1992; Low 1990; Ramson 1988). The main grubs described in the literature as edible root borers are the larvae of moths, such as the Witchetty Grub Moths (*Endoxyla leucomochla*, *E. biarpiti*) and the Giant Swift Moth (*Trictena argentata*, also known as Ghost Moth) (Tindale 1938b, 1966). The larvae of these moths are amongst the heaviest insects in the world. In the case of grubs boring into tree trunks, foods included the larvae of the Long-horned Beetle (*Bardistus cibarius*) and Wattle Goat Moth (*Endoxyla eucalypti*).

In eastern New South Wales, Aboriginal people collected edible grubs from Grass Tree (*Xanthorrhoea* species) trunks and Wattle (*Acacia* species) wood (Bennett 1860). Along the banks of the Murray River in south-eastern Australia, Tindale claimed that: 'Aborigines scrape the ground with digging scoops and smell the holes, thus detecting the high humidity maintained in occupied burrows; a conveniently hooked stick retrieves the grub' (Tindale 1953, p. 63). For these river people, the larvae of the *witjeti* (Witchetty Grub, *Xyleutes amphiplecta*), which lives on Saltbush (Chenopodiaceae) roots, served as both their food and fishing bait (Tindale 1966).

An analysis of contemporary placenames derived from Indigenous words reflects the importance of insects, particularly edible grubs. This Aboriginal Biocultural Knowledge is encoded in the naming of sites. For instance, museum researcher Norman Tindale noted that:

> … the Murray River irrigation town of Waikerie [in South Australia] preserves the memory of great feasts of *Trictena argentata* moths. They still emerge from the roots of the red gums [*Eucalyptus camaldulensis*] and fly every year on the night of the first big autumn rain in early- or mid-April (Tindale 1966, p. 182).

The term, *weikari*, was said to mean 'the rising', in reference to the moths emerging from the ground (Tindale [cited Clarke 2009c, p. 156]).

The skill of Aboriginal foragers in finding insect foods, particularly grubs, became legendary among European colonists. Colonist Peter Beveridge described the subsistence of Aboriginal people living in the river systems of south-eastern Australia, stating that apart from vegetable and vertebrate foods:

> Grubs also of all kinds and sizes are greatly appreciated by them, more especially the large one common to the gum-trees all over the Colony. The natives are very expert in discovering the shrubs and trees in which grubs are to be found, in fact they seldom err; yet to a casual observer, or even one with some acuteness, there is not the slightest perceptible difference in the appearance of a tree or shrub containing numerous grubs and those which do not contain any, but aboriginal observation is wonderfully keen in all matters pertaining to nature; even objects seemingly the most trivial fail to elude their ever ready perception (Beveridge 1883, pp. 36–7).

The adult forms of moths were also eaten. In the 1830s, colonist Edward J Eyre noted that:

> In the southern districts of the Colony at the Tindery Mountains a kind of moth (called 'Bougon' by the natives) [Bogong Moths, *Agrotis infusa*] congregated at a particular season of the year in such great numbers that the blacks flock from all quarters to catch and eat them. They are roasted and the wings rubbed off and then both look and taste very like a burnt almond. The blacks never look so fat or shiny as they do during the 'Bougon' season and even their wretched half starved dogs get into good condition then, in such profusion and so fattening are they (Eyre 1832–39 [1984, p. 60]).

In the Australian Alps, the annual swarming of 'bugong' (Bogong Moths) to aestivate in the rocky crevices of the Snowy Mountains in south-eastern New South Wales provided a superabundant Aboriginal food supply (Bennett 1834; Flood 1980; Froggatt 1903; Worsnop 1897). Here, the summer arrival of Bogong Moths attracted a large number of Aboriginal visitors to the region for feasting. From November to January the moths were collected by smoking them out with small fires, with a bark sheet or blanket placed below to collect them, or through the use of nets. The bodies of the moths, which are oily, were lightly cooked on hot stones, and either eaten directly or pounded into cakes for dry storage.

Termites and ants

A favourite food in the springtime was termites, recorded as 'white ants' and 'flying ants' (Eyre 1845). Egg-laying females were separated from the dirt by winnowing in a bark trough. In western Victoria, a colonist described the local people possessing the Aboriginal Biocultural Knowledge that when Marpeankurrk (Arcturus) 'is in the north at evening, the Bittur [termite larva] are coming into season, when she sets with the sun the Bittur are gone and (Cotchi) summer begins' (Stanbridge 1857, p. 138).

True ants (Formicidae) were also eaten. In eastern New South Wales, Aboriginal people gathered several kinds of ant 'larvae', from both ground and tree-living species (Beveridge 1883). In this region, Aboriginal foragers collected the edible workers by jumping upon nests, letting the swarming ants run up their legs, and then scraping them off (Campbell 1926). In central Victoria, ant eggs removed from a nest in a hollow tree were scraped into a kangaroo-skin bag, which was then shaken in order to separate the relatively heavy eggs from the loose bits of decayed wood and other debris (Blandowski 1855).

Honey

For Aboriginal hunter-gatherers, nests of the Australian stingless bees were highly sought after as sources of honey, comb, pollen and grubs – all often eaten mixed together (Akerman 1979). The most prolific honey-producing bees in northern Australia are *Tetragonula* (formerly *Trigona*) species (Dollin *et. al.* 1997). These bees are not present in temperate regions today, as it is suggested that they were displaced in the southern regions by the early introduction of the European Honey Bee (*Apis mellifera*) (Matthews 1976). Indigenous Australian bees, together with their nests and honey, are generally referred to in Aboriginal English as 'sugarbag' (Arthur 1996, p. 61). It has been remarked that they are 'so dear to the stockman's blackboy, who will neglect everything else to cut the coveted morsel out of the branch' (Froggatt 1903, p. 12).

Across south-eastern Australia, there is much uniformity in ethnographic accounts of how Aboriginal foragers located bee nests. For instance, in Eyre's account of honey gathering from the Murray River in South Australia it was reported that:

> It is procured pure from the hives of the native bees, found in cavities of rocks and the hollow branches of trees. The method of discovering the hive is ingenious. Having caught one of the honey bees, which in size exceeds very

little the common house fly, the native sticks a piece of feather or white down to it with gum, and then letting it go, sets off after it as fast as he can: keeping his eye steadily fixed upon the insect, he rushes along like a mad-man, tumbling over trees and bushes that lie in his way, but rarely losing sight of his object, until conducted to its well-filled store, he is amply paid for all his trouble. The honey is not so firm as that of the English bee, but is of very fine flavour and quality (Eyre 1845, vol. 2, pp. 273–4).

Similarly, from southern Victoria an explorer/botanist described the use of plants to help locate honey, and said that a bee 'was caught and marked by the boy with the feather-like seed of a composite [daisy] plant, and followed to its home in a neighbouring gum tree; this betraying the little industrious community of which it formed a member' (Bunce 1859, p. 77).

Shellfish

Crustaceans were another abundant food source available to Aboriginal people living in marine, estuarine and riverine environments. Along rivers and more permanent streams, Aboriginal women obtained various freshwater crayfish among underwater debris, sometimes using nets (Eyre 1845; Beveridge 1883). The species collected in this manner in south-eastern Australia included the Yabby (*Cherax albidus* and *C. destructor*) and the large Murray Lobster (*Euastacus armatus*). Aboriginal men also gathered Murray Lobster using a large bow-net, which was dragged close to the bottom of the shallows by two or three people (Eyre 1845).

The coastline was a rich zone for collecting a wide variety of molluscs (Clarke 2003a). The 'cobra grub' that Aboriginal foragers collected from old logs exposed to seawater along the New South Wales coast are not insects, as suggested by some (Campbell 1965), but molluscs also known as Shipworms (*Bankia australis*) (Dixon *et. al.* 1992; Rimmer *et. al.* 1983). Freshwater molluscs were a highly prized seasonal food source in central Victoria (Blandowski 1855). Reed rafts were used on Lake Alexandrina in the Lower Murray for travelling to mussel beds, where the women dived with net bags to gather the molluscs (Angas 1847b). For the Murray River of South Australia, Eyre wrote that:

> The women whose duty it is to collect these [mussels], go into the water with small nets (*lenko*) hung around their necks, and diving to the bottom pick up as many as they can, put them into their bags, and rise to the surface for fresh air, repeating the operation until their bags have been filled. They have the power of remaining for a long time under the water, and when they rise to the surface for air, the head and sometimes the mouth only is exposed (Eyre 1845, vol. 2, p. 267).

Aboriginal women kept mussel shells for use as spoons and cutting implements (Angas 1847b). In Victoria, missionary John Bulmer recorded that:

> Sometimes when a man was very hungry (*wilke wilkanu*) or in Gippsland tongue (*ganu ganook*) he would take a bag and dive to the bottom of the river and get mussels, it would take 3 or 4 dives to fill his bag, but as soon as it was

full he would quietly go ashore & roast his catch. He seemed to enjoy his repast (Bulmer 1855–1908 [1999, p. 61]).

The collecting of food such as crustaceans and shellfish has left an enduring pre-European human relic, in the form of middens, upon the landscape of temperate mainland Australia (Angas 1847b; Luebbers 1978; Cann *et. al.* 1991). Middens, comprised of cockles, mussels and other remains from cooking fires, were commonly found along the coast and on riverbanks.

Fish

There is a large fish fauna in the wetlands of south-eastern Australia (Lintermans 2007), so for Aboriginal groups living in the region fish was a major seasonal food source. At Lake Tyers in the Gippsland of eastern Victoria, Bulmer noted that:

> Fish formed a very important article of diet during the summer months. In Gippsland different fish were plentiful according to the season. At the beginning of summer they were able to spear the flounder [Bothidae and Pleuronectidae] and the fat mullet [Mugilidae]. Just before and after the lake opened to the sea (spring or winter) fish attempting to escape to sea were plentiful and many eels and mullet were obtained. The Aborigines generally appeared at these times in good condition, their well fed bodies being well greased with fish fat, they shone from head to foot (Bulmer 1855–1908 [1999, p. 59]).

From such an account, it is apparent that fish were a favourite food source and that much effort would have been expended in obtaining it.

Aboriginal groups living along the Murray River in northern Victoria were also heavily reliant upon fish, as here it was observed that:

> The Murray abounded with fish of many varieties; the *boorndo* (large cod fish [*Maccullochella peelii peelii*]), *boorndo boorndo* (a small cod [*M. macquariensis*]), the *tilyigu* or bream [*Nematalosa erebi*] and *bangnalla* a golden perch [*Macquaria ambigua*]. In the spring, a small very nicely flavoured fish called the *naambu* abounded in the billabongs and large quantities were caught using a net … When the river was in flood and many trees were partially submerged, the cod entered holes in the submerged trees to spawn. The fish was detected by noting commotion of the water and the Aborigines would dive with a small hand net which they placed over the aperture in the tree (Bulmer 1855–1908 [1999, p. 59]).

Similarly, it was reported that the 'Loddon tribe' in north central Victoria lived principally on Bulrush (*Typha* species) rhizomes, which they baked, but in 'January they collect in large numbers to enjoy the fishing season on the Murray' (Blandowski 1858, p. 136).

Ngarrindjeri people in the Lower Murray gave cultural significance to the fact that when their babies tried to speak, their first words were *mam*, which sounded like *mame*, being their word for fish (Taplin 1859–79). Adults apparently interpreted this as the infant's desire to eat

Remains of the Goolwa Cockle (*Donax deltoides*), in a coastal midden chiefly comprised of the shellfish remains from Aboriginal cooking fires. PA Clarke, Coorong, South Australia, 1990.

fish. Ceremonies were performed in order to maintain the high water levels for fishing. Howitt related that:

> There is a spot at Lake Victoria [= Lake Alexandrina], in the Narrinyeri [Ngarrindjeri] country, where when the water is, at long intervals, exceptionally low, it causes a tree-stump to become visible. This is in charge of a family, and it is the duty of one of the men to anoint it with grease and red ochre. The reason for this is that they believe that if it is not done the lake would dry up and the supply of fish be lessened. This duty is hereditary from father to son (Howitt 1904, pp. 399–400).

Techniques for capturing fish ranged from netting, spear fishing, trapping and opportunistic harvesting, with live storage in pens also practised (Clarke 2002). There are documented cases of Aboriginal use of fish poisons in the upstream reaches of the Murray River on the border of northern Victoria (Curr 1883). When large numbers of fish died for natural reasons in the river or lake, often due to seasonal change of water conditions, these were quickly gathered (Eyre 1845). Before the European construction of the barrages across the Murray Mouth separating the Coorong from Lake Alexandrina, the incoming salt water drove upstream species such as Murray Cod, which suspended fishing in the lakes until it receded (Taplin 1859–79).

It was claimed that before the arrival of Europeans, Aboriginal people in south-eastern Australia did not widely use the fishhook and line (Eyre 1845; Meyer 1846 [1879]); Smyth 1878; Davies 1881; Curr 1883; Worsnop 1897; Massola 1956). Nevertheless, the use of bone bi-points or fish gorges (*muduk*) and fishing lines has been recorded along the Murray River (Pretty 1977; Gerritsen 2001). Shell fishhooks have also been recovered from coastal shell middens in eastern Victoria (Mulvaney and Kamminga 1999). In the Gippsland area, bone hooks were reportedly used (Bulmer 1855–1908 [1999]).

Netting

Marine net fishing often involved the coordination of a large number of people. The colonist Thomas Worsnop provided an account of Aboriginal people sea-fishing in the 19th century:

> In Encounter Bay I have seen the natives fishing almost daily. Two parties of them, each provided with a large net, square in form, with a stick at either end, and rolled up, swam out a certain distance from the shore, and then spread themselves out into a semicircle. Every man would then give one of the sticks round which his piece of net was rolled to his right hand man, receiving another from his left hand neighbour, bringing the two nets together, thus making a great seine. They now swam in towards the shore, followed by others of their number, who were engaged in splashing the water and throwing stones, frightened the fish and prevented their escape from the nets (Worsnop 1897, pp. 90–1).

If nets were not available, then branches and bushes could be used to drive the fish up onto the beach (Angas 1847b).

Net fishing in freshwater creeks generally required fewer people than sea fishing, due to the more enclosed spaces (Clarke 2002). Nets were made from Sedge (*Cyperus* species) stem or Bulrush rhizome fibre (Clarke 2012). While sinkers and floats were not recorded as being used in association with any type of net fishing along the Murray River, in Gippsland fixed-position (set) nets had bark floats (*plearts*) and stone sinkers attached to them (Bulmer 1855–1908 [1999]). A settler's description of net fishing in the lagoons alongside the inland rivers of south-eastern Australia involved the use of reed bundle floats and clay sinkers to keep the net vertical in the water column (Beveridge 1883). In this case a man at either end would draw the net together, with Beveridge having stated that:

> On many occasions I have seen three, and four hundred-weight [150 to 200 kgs] of fish drawn from lagoons at single hauls, consisting of cod, perch, catfish [*Tandanus tandanus*], blackfish [river blackfish, *Gadopsis marmoratus*], and turtle. It is quite a sight to see them all tumbling and jumping about on the grass, codfish from 50 pounds [23 kgs] downwards, and perch, both gold and silver [*Bidyanus bidyanus*], from 10 pounds [4.5 kgs] down to 2 pounds [0.9 kgs]; the large mesh of the net prevents the landing of small fish, unless on very rare occasions (Beveridge 1883, p. 47).

Whether intended or not, techniques such as these would have given some protection for the fry of favoured game fish. Historical sources document the use of both dragnets and set nets in south-eastern Australia (Bulmer 1855–1908 [1999]).

Spearing

Fishers manipulated the available interface between shade and light in order to more easily spear fish. During the day, men stood motionless in the river and in shallow seawater, attracting the species such as Mulloway (*Argyrosomus japonicus*) and Snapper (*Pagrus auratus*) by their shadows. When close, the fish were stabbed with large double-pronged spears (Meyer 1846 [1879]; Smith 1930). According to Bulmer:

> On the Murray the Blacks used to dive with spear in hand and spear the fish underwater. This was relatively easy as the Murray Cod was so large, sometimes weighing as much as 120 lbs [54 kgs] (Bulmer 1855–1908 [1999, p. 59]).

Spears were also used in the Gippsland. It was recorded in comparison to bone hooks, that:

> The fish spear was a more useful instrument. I often saw them used very successfully particularly at night when a torch was used. Confused, the fish remained stationary long enough for the Aborigines to take aim (Bulmer 1855–1908 [1999, p. 61]).

Sound was also used when fishing. In the Lower Murray, fishermen frightened fish from their hiding places with a loud noise, created by thrusting one of their large fishing spears into the water (Unaipon 1924–25 [2001]). The compressed air caught between the prongs rose to the surface with a loud report. Another fishing tactic was to use cliff top lookouts as vantage points to see when fish schools had arrived (Clarke 2002).

Along the Murray River, a bark canoe was used as a platform to fish from at night (Angas 1847b; Edwards 1972; Meyer 1846 [1879]). A fire, contained by a clay hearth in the middle, served to attract the fish to be struck by spear or club and also served for cooking the catch. Rafts were made from Common Reed (*Phragmites australis*) stems and dried Grass Tree (*Xanthorrhoea* species) flower stalks. There were various rituals and prohibitions practised to maintain fish stocks (Clarke 2002). According to Bulmer, 'There was sometimes a shortage of fish on the Murray, then they used to go in their canoes and sing while they paddled their canoes as if they were invoking to the god of the river to help them' (Bulmer 1855–1908 [1999, p. 59]). On Victorian reaches of the Murray River, women also reportedly fished from canoes (Bulmer 1855–1908 [1999]). Mobility across water was important in order for fishers to reach areas favoured by particular species of fish.

Fish trapping

The landscape was modified for fishing by the building of stone and wood fish-traps, which are sometimes described as weirs in the literature (Eyre 1845; Smith 1930; Tindale 1974). In south-east New South Wales, anthropologist Robert H Mathews recorded that:

> Catching pens or fish-traps, *ngullaungang*, are made across narrow, shallow inlets on the sea coast or along the course of rivers. These are made by tying together bundles of tea-tree [*Melaleuca* species], and laying them close tighter

like a wall across a creek or narrow shallow arm of the sea. These walls or barricades are slightly above the surface of the water. A gap or gateway is left in mid stream so that the fish can pass through, and when a sufficient number are enclosed, the gateway is blocked up by other bundles of tea-tree, which have been prepared beforehand (Mathews 1904a, p. 253).

According to the above account, a portion of a dead man's skin kept in a bag was used to ritually lure the fish into the enclosure.

In parts of south-eastern Australia today, there remains the physical evidence of Indigenous earthworks made for fishing purposes, some of which are of considerable scale, covering many hectares (Lourandos 1997; Mulvaney and Kamminga 1999; Worsnop 1897). In the Lower Murray and south-east of South Australia, the Aboriginal construction of long trenches to concentrate fish saved much labour expenditure (Clarke 2002; Coutts *et. al.* 1978). With similar but more extensive works at Lake Condah in south-western Victoria, Aboriginal people used trenches to catch Short-finned Eels in wicker baskets, and they then smoked their catch in the hollowed trunks of eucalypt trees (Builth *et al.* 2008; Coutts *et. al.* 1978; Gunditjmara People and Wettenhall 2010; Jones 2011; McNiven *et al.* 2012, 2015; Rose *et al.* 2016). Channels, known as *vam*, were utilised for eel fishing at Mount William Swamp (Robinson 1841 [cited Clark 2014b]).

Rules applied to those who could visit key fishing sites, as the weirs were owned by particular clans (Builth 2005). The use of fish-traps, weirs and trenches enabled Aboriginal people to seasonally congregate in larger numbers than would otherwise have been possible. Not all of the Aboriginal groups that Europeans encountered in temperate Australia fished, as the Tasmanians stopped eating fish some thousands of years ago (Horton 2000; Jones 1978). This dietary change is believed to be the result of a refinement in hunting and gathering strategies that made fish redundant to the Tasmanian diet. In contrast, fish remained central to Aboriginal subsistence in south-eastern Australia (Lawrence 1968).

Birds

Temperate Australia was a region rich in a variety of wild fowl, particularly waterbirds. The large expanse of lakes and lagoons here enabled the Aboriginal people to depend upon avian food to a much greater extent than would have been possible in any of the more arid regions (Clarke 2016b). Aboriginal hunters had many methods for capturing such a diverse range of species. The hunting of birds was highly seasonal due to factors such as migration and breeding behaviour. Some waterfowl, such as ducks and swans, were also much more easily caught when moulting (Bulmer 1855–1908 [1999]; Taplin 1859–79).

Eggs

Eggs were particularly important seasonal foods, and for this reason there were restrictions on the collecting of them in certain places. Among the Ganai (Kurnai) people of Gippsland in Victoria, a patrilineal clan possessed 'exclusive ownership by inheritance of the swan's eggs at the breeding place on Lake Kurlip' (Fison and Howitt 1880, p. 296). Along the Coorong in South Australia, Tindale claimed that:

> In September eggs of swans [Black Swans, *Cygnus atratus*], pelicans and ducks became important and excursions were made to places where emus had begun to lay their giant eggs. It was also the beginning of the season for exploiting mallee fowl [*Leipoa ocellata*] eggs, their incubating mounds being robbed but the birds left unmolested (Tindale 1981, p. 1879).

The peoples of this area could extend the seasonal use of eggs from certain species that produce thick shells by burying them in moist sand (Clarke 2016b). Since birds are widespread, a variety of bird eggs and hatchlings were collected for food in the wetlands as well in the dry scrublands and the forest canopies.

Snaring

Precise knowledge of the habits of gamebird species was particularly important for techniques such as snaring, as often the hunter was not around to observe whether the snare was in the correct position (Smyth 1878; Worsnop 1897; Smith 1930). The use of snaring-rods, which are long poles with a noose attached to the end, was a variation of the snare technique that was common in south-eastern Australia (Meyer 1846 [1879]; Angas 1847b; Worsnop 1897). A method for catching Black Swans involved the hunter, armed with a swan-wing decoy and snaring-rod, hiding among bulrushes and reeds by the margin of a lagoon (Berndt *et al.* 1993; Smith 1930). The splashing of the wing, imitating a bird in distress, attracted distant swans. The loop of the snaring-rod was quickly extended over the bird's head, the rod pulled in and the bird killed. The inquisitiveness of other species was also taken advantage of, with a hunter able to 'decoy pelicans within his reach by imitating the jumping of fish by throwing mussel shells or splashing the water with his fingers' (Smith 1909, p. 10).

Another bird hunting strategy was to place upright sticks a short distance into the lake in areas where shags and cormorants were known to frequent (Angas 1847b). The hunters would swim out with weed-covered heads, snaring the birds with rods as they landed to roost. In Victoria, a similar device was used to capture the Plains Turkey (Australian Bustard, *Ardeotis australis*) (Bulmer 1855–1908 [1999]). Ngarrindjeri hunters in the Lower Murray took advantage of their natural inquisitiveness, luring these large birds away from their daytime shade cover with a butterfly decoy, comprised of white feather plumes that were either swan down or pelican breast feathers mounted on sinew tied to a stick (Berndt *et al.* 1993). During the summer, the Plains Turkeys were more easily caught as they were at this time preoccupied with catching butterflies and other insects.

Netting

Nets were widely used across south-eastern Australia to catch birds, particularly flocks of waterfowl (Eyre 1845; Krefft 1862–65; Beveridge 1883; Worsnop 1897; Smith 1930, p. 223). The nets were strung between two trees in the flight path of the birds. Ducks were flushed and made to fly low with thrown bark discs or boomerangs, accompanied by shrill whistles from hunters to imitate hawks or falcons (Beveridge 1883). This kept the ducks flying low and into the outstretched nets, with many birds caught at once in this manner. Along the backcountry of the Murray River Valley, strong nets were used for hunting Emus, and according to Bulmer:

The net was spread between two trees not far from a watering place. The blacks formed themselves into a large circle and as the emu came in sight it saw men on every side for they were very gradually making the circle smaller. The only clear place was the vicinity of the net, to this it made its way only to get entangled, for the net came down and the emu was helpless. It was soon despatched by the clubs of the men (Bulmer 1855–1908 [1999, p. 64]).

Stalking

Catching birds settled on the surface of an open section of water involved the hunter swimming. Wearing a hat made of rushes, reeds or bulrushes, he would swim out to the waterfowl and pull them under (Bulmer 1855–1908 [1999]; Worsnop 1897; Smith 1930). A flock of Grey Teal Duck (*Anas gracilis*) on the water could be enticed towards the bank by waving the flower tops of several reeds tied together, with the hunter hidden from view (Museum Board 1887; Smith 1930). The reeds apparently looked like the tail of a wild dog (Dingo, *Canis lupus dingo*). When close, the hunter stood up to throw clubs and spears at the birds in the confusion as they took flight. Another method required two boomerangs: after throwing the first to make the birds stay down, the hunter threw another, but much lower to sever the head of any duck it hit (Clarke 2016b; Smith 1909, 1930). Large birds could also be taken by spear (Eyre 1845).

Emus, being large flightless animals, were generally caught in the same way as kangaroos, by stalking with bough shields or by trapping in concealed pits (Smith 1930). An alternative method was to lie in wait next to an Emu pad leading to water. When an Emu approached in the late afternoon, the hunter would thrust out his hand, catching the bird by one of its legs while he clubbed it to death (Smith 1930). In the Gippsland, Emus were generally hunted with spears propelled by spear-throwers (Bulmer 1855–1908 [1999]). Emu hunting was heavily seasonal due to several factors. In Victoria, Bulmer reported that:

> Except when fat, emu flesh was coarse and tough and not much used. The Aborigines preferred to hunt the emu at the beginning of summer after it had been feeding on the *belka*, a plant [possibly Yam Daisy, *Microseris lanceolata*] with a flower resembling that of the dandelion. According to their epic of creation *perrelko barlkama thalgau* – when the emu eats the *belka* it would be strong and fat. I think the fat was the favourite part. It was considered by the Murray people to be a delicacy it was cut into strips and sent as a present to their friends (Bulmer 1855–1908 [1999, p. 64]).

Reptiles

Aboriginal people in south-eastern Australia commonly ate reptiles, particularly larger species of skink and goanna (Bulmer 1855–1908 [1999]). Snakes, however, were often avoided, being particularly hazardous during Bulrush root-collecting expeditions in the swamps (Angas 1847b). Aboriginal people tracked freshwater tortoises (such as the Eastern

Long-necked Turtle, *Chelodina longicollis*) as they moved inland to lay their eggs, and along the river, men generally caught them by diving (Angas 1847b; Blandowski 1855; Eyre 1845; Smyth 1878). For cooking, the gathered tortoise eggs were covered with foliage and then fried in hot ashes.

Frogs were either dug out of the ground by women using a digging stick, or caught in the swamps (Eyre 1845; Bulmer 1855–1908 [1999]; Krefft 1862–65; Smyth 1878; Worsnop 1897). Tadpoles were also eaten, with a South Australian colonist having remarked that: 'I have personally seen the native women of the Encounter Bay tribe at Victor Harbor catch tadpoles from the claypans with a very fine net and cook them in a large mutton-fish [*Haliotis* species] shell' (Worsnop 1897, p. 83).

Terrestrial mammals

Large mammal meat was highly sought after by Aboriginal people across Australia, both as food and for its symbolic value (Clarke 2003a). Kangaroos and wallabies were stalked just before dusk, a time when the animals became more active through feeding (Smith 1930). Hunters used a shield of branches to hide, allowing them to move within spear-throwing distance of their game. A good hunter could strike the kangaroo or wallaby when its head was down feeding. In this way there was a good chance that the dying beast would not frighten away other animals. Rituals were also utilised before hunting. For instance, it was recorded that: 'Near Dandenong there is a rock on which the Ngarukwillam clansmen of the Wurunjerri [Wurundjeri] tribe used to place leafy boughs when going out hunting kangaroos, to ensure a good catch' (Howitt 1904, p. 400). In Aboriginal Australia, hunters typically smoked themselves before setting out to kill terrestrial game in order to remove spirits that might be following them and disturbing game (Clarke 2012). Hunters typically rubbed themselves with a variety of plants, as well as mud and ashes, to help disguise their scent from game animals (Basedow 1907; Clarke 2012; Roth 1901).

Drives

The large-scale hunting of big animals, such as kangaroos and Emus, involved the cooperative efforts of groups of people. A party of men would go out early in the morning, armed with barbed spears and positioning themselves in a large semicircle (Beveridge 1883; Bulmer 1855–1908 [1999]; Eyre 1845; Smyth 1878). Women, children and other men started driving game in the direction of people lying in wait. Often this involved firing the countryside and using the topography of the landscape to the hunter's advantage (see Chapter 7). Gradually the space between driving party members was reduced, the game finally cornered and killed with clubs. If fewer people were available, then large nets would be placed across a well-worn animal pad, and then game driven into it. Sometimes brush fences were used to drive kangaroos and wallabies towards a point or a concealed pit where they could be more easily killed (Eyre 1845; Smith 1930). Cliff edges were also often used as a termination point (Robinson 1842 [cited Clark 2014b]; Taplin 1859–79). In cornering and running down the game, Dingoes were an aid to hunting (Cahir and Clark 2013; Dawson 1881).

Tree climbing

In the forest canopy, various arboreal marsupials were hunted. The scaling of trees with smooth branchless lower trunks could be achieved with the aid of a pointed climbing-stick to chop toeholds or by the use of a climbing band made from Stringybark (*yangoro*, Messmate *Eucalyptus obliqua*) strips (Bulmer 1855–1908 [1999]; Clarke 2012). Hollow trees where possums lived were located by examining tree trunks for fresh scratch marks and small pieces of fur caught on the bark (Eyre 1845; Meyer 1846 [1879]; Smith 1930). The skins of possums were eagerly sought after for use in the making of warm winter clothing (Eyre 1845). The relatively slow moving Koala or 'native bear', was easily taken from trees or when walking on the ground, and the 'natives may not skin the bear. He is roasted whole in his skin. The flesh is said to taste like pork' (Smyth 1878, vol. 1, p. 191).

Digging

Burrowing mammals, particularly Short-beaked Echidnas (*Tachyglossus aculeatus*), Bandicoots (such as the Southern Brown Bandicoot, *Isoodon obesulus*) and various large Rats (Muridae), were either dug up or smoked out of their holes (Angas 1847b; Bulmer 1855–1908 [1999]). In the Murray River region of South Australia, Eyre recorded that:

> Rats are also dug out of the ground, but they are procured in the greatest numbers and with the utmost facility when the approach of the floods in the river flats compels them to evacuate their domiciles. A variety is procured among the scrubs under a singular pile or nest which they make of sticks, in the shape of a hay-cock, three or four feet [90 cm to 1.2 m] high and many feet in circumference. A great many occupy the same pile and are killed with sticks as they run out (Eyre 1845, vol. 2, p. 268).

The above quote refers to a now locally extinct Stick-nest Rat (*Leporillus* species) and demonstrates the seasonal nature of many hunting practices. Wombats (such as the Southern Hairy-nosed Wombat, *Lasiorhinus latifrons* and Common Wombat, *Vombatus ursinus*) could also be dug up or smoked out of their burrows (Angas 1847b; Blandowski 1855; Bulmer 1855–1908 [1999]). Another method of capturing wombats, from the Lower Murray, involved the hunter hiding in bushes during the late afternoon, from where he could observe from which hole a wombat emerged (Smith 1930). When the animal moved on, the hole would be blocked about a metre in. The hunter would then frighten the wombat back to his hole, trapping him between the exit and the block.

Marine mammals

Aboriginal contact with marine mammals happened when they were resting in coastal colonies, stranded on the beach or penetrating inland waterways. There has been an argument that Indigenous people were able to access most offshore islands using watercraft before European settlement (Draper 2015), although it is unlikely that Aboriginal people in south-eastern Australia were able to use their fragile watercraft, built for calm inland water conditions, to kill even small whales (Clarke 1994; see Chapter 8). This situation is in contrast to that for Aboriginal hunters in northern Australia, who had more access to the sea through their use of robust dugouts and outriggers gained through contact with Macassans and Torres Strait Islanders (Haddon 1913; Tindale 1926; Baker 1988). For marine animals stuck in shallow water or stranded on land, killing was a more straightforward task. At Encounter Bay in South Australia, Ramindjeri people swam or travelled on rafts to West Island, which is less than a kilometre off the coast south-west of Victor Harbor, in order to kill Australian Sea Lions (*Neophoca cinerea*) and New Zealand Fur Seals (*Arctocephalus forsteri*) (Tindale 1941). They also travelled to Pullen Island on rafts in calm weather for sealing. More distant islands, such as Kangaroo Island, were places beyond the range of Aboriginal hunters immediately before European settlement (Clarke 1996, 1998a). There is no evidence to suggest that Aboriginal people used watercraft to actively hunt whales or dolphins in south-eastern Australia.

Sealing

Across Victoria, the Australian Fur Seal (*Arctocephalus pusillus*), Leopard Seal (*Hydrurga leptonyx*) and the Australian Sea Lion were frequently implicated in the inland 'sightings' of the mythical bunyip (see Chapter 3), although they were often hunted on the coast. At Cape Otway in south-western Victoria, seal bones were reported as being present, along with other bones and shells, in Aboriginal middens (Wilkinson 1865 [cited Howitt 1904]). In the Gippsland area, the seal (*ngaliwan*, probably Australian Fur Seal) was killed when found on the beach (Bulmer 1855–1908 [1999]). In the south-east of South Australia, the New Zealand Fur Seal and the Australian Sea Lion were hunted (Tindale 1941). Due to Aboriginal hunting activities in the pre-European period, many of the main breeding grounds of seals were on offshore islands, such as those in Bass Strait, which were largely places beyond the range of Aboriginal hunters immediately before European settlement. In the early historic period the labour and skills of Aboriginal women were heavily utilised by European sealers (Clarke 1996, 1998a).

The stranding of whales and dolphins

Aboriginal foragers took advantage of the stranding of whales and dolphins whenever they occurred (Clarke 1994, 2001). In South Australia, whales were occasionally stranded along the Lower Murray coast, with one early observer suggesting that they are 'possibly flurried by getting into the volume of fresh water of the Murray River' (Howitt 1904, p. 132). In the Gippsland area it was claimed that: 'The whale (*kaandha* [probably several species]) and the porpoise (*kornon*) [probably several species of dolphin and porpoise] are only procured when stranded. No efforts are made to catch them' (Bulmer 1855–1908 [1999, p. 59]). The seas off southern Australia are part of the migration route of the Southern Right Whale (*Eubalaena australis*) across the Southern Ocean (Carroll *et al.* 2011). The occasional stranding of whales were events that attracted many Aboriginal bands to the coast for feasting, and while Europeans stated that whales were not previously hunted, those Aboriginal people concerned believed that they had a more active role in some beaching incidents with their 'strong men' singing to cause a stranding (Tindale 1974, pp. 18, 23–4, 80).

It appears that in most Aboriginal languages, all the large whale species were classified in the same category. Records from Encounter Bay in South Australia suggest that whales were generally called *kondoli* (Clarke 2001), although other equivalent terms existed. A word that referred to the whales blowing water was recorded in South Australia, and it was recorded by Tindale that here:

> The whales which frequented the shores off the mouth of the Murray River and the rocks at Encounter Bay where they often came close in shore were called *winkulare*, literally the 'whistlers' or 'blowers' from the intransitive verb *winkulun* 'whistle' because of their 'blowing'. Magical spells designed to entice the whales ashore were practised by the 'clever' men (Tindale [cited Clarke 2001, p. 20]).

For the lower south-east of South Australia, there is a transcription of a whale song in the Booandik (Bunganditj) language. This was translated as 'The whale is come, And thrown up on land' (Smith 1880, p. 139), with the term for whale being *konterbul*. These songlines were repeated over and over, and perceived as inducing a stranding. At Twofold Bay in south-eastern New South Wales, local Aboriginal people made a distinction between the Humpback Whale (*Megaptera novaeangliae*) and the smaller Orca (Killer Whale, *Orcinus orca*) that hunted them, with the Orca known as *beowa* (Anonymous 1912).

Members of the Kondilindjera ('whale people') at Encounter Bay believed that some of them could 'sing' whales in towards the shore or out, by standing on a rock and singing some 'wordless chant' (Tindale [cited Clarke 2001, p. 20]). There is a whale song recorded in the Ramindjeri dialect spoken at the northern end of Encounter Bay that was intended to protect whales that were close inshore (Tindale 1931–34, 1937). This was sung by a man of the *kondoli ngatji* ('whale totemic clan'), in order to assist a female whale and calf escape the shallow waters of Encounter Bay, and thereby thwart 'evil-minded' people who desired the whales to be stranded so that *kraipunuk* or oil could be collected for 'spear poison' (Tindale 1937, p. 112). Tindale stated that:

> A man who had the whale as totem would not eat whale. If a whale became stranded on the coast, he would give permission for others to use it as food, but he himself would merely rub himself with the oil which it produced. If he saw a whale floundering off the shore, he would sing a magical song, telling the whale to avoid the shallows and escape to the sea (Tindale 1936b, p. 23).

Knowledgeable men in eastern Victoria reputedly knew when a whale stranding had occurred. Howitt said that:

> Besides learning news about absent friends and possible enemies from the ghosts, the *Birraarks* ['spirit-medium' men] were material benefactors to their tribesmen, as, for instance, when the *Mrarts* [ghosts] informed them of a whale stranding on the shore, for it was in such cases thought that the whales were killed by the *Mrarts* and sent ashore for the Kurnai. At such times

Aboriginal camp incorporating whale bones. Whaling station, Victor Harbor, South Australia. George French Angas watercolour, *South Australia Illustrated*, 1844. Image courtesy of the National Library of Australia.

messengers were sent out, and the people collected to feast on the food sent them. No doubt the *Birraark* was at such times not forgotten (Howitt 1904, pp. 391–2).

Whale materials were seen as containing special powers. There is a detailed account of how a man named Tankli among the Ganai (Kurnai) had become a *Mulla-mullung* or 'clever man', which involved older men tying him with a cord made from whale sinew and being ritually swung through the air over the sea at Corner Inlet and landing at Yiruk, known today as Wilsons Promontory (Howitt 1904). The symbolic importance of whales across Aboriginal Australia is reflected in the rock art of the Sydney area, which depicts people inside whales (Stanbury and Clegg 1990). It has been suggested that this might be explained by the eastern seaboard practice of sick people lying inside the body of a stranded whale.

Normally, the coastal districts were sparsely populated during the wintertime (Clarke 1994; Smyth 1878, vol. 1; Tindale 1938a; 1974). The cutting of the flesh and blubber was performed with flint knives, *maki*. Aboriginal people cooked the fat before it was eaten and the oil was rubbed on their bodies for protection against the weather. The *witjeri*, leaves of the Pigface (*Carpobrotus rossii*) that grows along the foreshore, were used as a relish with the whale meat. Ceremonies were held during the nights of the feast, and the site of the stranding was not left until all the blubber was eaten.

Aboriginal people used the whale earbones as drinking containers and water storage vessels (Clarke 2001; Hemming *et al.* 2000). In the Rivoli Bay area of the south-east of South Australia, whales also stranded, providing big feasts for the local people. The Indigenous name for this place was Weirintjam or Wilitjam, which was said to mean 'place of whales' from the Potaruwutj word *weirintj*, a whale (Tindale c.1931–c.1991). Also in the south-east of South Australia, it was reported that: 'A dead whale cast ashore occasionally provided them with a feast they delighted to recall to mind in later years' (Stewart c.1870–c.1883 [1977, p. 62]).

Discussion

Overall, the Aboriginal diet of south-eastern Australia was probably based on the sources of food that were constantly available, such as roots and rhizomes (Lawrence 1968). When seasonally available, sources such as fish and waterfowl would have led to less reliance on vegetable foods. The high cultural value of sources, such as kangaroo and Emu meat, meant that considerable efforts would be expended to procure it (Clarke 2003a; Dawson 1881). Insects were also a favoured food source (Campbell 1926; Tindale 1966), although modern Western Europeans perceive them as not suitable for consumption (Yen 2010, 2015). Emergency ('hard time') food sources would have been available during harsh climate episodes, such as droughts and floods, as well as providing local groups the resources to feed a large influx of people arriving for ceremony (Cane 1987; Clarke 2003a, 2007a). Aboriginal Biocultural Knowledge was crucial during periods of pressure on local food supplies, as many of the emergency food sources were difficult to collect and prepare.

In the past, scientists believed that the Australian flora and fauna did not possess any species that would have allowed the development of societies of horticulturalists or pastoral nomads. In 1868, a scientist spoke of the 'vegetable poverty of Australia, in so far as human food is concerned' (Crawfurd 1868, p. 113). In the late 19th century, German geographer and ethnographer Friedrich Ratzel blamed the limitations arising from the natural resources of Australia, which included the plants, animals and water sources, for preventing the development of any form of agriculture among the Aboriginal people (Ratzel 1896). He considered that the lack of Indigenous human food production in Australia was due to 'the pressure of a climate particularly untrustworthy in respect to moisture' (Ratzel 1896, vol. 1, p. 342). With little understanding of the role of Aboriginal Biocultural Knowledge in Australian societies, these scholars saw Aboriginal foragers as slaves to the environment.

Through much of the 20th century, the heavily environmental deterministic models of human/environment relationships in Aboriginal Australia prevailed. In 1939, South Australian-based scientist John B Cleland stated that in Australia:

> ... there is no mammal that could be a source of milk, and the natural pastures would be a much more suitable field for the multiplication of kangaroos and wallabies than any process of domestication and control could be. In other words, the Tasmanian and Australian natives reached a land that gave them no opportunity whatever to practise the arts of plant cultivation or of animal

husbandry, even had they possessed a knowledge of such before their arrival (Cleland 1939, p. 4).

In an ecological paper published in 1981, museum researcher Norman B Tindale spoke of a dilemma with the alleged primitiveness of Aboriginal Australians, who were still nonetheless able to thrive in such a challenging landscape. He remarked that: 'There is increasing appeal in an understanding of how fellow man, armed with little more than some primitive stone and wooden implements, and a wholesome philosophy, is able to grapple with his oft hostile and always untamed environment' (Tindale 1981, p. 1855).

More recent research in south-eastern Australia, with its comparatively resource-rich environment, has overturned the above views of an impoverished Australian landscape. In south-western Victoria, archaeologists have argued that local Indigenous people, like the Gunditjmara people, were 'transegalitarian' forager societies that exhibited elements of cultural and social complexity that conflict with the highly idealised views of egalitarian and highly mobile hunter-gatherers (Richards 2011). Such transegalitarian societies can only exist in places where economic surpluses are regularly and seasonally produced, and where 'the magnitude of socioeconomic complexity is dependent on the richness and nature of the resource base and in particular … it is dependent on the ability to produce surpluses' (Owens and Hayden 1997, p. 123). For the Gunditjmara people, a resource from which they could produce a surplus was the Short-finned Eel, and it has been argued that through the construction of a series of trenches, further habitats for eels were made in order to make the population more stable and for making the harvesting of them easier (Crook *et al.* 2014; Head 1989; Jordan 2012; Lourandos 1980; McNiven *et al.* 2012). Such developments were based upon the Aboriginal Biocultural Knowledge of eels.

Chapter 6
Water

Ian D. Clark

Introduction

An adequate intake of drinking water is a fundamental human physiological imperative. The ability of Aboriginal peoples to survive and flourish across Australia in times of severe drought has captured the popular imagination for many decades (Bayly 1999). Traditional knowledge regarding the location of water, including how to obtain it in the absence of surface sources, was important for the survival of Aboriginal people. In semi-arid and arid regions, the importance of native water 'maps' in the form of myth narratives that contained the names and locations of water – such as wells, rock-holes, claypans, and waterholes – cannot be understated. Knowing which birds and insects were water-finders or 'diviners' and which trees were sources of water, was another fundamental part of Indigenous Biocultural Knowledge.

White (2009, p. 1) has noted that the 'more important water supplies usually have a totemic significance and play a very important part in Aboriginal ceremonial and social life and were often central trading locations'. There are many Creation stories related to water, especially the source of water and the origin of large rivers such as the Murray River. The Ganai people at Lake Tyers told the story of how at one time the Earth was waterless and experiencing an extraordinary drought, as all the waters were contained in the body of a huge frog (possibly the Giant Burrowing Frog, *Heleioporus australiacus*) (Smyth 1878, vol. 1). At a meeting of the wisest of the animals it was decided that if the monster frog could be made to laugh (Kramban), then the waters would run from its mouth and there would be plenty of water for all parts of the Earth. Several animals danced and capered before the frog but it remained solemn. It was only when No-yang (the eel) began to wriggle and distort himself that the frog's jaws opened and he began to laugh. All the waters came out of his mouth and there was a flood (Koorpa) and huge numbers drowned. The Pelican (Booran, *Pelecanus conspicillatus*) saved the black people by cutting a large canoe (Gre). By sailing among the islands that appeared in the great waters, he took the people up into his canoe and kept them alive.

About 8000 B.P. a major hydrological change occurred to the Murray River which saw it divert south of Barmah through the Barmah sand hills and join the Goulburn River near the centre of the old Lake Kanyapella (Lyons 1988, p. 171). According to an Aboriginal legend about this change, the Aboriginal people living in the area could see that the water was mounting up behind the Barmah sand hills and reaching the tops of trees. This so concerned them that they dug a small channel through the sand line with their digging sticks. This allowed the Murray to escape and meet the Goulburn River (W. Atkinson pers. comm. in Lyons 1988, p. 172).

In the early 20th century, Natune, an Aboriginal man from Moorundie, provided a Creation account of the Murray River that involved the wanderings of Nurela, who was described as a blind woman being accompanied by two young children as her guides. At Lake Victoria, near the borders of New South Wales, Victoria and South Australia, she created the River through driving back the sea, but Nurela then became lost and travelled like a 'drunken bee'; with her meandering course causing the River to become very long. This lengthening of her journey was fortunate for Aboriginal foragers who came later, as it increased the number of hunting and fishing grounds, with a lagoon at each river elbow. Fossils jutting out of cliffs along the River were said to be the remains of fish killed and eaten by Nurela and her children (Bellchambers 1931).

There is an extensive literature based primarily on early historical accounts that describe the ability of Aboriginal people to live in regions where there was a paucity of drinking water (Noble and Kimber 1997). As early as 1850, a correspondent named 'Bibimus', which means 'We drink', was alarmed at the number of travellers who perished in the Mallee for want of water (Bibimus 1850, p. 2). 'Now it would startle incredulous folks to tell them that in the Mallee Scrub they are treading on never failing spring of water, and obtainable at all times with the greatest facility.' He noted that Aboriginal people obtained water in the Mallee from trees known as 'Weir Cuttiur', which contain water, and he claimed that in the month of February he had obtained two quarts [2.27 litres] of water in less than half an hour.

At the turn of the 20th century, there was considerable public interest in the sagacity of Aboriginal people in obtaining water. Many articles in both metropolitan and rural newspapers extolled the virtue of knowing about Aboriginal methods of extracting water from plants. For example, the method outlined by Bibimus was repeated by Anonymous (1893), who asserted that travellers in the Mallee should be acquainted with the Aboriginal method of quenching their thirst when waterholes are empty.

Reading the signs: animals and insects as indicators of the presence of water

Aboriginal people were adept at reading the landscape, and they knew that the presence of certain birds, animals, insects and plants indicated that water was nearby. AT Magarey discussed Aboriginal water-finding skills in semi-arid regions in a lecture he gave to the South Australian branch of the Geographical Society of Australasia in June 1895:

> The mere fact that kangaroos, wallabies, wombats, and dingoes abounded in any district did not indicate the presence of water, as, like the camel, these animals have the faculty of living on the moisture derived from their food for some days. Swarms of small ants occasionally indicated the presence of water, but moisture was generally prevalent in the ground over which the red hornet, as it was known by bushmen, hovered, and in such places if holes were made

probably small supplies of water would be obtained … as far as birds were concerned such varieties as the cockatoo, crow, emu, and laughing-jack did not indicate the near presence of water, although the diamond finches and pigeons did (Magarey 1895b, p. 8).

Eleanor Clowes was another who was effusive in her praise of the water-finding skills of Aboriginal people:

> He is the most expert water-finder in the world, knowing at a glance, by the way vegetation grows, where water will be found, and at what a depth[,] and in what places and where, after a heavy dew, he may be able to collect enough to fill his water-bag, or the shell or skull, with the orifice sealed up, which serves the same purpose; collecting, even in the heart of the desert, where any white man would die of thirst, a sufficiency of water from the long tap-root of the gum-scrub (Clowes 1911, pp. 300–1).

Hydronyms as clues in the cultural landscape

Hydronyms or the names of rivers, lakes, springs and other water bodies, are an important means by which Aboriginal people communicate the location and suitability of water sources. Placenames also provide insights into the Aboriginal cultural landscape. 'Gheringal, or Kirrngkarl, was the name of a lake near Derrinallum, in south-west Victoria, and a spring in the Parish of Colongulac, which means 'dog urine', which may mean the water is brackish and undrinkable' (Clark and Heydon 2002). It is also possible that the placenames relate to a mythical story involving the actions of a dog. Hydronyms that reference water saltiness include Boninjeb, or Pine Lake, south-east of Horsham, meaning 'salt water'; Barriyalug, referring to Mt Emu Creek, meaning 'salty creek'; Korraynamitj or Lake Corangamite meaning 'bitter water'; and numerous waterholes along the Hopkins River were known as Lapeeyt or Lapeeyt Lapeeyt both indicating salt water (Clark and Heydon 2002). Woombeyer referring to Kirkstall near Koroit, means 'bad smelling water'.

Hydronyms are rich in Aboriginal Biocultural Knowledge, and are too extensive to go into detail here, however, a quick survey will demonstrate their value to this study (all the examples are taken from Clark and Heydon 2002). The backwater of the Nicholson River in East Gippsland is Yowen-burrun, signifying the edible root of the water plant; Yarram-yarram (Beaufort) means waterholes or many crayfish; Witjibar, or the Richardson River, means 'basket grass river'; and Migunang wirab (McKenzie Falls) means 'the blackfish cannot get any higher up'.

Some hydronyms refer to the actions of ancestral heroes, such as Mudyin gadjin (the confluence of the Loddon River) meaning '(he) picked up water' which refers to the actions of a mythological ancestor; Red Man Bluff or Ngarra mananinja gadjin 'having water in one's hand' has obvious references to the actions of ancestors, and may be part of the dreaming story of the *Buledji Brambimbul*, the two Bram brothers who created the Gariwerd (Grampians) landscape. A waterhole on the Wimmera River, south of Jeparit, was known as

Barengji-djul 'river-waterbags' – because waterbags made of wallaby skin used to be filled there. Wergaia legend tells of how a large Kookaburra sat there and laughed so loudly that all the waterbags fell over and spilled (see Hercus 1986, p. 199).

Rock-holes

Aboriginal people also took advantage of sources of water such as gnammas or rock-holes, and weathering pits and rock basins, which are water-holding depressions on the surfaces of rock formations, especially granites (Magarey 1895b). They vary in size from several centimetres to more than 10 m and from a few millimetres to several metres deep. Pit gnammas were an invaluable source of water for Aboriginal groups and they believed they were dug by ancestral beings (Timms and Rankin 2016). Aboriginal people had many techniques of securing food and water at rock-holes, such as placing large rocks over the holes as caps, which slowed evaporation and kept the water clear of animal fouling (Bindon 1997).

Water from plants

The need to obtain drinking water from plant roots in south-eastern Australia was probably confined to the Murray Mallee and Mallee areas of the Eyre Peninsula where surface supplies of freshwater were scarce (Clarke 1986a). Elsewhere, water from springs and soaks appears to be available year-round, other than, perhaps, in periods of prolonged drought. Magarey (1895b) has given us a detailed description of the trees that were used to supply root water, and Aboriginal methods of extraction (also see Smyth 1886, p. 2; Howitt 1904, p. 51).

Botanist Joseph Maiden, in an extensive study of the flora of New South Wales, discussed Aboriginal methods of obtaining water from Mallee trees. He stressed that no adult in Australia should be ignorant of this technique, as there is 'no doubt that a knowledge of this method of obtaining water would have been the means of saving the lives of many persons who have suffered one of the most terrible of all deaths - death from thirst' (Anonymous 1914a, p. 11).

> The blacks also knew particular hollow trees in certain localities, where supplies of water were found in the fork, varying in quantity from a pannikin full to a gallon. The water was obtained from these hollows by sucking it through grass stems, or by throwing in balls of grass and then sucking the moisture out. Another source of supply was in the stems of young trees, which, if turned up on end, would give forth small quantities of water, which could be drained into any vessels available. The trees from whose roots supplies of water could be obtained were then described. The water mallee, of which there were six varieties, was particularly referred to in this connection, and the manner in which the blacks scraped away the ground from the roots, and cut them into lengths, and then let the moisture drip out was mentioned. These roots were usually from two to nine inches [5 to 23 cm] below the surface of the ground, and a healthy mallee would give enough water to enable two or three men to quench their thirst. The currajong, the needle bush, the casuarina, which was

known as the desert oak, and also the acacia, had varying quantities of water stored up in their roots and stems (Magarey 1895b, p. 8).

Analysis of the category of 'water-trees' includes eucalypt species of Mallee (such as *Eucalyptus dumosa*, *E. gracilis*, *E. incrassata*, *E. oleosa*, and *E. fasciculosa*) that are common in the semi-arid parts of south-eastern Australia (Clarke 1986a). The availability of root water permitted Aboriginal people to enter areas lacking surface water – the Ngarkat people of the Murray Mallee relied heavily on root water and only needed to travel to the Murray River at times of severe drought. Local clans did not permit the Ngarkat to camp near the water; rather, they would have to use a particular track to the Murray River where they would collect their water in skin bags and then return to their camps in the Mallee (Clarke 2009c; Tindale 1974). Howitt (1904, p. 51) noted that at times of drought the Berriait people were forced to go to the rivers for water, 'and as these were occupied by other tribes such as the Barkinji and the Wonghibon, they had to fight their way in strong parties'.

Dr Benjamin W Gummow, a medical practitioner at Swan Hill, and Honorary Correspondent to the Aboriginal Board in Victoria, in correspondence to R Brough Smyth, dated 9 April 1872, discussed the value of knowing how to obtain water in semi-arid regions:

> … it frequently happens to the natives, when out in the Mallee country, that the waterholes on which they had counted on obtaining a supply of water, have dried up, but they are never, therefore, at a loss. They select in the small broken plains some Mallee trees, which are generally found surrounding them. The right kind of trees can always be recognised by a comparative density of their foliage. A circle a few inches deep is dug with a tomahawk around the base of the tree; the roots, which run horizontally, are soon discovered. They are divided from the tree and torn up, many of them being several feet in length. They are then cut into pieces, each about nine inches [23 cm] long, and placed on end in a receiver, and beautifully good, clear, well-tasted water is obtained, to the amount of a quart [1 litre] or more, in half an hour. This method of procuring water is not confined to the Mallee only. The roots of several other trees yield water. A knowledge of this means of getting water, and of the trees which yield it, says Dr Grummow [sic], would have saved the lives of very many white men, whose bleached skeletons, lying on the arid plains, alone testify to their once having existed (Smyth 1878, vol. 1, p. 220).

William Lockhart Morton, a settler at Morton Plains Station near Birchip in the 1840s and 1850s, conducted experiments to see how much water could be obtained from the water-yielding Hakea (*Hakea leucoptera*): 'the first root; about half an inch [1.2 cm] in diameter, and 6ft. to 8ft. [183 to 244 cm] long, 5 yielded quickly, and in large drops, about a wineglassful of really excellent water' (Anonymous 1914a, p. 11).

Other sources of water included the Kurrajong and the native vine:

> Mr. Bennett refers to the currajong as another water tree. This is the well-known Brachychiton populneuum [sic] [*Brachychiton populneus*], a handsome tree which is often found growing in Australian shrubberies, and which is

indigenous to Victoria, New South Wales and Queensland. In the coastal districts of New South Wales is a native vine – Vitis hypoglauca – which has thick lianes [lianas], or stems, which hang pendant from the trees. Dr. George Bennett records that Mr. Bidwill's life (he was director of the Sydney Botanic Gardens for a short period in 1847) was saved when he was lost in the bush by the water he was able to procure by incising one of these vines (Anonymous 1914a, p. 11).

German traveller Friedrich Gerstaecker has left a detailed description of the succulent plant commonly known as Pigface (*Carpobrotus rossii*) as a source of refreshment:

> But here on the Edward, commences the best country for sheep farming in the whole of Australia. The reason of this is the existence here of a small kind of shrub, 'salt-bush,' it is called, which possesses, in a greater or less degree, a saline taste, and juicy leaves which sheep are said to be fond of feeding upon. Besides this, a very juicy plant, a kind of cactus, with short, thick, three-cornered leaves, or leaf-like stems, filled with water, grew everywhere. In the most dry places, where no rain had fallen for a very long time, I found this plant ('pig's-face,' as the shepherds have christened it in their simple way), in abundance, running along the ground like a vine, and throwing out leaves upon leaves, like a cactus. There are several kinds of this pig's face; some are bitter and hard, others saline, and one nearly sweet, with a pleasant taste. I have eaten many a meal of these plants with considerable relish. The blacks eat it also in great quantities … (Gerstaecker 1853, p. 427).

Pigface leaves and stems were carried as a source of water when Aboriginal groups were crossing dry plains (Tindale [cited Clarke 2009c]).

Owarine

Robinson observed a unique method used by the east Kulin people of central Victoria to find water in the ground, called *owarine* by his Boonwurrung and Woiwurrung guides: '… there was no water and the natives with torch of bark went and beat the ground and dug where it was hollow and got a little dinky worth; they call this mode of getting water *owarine*' (Robinson 17 May 1844 in Clark 2014a, p. 557).

Waterholes beside dry lakes

Water was obtained in the beds of dry lakes by digging holes. Aboriginal Protector George Augustus Robinson saw this method in action at Lake Bolac in 1841 – despite the lake being dry and the region in the grip of a severe drought, water was readily obtainable:

> No water, at least our native friends could not find any. All the natives of the lake evidently got their water by digging, as numerous holes were seen where they had been digging (Robinson 31 March 1841 in Clark 2014a, p. 282).

Robinson observed: 'Numerous holes where the natives had dug for water were met with in all parts of the sand, and dead eels strewed the sand and banks' (Robinson 1 April 1841 in Clark 2014a, p. 282).

Waterholes dug to retain rainwater

In 1840, when riding across the plains near the Dandenong Ranges south-east of Melbourne, Robinson noted several small holes in the ground made by Aboriginal people. He subsequently learned that the holes were dug to serve as caches to retain rainwater:

> I passed some small holes in the plains frequently and ask my native companion what occasioned them, said black fellow made them this I was aware they did but hardly thought they would do so when close to well-watered rivulet. I afterwards saw one or two holes of the same description made at the native hut where they were encamped at Narre.narre.warreen. It was rainy weather and this was an expedient [sic] to retain the rain water and a good one when we consider that they had no vessels either to fetch it in or keep it (Robinson 11 September 1840 in Clark 2014a, p. 209).

Adaptations in times of water scarcity

During times of water scarcity, Aboriginal people commonly used animal fat as a cleanser to substitute for water. Dawson considered this was an excellent substitute for water, though he acknowledged that many Europeans found the practice repulsive:

> In every respect the aborigines are as cleanly in their persons and habits as natural circumstances admit; and, although the universal custom of anointing their bodies with oily fat may be repulsive to highly-civilized communities, it is an excellent substitute for cleansing with water, and must have arisen, not only from the comfort it affords to the skin in various ways, but also from the difficulty of obtaining water in most parts of the country, even to satisfy thirst (Dawson 1881, p. 12).

Smyth (1886, p. 2) noted that when unable to obtain water, Aboriginal people would cover their bodies with moist earth.

Another way of obtaining water was through finding Water-holding Frogs (*Cyclorana platycephala* and *C. australis*), burrowed beneath the surfaces of claypans. They hold water in their bladders, and Aboriginal people would squeeze the body water out of them. The location of frogs was identified by reading observations on the ground surface, tapping the surface with the butt of a spear to detect their aestivation chambers, or by stamping on the ground and listening for their faint croaking (Bayly 1999, p. 23).

Thirst quenchers

Aboriginal people also used plants to quench their thirst in the absence of water. For example, the needle-like stems of Sheoak *Casuarina* and *Allocasuarina* were chewed to quench one's thirst (Nash 2004, p. 12). Another thirst quencher was the fruit of Snotty Gobbles or Mistletoe (*Amyema* species), a parasitic plant that grows on the branches of trees and is spread by Mistletoebirds (*Dicaeum hirundinaceum*). Aboriginal people chewed the fruit, much like chewing gum, to produce saliva to quench their thirst (Williams and Sides 2008, p. 86).

Dew water

Another source of water was collected in the form of dew droplets. Magarey explained that Aboriginal people were able to travel long distances by obtaining dew water (Magarey 1895b, p. 8). He 'described how Aborigines would go out before sunrise and collect dew-drops with a sweep of the arm or movement of a stick' (Bayly 1999, p. 23).

> A more efficient method was to form a ball of grass and use this as a sponge to gather the dew lying on grass. The water would then be squeezed out of the ball into a coolamon or pitchi. Another method was to hold a water container under the dew-drenched twigs of a shrub or tree which was tapped or shaken. The leaves of the sandalwood tree (*Santalum spicatum*) were said to be particularly useful for gathering early morning dew. Aborigines in central Australia collected dew from the desert heath myrtle, *Thryptomene maisonneuvei*, to make a sweet drink (Bayly 1999, p. 23).

Strainers and filters

Water sources can become foul late in the hot season, particularly during prolonged droughts. In response, Aboriginal people used plant parts to strain water to remove any impurities (Clarke 1986a, 2015c). Dawson discussed the use of the cob or flower-cone of the Silver Banksia (*Banksia marginata*) as a crude form of filtration of contaminated water that also served to sweeten its taste:

> When obliged to drink from muddy pools full of animalculae [microscopic organisms such as amoeba or paramecium], they put a full-blown cone of the banksia tree into their mouths, and drink through it, which gives a fine flavour to the water, and excludes impurities. The name of the cone, when used for this purpose, is tatteen mirng neung weeriitch gnat — 'drink eye banksia tree belonging to' (Dawson 1881, p. 22).

Smyth confirmed that Aboriginal people used the 'blossoms of trees to flavour and sweeten his drinks' (Smyth 1886, p. 2). Clarke (2015c) considers it is likely that the same technique would have been used in south-east South Australia.

European explorers sometimes availed themselves of Aboriginal survival techniques to obtain potable water (McLaren 1996). Major Thomas Mitchell recorded:

> Modes of drinking *au naturel*. The water, when we encamped, was hot and muddy, but the blacks knew well how to obtain a cool and clean draught, by first scratching a hole in the soft sand beside the pool, thus making a filter, in

which the water rose cooled but muddy. They next threw into this some tufts of long grass, through which they sucked the cooler water thus purified also from the sand or gravel. I was very glad to follow the example, and I found the sweet fragrance of the grass an agreeable addition to the luxury of drinking (Mitchell 9 December 1831 in Clarke 2012, p. 63).

Village sites and campsites at sources of permanent water

The need for Aboriginal people to gain permission to traverse the country of another clan, and to access its resources, was confirmed by Chief Protector George Augustus Robinson during his 1843 expedition to the Murray River district and the Western District of Port Phillip. On this expedition he was accompanied by two members of the Native Police Corps. In the country to the east of the Grampian Ranges, Robinson was informed that members of the Weeripquart balug, the Djabwurrung clan who owned the locality, had told a squatter that he could not stay there. When the squatter, Captain Rupert Allan, was beginning to establish a station near their waterholes, the Weeripquart people told him he could not remain there as it was their country and the water belonged to them. If their country was taken away, they were not in a position to go to another 'Blackfellows country for they would be killed' (see Robinson 16 April 1843 in Clark 2014a, p. 503).

It was common practice for squatters to situate their home stations and outstations near a ready supply of water, often waterholes in rivers and streams. At least one of these waterholes would have been reserved for the domestic use of the Europeans, and another for the use of stock. These were considered out of bounds to local Aboriginal clans-people, and if clans attempted to use waterholes selected by Europeans for their solitary use, there were obvious repercussions. The diary of Henry Mundy records an instance of conflict over the use of a waterhole at Caramut Station, near Caramut, in western Victoria, in the mid-1840s. A large group of Aborigines had gathered at a waterhole on his property, so Mundy went to talk to them. As he approached he noticed that many were strangers to him. Mundy was particularly concerned as 'the blacks [had] squatted around the good waterhole, and nobody knows what nastiness they'll chuck in it'. Ned, one of Mundy's employees, 'went to the mob of blacks as he told us with the intention to hunt them straight away. They were sitting around fires roasting oppossums, some splashing in the water, some lubras dipping their mangy dogs in our drinking water. This rose Ned's ire, yelling out pull away, pull away out of this you stinking varmin, kicking one, pushing another, punching them on the head with his fist, kicked their fires about, pull away, pull away' (Hughes 2003, pp. 100–2).

In May 1841, Robinson came upon an Aboriginal village of 13 huts beside a waterhole near Mt Napier in western Victoria. Three of the huts in the village had recently been occupied, as evidenced by the tools and weapons Robinson found in them.

> This place, previous to its occupation by white men, was a favorite resort and as this was the only permanent [sic] supply of water, a village had been formed. I counted 13 large huts built in form of a cupola. When seen at a distance they have the appearance of mounds of earth. They are built of large sticks closely

packed together and covered with turf, grass side inwards. There are several variations. Those like a cupola are sometimes double and have two entrances; others again are like a niech [sic]. Then there are some made of boughs and grass. And last are the common screens. The permanent huts are those in form of a cupola. Three of these huts had been occupied a day or two previous to my visit. A shield or, in the language of the natives, por.ral, as also a bucket or po.pare.re, and a shield of boughs for catching birds were left at the huts. The por.ral my natives took away. They wanted the bucket but I would not allow it (Robinson 10 May 1841 in Clark 2014a, p. 326).

Drought conditions blamed on the arrival of Europeans

In the late 1830s and early 1840s, south-eastern Australia experienced a significant drought event that ran from 1836 until 1845 (Helman 2009). The Wathawurrung people of the Ballarat district believed there was a causality between the drought and the arrival of the European settlers in the region. This was explained by Katherine Kirkland who squatted at Trawalla Station, west of Ballarat, in the early 1840s:

> We passed an immense salt lake, which is gradually drying up; its circumference is forty miles [64 km]. Many lakes, both salt and fresh, have dried up lately. The natives say it is the white people coming that drives away the water: they say, 'Plenty mobeek long time, combarley white fellow mobeet gigot' – in English, 'Plenty water for a long time, but when the white people come, the water goes away' (Kirkland 1845, p. 8).

Rainmakers or charmers

In the face of drought conditions or an abundance of precipitation, Aboriginal groups utilised the powers of rainmakers or charmers. William Thomas noted that in the Melbourne district when a continuance of rain was desired, the charmer would sing: 'Won-ner-rer Nger-wein Barm-we-are Won-ner-rer Tin-der-buk Koo-de-are Nger-wein Koo-de-are Tin-der-buk Kar-row-lin'.

> During the time that this is sung the charmer sits in his mia-mia, and with a piece of thin bark, about a foot or eighteen inches [30 to 46 cm] long, continues throwing hot dust from the fire into the air, alternately mumbling and singing the above song; in fact, all their charmings are in mumbling language, not known to the rest of the blacks (Thomas in Bride 1898, p. 91).

Thomas recalled Bobbinary, a celebrated Boonwurrung charmer-away of rain:

> I have known this man to be kept singing for hours. The blacks say, when Bobbinary was a child that it had been raining for some days, and 'blackfellows all sad, their bellies tied up to keep off hunger; that the child Bobbinary began to sing, and that sun immediately came out, and no more rain. That ever since then he has been able to send rain away' (Thomas in Bride 1898, pp. 91–2).

Howitt (1904, p. 397) noted that rainmakers and 'weather-changers are important persons in most parts of Australia, but especially in those parts of the continent which are subject to frequently recurring periods of drought'. In relation to the Kurnai people of Gippsland in eastern Victoria, he noted:

> Among the Kurnai there were rain-makers, and also those who caused rain and storms to cease. The former were to be found in each clan, and the methods used for producing rain by the Bunjil-willung,[1] or rain-men, were to fill the mouth with water and then squirt it in the direction appropriate to the particular clan, and each one sang his especial rain-song. The Brayaka squirted water, and sang, towards the south-west (Krauun); the Brayaka and Tatun-galung did this in the same direction; the Brabralung and the Krauatungalung squirted water towards the direction of the south-east, the east winds (Belling) being from their rainy quarter. From these several directions the rain came in Gippsland; and when, for instance, a south-westerly rain came to the Brabralung they said that it was the Brayaka who sent it, and so on with the others. These rain-makers could also bring thunder, and it was said of them, as of the other medicine-men, that they obtained their songs in dreams. I have before spoken of one of the Brayaka Head-men who was credited with the power of calling up the furious west winds, whence he derived his name of Bunjil-kraura. His song by which he stopped the gales which prevented his tribes-people from climbing the tall trees in the western forest, ran thus from Kutbun to bear or carry, Wang a bond, or something tied, and Kraura, the west wind. I did not hear the song by which he caused the western gales to arise, but I have no doubt that it was of the same character. When these gales came, he was propitiated by presents to send them away (Howitt 1904, pp. 397–8).

Howitt (1904, p. 398) also recounted the Wotjobaluk practice of rainmaking in north-west Victoria, and highlighted the time when the rainmaker agreed to make rain for a settler during a severe drought:

> To produce rain he took a bunch of his own hair which he carried about with him for the purpose. Soaking it in water, he then sucked the water out and squirted it to the westward. Or he twirled it round his head, so that the water passed out like rain. In this somewhat arid district the office was much thought of, and an instance came under my notice in which the rain-maker scored a success from a white man, in a severe time of drought. He and others of his tribe were camped at Morton's plains, and on his boasting of his power to produce rain, the then owner of the station said to him: 'I will give you a bag of flour, some tea, and half a bullock, if you will fill my tank before to-morrow night.' The tank was a large excavation just finished. This was early in the morning, and the rain-maker set to work at once, saying, 'All right, me

---

1 Willung is rain. The Kurnai say that the frogs when croaking in chorus in the swamps are 'singing for rain,' and that the big sonorous bull-frogs are the Bunjil-willung.

make him plenty rain come.' The next day there was a tremendous thunderstorm, the rain fell over all the run and the tank was filled. Then the rain-maker went to the owner of the run saying, 'You see, plenty rain come.' It was to the honour of the white man that he made him happy with the gifts which he had promised him.

Limiting water spillage

Layers of plant material were commonly used in Aboriginal Australia to help conserve water supplies (Clarke 2012). A sprig of fresh leaves was used to keep water cool and to limit splashing when carrying water back to camp in deep wooden containers. In the Coorong region of South Australia human skull vessels were covered with dry grass to prevent spillage.

Water extractors

While the pre-European landscape of south-eastern Australia was rich with lakes, swamps and rivers, there were foraging zones where drinking water was scarce during summer. Away from the wetlands, explorers, such as Thomas Burr (1844), recorded the existence of numerous wells. When extracting water, Aboriginal people used drinking straws that were made from hollow reeds, probably cut from the Common Reed (*Phragmites australis* (Cav.) Trin. ex Steud.) that has bamboo-like stems. This is a plant use that was recorded in Bunganditj mythology, with a giant man Brit-ngeal being in the habit of drawing his water from a cave pool via a long drinking reed (Smith 1880, p. 23). In the Murray River region of South Australia and north-west Victoria, initiation novices were made to drink through reed straws in order to avoid direct contact with water (Clarke 1999b; Mathews 1904a; Taplin 1874 [1879]).

Aboriginal people used a variety of water-extracting tools, such as sponges, straws and water plungers. These were essential to extract potable water from waterholes, water buckets and deep recesses such as rock fissures and crevices, and in the cavities of large trees. Sponges were used to soak up liquid as an aid for drinking. In western Victoria, a ball of grass tied to the end of a spear served as a sponge to soak up deep, hard-to-access drinking water (Clarke 2012). Robinson, in his many travels through south-eastern Australia, often witnessed the use of water extractors:

> At a clump or copse of tea tree saw a native well, about 2½ deep and water. The hole was about a foot wide at top. To get the water out of such holes the drinking reed is indispensable. Besides which, the water can be obtained without disturbing the sediment. Grass is laid in the well and the water filters through it (Robinson 10 May 1841 in Clark 2014a, p. 326).

> Some tribes when water is scarce construct reservoirs or tanks and others inhale it through long reed tubes between fissures and crevices from below the surface. I have indeed seen it sought for and found in deep cavities of large trees (Robinson 1845 in Clark 2014b, p. 486).

Dawson noted that water was obtainable from yabby holes using a reed straw:

> In summer, when the surface of the ground is parched, and the marshes dried up, the natives carry a long reed perforated from end to end, which they push down the holes made by crabs in swamps, and suck up the water (Dawson 1881, p. 22).

Reed pipes or straws had different names across south-eastern Australia, for example, *tchope tchope* (Djabwurrung); *tome dome* (Wathawurrung), and *ngallome* (Gulidjan) (Robinson in Clark 2014b, p. 238, 286). The lengths of the reed varied according to their intended use; Robinson noted the following lengths: less than a foot long (30.5 cm) to drink from a bark bucket (Robinson 21 April 1841 in Clark 2014a, p. 298); and two and a half feet long (76 cm) to drink from waterholes (Robinson 13 April 1841 in Clark 2014a, p. 293).

In August 1841, Robinson was presented with a *tchope tchope* by the clan head of the Bulugbara clan (Djabwurrung) that belonged to Lake Bolac (Robinson 2 August 1841 in Clark 2014a, p. 414). The gifting of reed straws often took place at cultural ceremonies when Aboriginal people were granting strangers temporary access to their country and its resources – the straw symbolised access to water. Straws had an important ritual use in the Lower Murray region of South Australia, where Ngarrindjeri male initiates had to avoid contact with all water, and could only drink through a hollow reed (Clarke 2012).

Water storage/vessels

Other than waterbags made from animal skins, Aboriginal people developed methods of storing and carrying water using buckets made from bark and from the gnarls of eucalypt trees. At the confluence of the Ovens River, Robinson came upon an Aboriginal camp and saw 'several small vessels for holding water. These were the bark from the excresence of a tree, stunted box kind thus: and shows the ingenuity of these singular people. The excresence cut out in form of a diamond' (Robinson 29 April 1840 in Clark 2014a, pp. 159–60). Curr called these vessels 'calabashes', and asserted they were used for the purpose of holding or carrying water. He believed they contained from one or two gallons [4.5 to 9 litres]. 'They were made out of the knots or excrescences which are common to both box and gum tree' (Curr 1883[1965], p. 133).

Smyth has a detailed description of these vessels, called *tarnuk bullito* or *tarnuk bullarto* (*tarnuk* – bark bucket or basin; *bullito* – big, hence 'big bucket' or 'big basin'):

> The vessels used for holding and carrying water by the Aborigines of Victoria were commonly made of the gnarls of gum-trees, or the bark covering the gnarls, or of a portion of the limb of some tree. The large tub – Tarnuk bullito or Tarnuk Bullarto – was either a hollowed log or a large gnarl hollowed by fire and gouging (Smyth 1878, vol. 1, p. 346).

Smyth noted that the *tarnuk bullito* were not carried from camp to camp as they were too heavy to carry:

> The large tub nearly in the centre of the Fig. 163 is the *Tarnuk bullito*. It is a large hollowed gnarl. The marks of the fire which was kindled in it to burn out

Examples of Tarnuks. Source Smyth 1878, vol. 1, p. 347.

the interior are still clearly perceptible, though it has been hacked and gouged for the purpose of increasing the capacity. It is a very heavy vessel. This is rather an unusual form of the *Tarnuk*. Such vessels were ordinarily made of the naturally bent limb of a tree, or of an uprooted tree. The limb or tree was placed in a hollow excavated in the ground, and a large cavity was formed in it by burning and gouging. The *Tarnuk bullito* was not carried from camp to camp. It was too heavy for carriage, and one could always be made at each camping ground, if the old one left by the tribe on the last visit was decayed or damaged (Smyth 1878, vol. 1, p. 347).

Tarnuks were also used to make a sweet beverage from the blossoms of the honeysuckle and box.

The *Tarnuk bullito* was used for pounding and macerating the blossoms of the honeysuckle and box, from which a beverage was obtained - sweet - somewhat like sugar and water, but with a flavor of its own (Smyth 1878, vol. 1, p. 347).

A smaller wooden bucket, simply called a *tarnuk*, was made from the hollowed gnarls of gum trees. Smyth has given us a detailed description:

> The *Tarnuk* in all the specimens I have seen is the hollowed gnarl of a gum-tree. Unlike the *Tarnuk bullito*, however, it is made very thin, and the interior is smooth. It was smoothed, no doubt, by laborious scraping. It is light, and, even when full of water, would not be a very heavy burden. The bark covering the gnarl, but most often the layer of wood next to the bark, was used for these vessels. Those made of such wood are, I believe, the lightest, as they are certainly the best. The twine for carrying the vessel was made of the fibre of the stringybark or some other vegetable fibre, and was passed through holes pierced on each side of the *Tarnuk*. The gnarled tree shown in the drawing is not an unfair representation of the mode of growth of some of the eucalypti, and it was from such knobs and gnarls as are there depicted that the natives found materials for the *Tarnuks*. On the River Powlett, in Gippsland, and elsewhere, the gnarled trees are seen stripped of their bark, and the larger excrescences have been cut off with the stone tomahawk for the purpose of making water vessels (Smyth 1878, vol. 1, pp. 347–8).

Smyth explained that filling and carrying the *tarnuks* was a gendered activity, and was the responsibility of women:

> The two buckets—one with a string for carrying it—on the left-hand side of the figure, and the other on the right—are the *Tarnuk* proper. This vessel was used for carrying water from place to place when journeying, and for keeping water in when encamped. The women always carry these buckets, and fill them with fresh water when they reach a creek or water-hole. They are indispensable to a tribe that is wandering through forests or over plains where water may not be met with at every place of encampment (Smyth 1878, vol. 1, p. 348).

Bark buckets

Water buckets were also constructed using bark from the wattle or acacia tree. Robinson, Dawson, and Smyth have given detailed accounts of the construction of the bark water buckets used in Victoria:

> po.pare.re or bark bucket made from the bark of the wattle tree, 11 or 12 inches [28 to 30 cm] deep, 12 inches [28 cm] long. Is made by taking a sheet of bark 22 or 24 inches [55 to 60 cm] long and a foot [28 cm] broad then folded and sewed up at each end with grass and the seam gummed a grass rope handle with a loop is added. A little grass is laid loose on the bottom inside and a small reed, foot [28 cm] in length, through which the water is employed in holes or imbibed, and the reed being put on the loose grass at the bottom serves as a

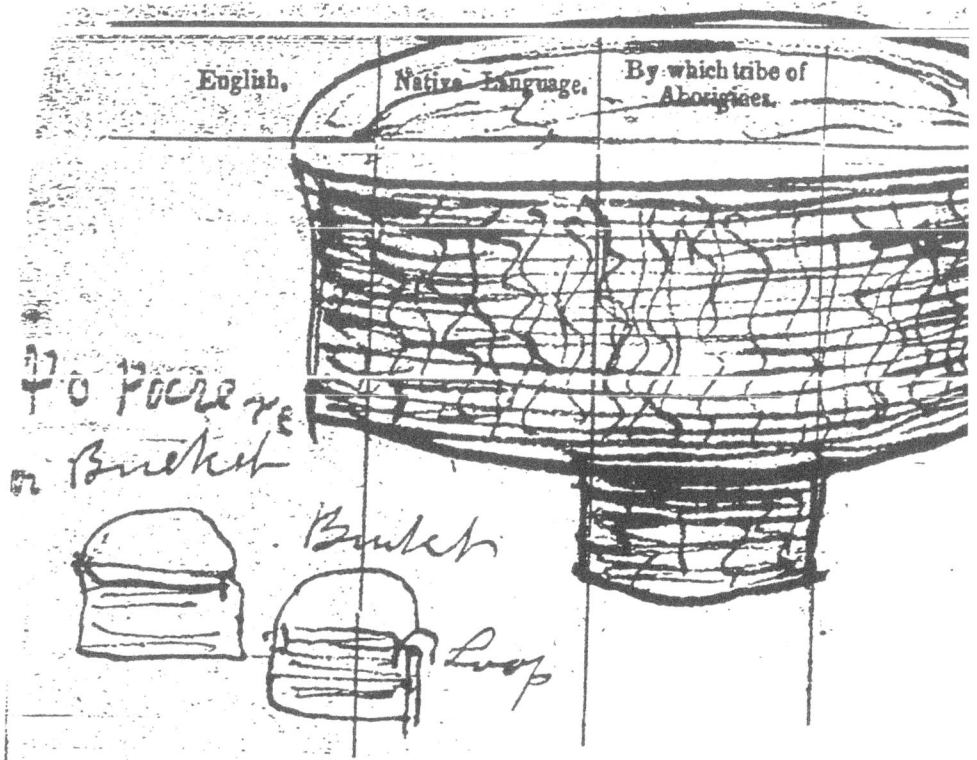

Po.pare.re or bucket. Source: Robinson Papers in Clark 2001a, p. 91

filter to prevent the impurities from passing through (Robinson in Clark 2014b, p. 241).

Another vessel, named 'popaeaer yuu', is used for carrying water, and is formed of a sheet of fresh acacia bark, about twenty inches [50 cm] long by twelve [28 cm] broad, bent double and sewed up at each side with kangaroo tail sinews, and the seams made water-tight with an excellent cement, composed of wattle gum and wood ashes, mixed in hot water. After the bucket is made it is hung up to dry, and the contraction of the inner bark causes the vessel to assume a circular shape, which it retains ever after. It is carried by means of a band of twisted wattle bark fixed across its mouth (Dawson 1881, pp. 14–15).

Smyth confirmed that these bark water buckets were used in Gippsland. He noted the ends of the bucket were 'tied exactly in the same way as they tie the ends of a canoe. This vessel is called *Gil-ang*' (Smyth 1878, vol. 1, p. 349). Smyth recorded that the buckets were constructed using green bark:

> … vessels for holding water are generally made of green bark. Pieces are cut into various shapes, laid on the fire or in hot ashes until they are soft and the edges begin to contract, and then they are easily wrought into the forms desired by

the natives. When the bark is heated, it can be drawn into many shapes without breaking it or causing it to crack (Smyth 1878, v. 1, p. 349).

Dawson discussed a second type of water container, a much longer bucket, in the shape of a canoe, called a *torrong* in western Victoria. He also noted the practice of adding banksia cone into the water to sweeten it:

> For keeping a supply of water in dry weather, a vessel called 'torrong' — 'boat' — is made of a sheet of bark stripped from the bend of a gum tree, about four or five feet [122 to 152 cm] long, one foot [28 cm] deep, and one [28 cm] wide, in the shape of a canoe. To prevent dogs drinking from it, it is supported several feet from the ground on forked posts sunk in the earth. A wooden torrong is often used in the same way, and is formed from a bend of a gum tree, hollowed out large enough to hold from five to six gallons [23 to 27 litres]. As the water which they use is frequently ill-tasted, they put some cones of the banksia into the torrong, in order to give a pleasant flavour to its contents (Dawson 1881, p. 15).

Smyth noted that when it was not possible to find a tree suitable to make a *tarnuk bullito*, a different kind of bucket was made, in the shape of a canoe:

> When it was difficult to get a limb of a tree, or a tree suitable for a *Tarnuk bullito*, the natives cut a thick piece of bark from off the curved limb of a gum-tree, heated it in ashes, and bent it so as nearly to resemble the shape of a canoe, and stopped the ends with clay. This was a temporary expedient most often resorted to on hurried journeys. The bark of the *Eucalyptus viminalis* was preferred for the purpose (Smyth 1878, vol. 1, p. 347).

Waterbags made from animal skins

Waterbags were made from animal skins (Magarey 1895b). The skins were turned inside out (Robinson 26 November 1842 in Clark 2014a, p. 483), before stitching up the limbs – this made them more waterproof, as the oils in the fur repel water and skin, as a membrane is permeable only one way. Small waterbags were made using the pouch of a female kangaroo:

> A small water-bag, called 'paanuung' is formed of the pouch of the kangaroo, which, when fresh, is stuffed with withered grass till it is dry. A strip of skin is fixed across its mouth for a handle (Dawson 1881, p. 16).

When needing to carry water a longer distance, a larger bag made from wallaby skin was used:

> For carrying water to a distance a bag called 'kowapp' is used. It is made of the skin of a male brush or wallaby kangaroo, cut off at the neck and stripped downwards from the body and legs, and made water-tight by ligatures. The neck forms the mouth of the bag. This vessel is carried on the shoulders by the forelegs (Dawson 1881, p. 16).

Robinson in reflecting on a visit to the Mallee district in the Port Phillip district in 1845, discussed Aboriginal methods of carrying water, and strategies adopted in times of drought:

> These natives like other tribes of the north convey water in skins; they are simple and easily prepared: the skin of an opossum or wallaby is merely stripped from off the animal, whole and entire. When it is fit for use; the fur is retained and turned inwards; the truth of the apothgm [sic] [apothegm, aphorism] 'necessity the parent of invention' is strikingly verified in these opossum skin water vessels, yet with this example, and the leather vessels of India, the equestrian traveller still adheres to his colonial and wonted predilection of vessels of wood (i.e. water kegs) than a more unsuited, precarious, and inconvenient mode of conveying water, especially on horses over the heated and arid plains of the interior cannot well be imagined. There are however few spots but have had its aborigines. A striking instance the man found by Captain Sturt on his late exploration and who had crossed at a season of great drought the wide and searching desert when fallen in with yet I am aware that by some travellers the contrary has been asserted, but who it will be conceded had few opportunities in their discursive journies [sic] of becoming acquainted with the aboriginal character. The truth is that the natives of the sterile country leave it in times of drought, and retire to the more favoured lands of their friendly neighbours (Robinson 1845 in Clark 2014b, p. 486).

Smyth believed the skin of the native cat was preferred. Presumably he is referring to native Quolls (genus *Dasyurus*):

> It is taken off with the greatest care, the incision and the skin which covered the feet, &c., are carefully sewn up and made water-tight, and the neck is left open. This vessel is carried with a string, formed into a loop and passed over the head, the skin of water hanging at the back. These vessels resemble the water-skins used by the ancient Egyptians (Smyth 1878, vol. 1, p. 348).

Water-carrying bags were also made from the entrails from kangaroos (Clarke 2009c; Hahn 1838–1839 [1964]).

Drinking vessels

Across south-east Australia, an assortment of vessels were used to drink water, ranging from wooden vessels, shells, and in south-east South Australia, human skulls were used as cups. Smyth noted that amongst the Woiwurrung of the Yarra River, a wooden drinking vessel shaped like a shoe was known as *No-been-tarno*: 'The shoe-shaped vessel shown in the figure in the foreground was used as a drinking vessel—the water being taken either out of the *Tarnuk* or out of a creek' (Smyth 1878, vol. 1, p. 347).

When they were procurable, Aboriginal people used shells as drinking vessels (Smyth 1878, vol. 1, p. 349). In examining the contents of an Aboriginal camp Robinson found among the utensils 'the Haliotus or ear shell which they use as drinking cups, called munjer' (Robinson 21 April 1841 in Clark 2014a, p. 298). Presumably Robinson is referring to the Common Ear Shell or Abalone (*Haliotis rubra*).

Some Aboriginal groups in south-east Australia used human skulls as drinking cups. When Chief Protector Robinson visited South Australia in July 1846 he was struck by the difference in some of the customs of the Aboriginal people from Adelaide and Encounter Bay compared with those of the Port Phillip district (what is now Victoria). In particular, he was surprised to see human skulls being used as drinking vessels (Robinson 9 July 1846 in Clark 2014a, p. 719).

> Among many of the tribes may be seen a strange sort of ornament or other utensil—namely, a drinking cup made of a human skull. It is slung on cords and carried by them, and the owner takes it wherever he or she goes. These ghastly utensils are made from the skulls of the nearest and dearest relatives; and when an Australian mother dies, it is thought right that her daughter should form the skull of her mother into a drinking vessel. The preparation is simple enough. The lower-jaw is removed, the brains are extracted, and the whole of the skull thoroughly cleaned. A rope handle, made of bulrush fibre, is then attached to it, and it is considered fit for use. It is filled with water through the vertebral aperture, into which a wisp of grass is always stuffed, so as to prevent the water from being spilled. Although they consider that to convert the skull of a parent into a drinking vessel, and to carry it about with them, is an important branch of filial duty, they seem to have no very deep feelings on the subject. In fact, a native named Wooloo sold his mother's skull for a small piece of tobacco (Wood 1870, vol. 2, p. 86).

Eyre also refers to the use of skulls as drinking cups. The sutures are closed with wax or gum (Smyth 1878, vol. 1, p. 349). Tindale (1938a) noted that human skull vessels were known as *merikin*, by the Tangane people of the Coorong, and were made waterproof by being plastered with a mixture of red ochre and whale or emu oil.

Conclusion

In south-eastern Australia, the loss of language and knowledge of Aboriginal myth has led to a loss of Indigenous Biocultural Knowledge. Nevertheless, although fragmented, the corpus of Aboriginal knowledge in the study area concerning water remains significant, as this survey has shown. Aboriginal people were adept at water management and in certain regions, such as western Victoria, extensive trenches were constructed in the ground to concentrate fish and eels. Stone and wooden weirs of various sizes were placed in rivers and streams and in channels to facilitate the capture of fish and eels using baskets. Poisons were also added to waterholes to catch quantities of fish. These diverse methods of water management are discussed in greater detail in Chapter 5. Throughout south-eastern Australia, there are Aboriginal accounts of water spirits, generally called 'bunyips' in the popular literature, and these are discussed in Chapter 3.

Chapter 7

Fire in Aboriginal south-eastern Australia

Fred Cahir and Sarah McMaster

Introduction

In south-eastern Australia, climatic, topographic and vegetative characteristics have combined to produce a landscape in which fires occur with some regularity (Cruz *et al.* 2012). The historic record suggests that the frequency of fire in this area was a common source of interest (and concern) for European explorers and colonists first encountering the area. In most instances, the Europeans attributed fires to Aboriginal people, and theorised the potential motivations and function of burning practices. This chapter will introduce archival material that illuminates these colonial perspectives on Aboriginal fire use in south-eastern Australia, and will discuss the ethnographic observations that were made of 'customary' uses of fire. This will include the association that fire was seen to have with religious, mortuary, hunting and communication practices. This chapter will begin by discussing the impacts that explorers and colonists had on the customary fire practices of Aboriginal people, recognising that while the historical record contains invaluable material about Aboriginal burning practices, its descriptions of Aboriginal burning practices must be assessed carefully.

As Indigenous and European cultures encountered one another on the colonial frontier, it is also interesting to consider how fire, a simultaneously essential and dangerous force, featured in the historical relationships between Aboriginal people and the newly arrived Europeans. To explore this question, this chapter will discuss the way that Aboriginal knowledge of fire was transferred cross-culturally, and question how this may have influenced the lives of both Indigenous and non-Indigenous people. By drawing from the historical records, it will examine the effects that fire, including both its purposeful use and its accidental occurrence, had on frontier relationships.

The historical record

The Aboriginal fire-practices observed by the European colonists were likely different to those that existed before colonisation. This was in large part because the depopulation, caused by introduced diseases and other factors such as restricted access, left fewer people present to enact fire practices. On 20 February 1841 Robinson reported that Jamieson, a squatter, had told him that: 'they never allowed the blacks to come on their station', and that if they did, they were driven off. Upon asking whether the 'blacks' had 'done any mischief,'

Robinson was told, 'no'; the Aborigines had 'only burnt off the grass … with a fire stick' (Clark 2014b, p. 95). In addition to fire practices being disrupted by dispossession from land, it is also evident that practices were altered by the very presence of colonists. For instance, it can be difficult to discern from descriptions in the historical record which fires Aboriginal people created in order to threaten or ward off the squatters and which fires were lit as part of their traditional mode of managing the grasslands landscape. Some, such as explorers Hume and Hovell, concluded that the 'numerous fires which were being made around them', by the Wadawurrung clans near present-day Lara (west of Melbourne), were signal fires (Bland 1965, p. 70). They became uneasy that their presence was being communicated to other Aboriginal peoples in the district. Many other colonists were unaware, or did not record their opinion about the intention of the fires they ascribed to the actions of Aboriginal people. Robert Wrede journeyed overland from New South Wales to Victoria and from there sailed by ship to Adelaide. His description illustrates the typically imprecise nature of many observations of fire in the area: '7 April 1838 [at Yass] we passed through the bush burning with fury on all sides of us …11 February 1839 we were close in shore, past Portland Bay … Saw plenty of natives' fires' (Nicholls and Wrede 2012, pp. 56, 162).

In many instances, historical records offer generalised and speculative descriptions of fire. Exploring by boat around what was later to become Port Albert, Brodribb found the land to be 'well grassed, although thickly timbered' (Brodribb and Bennett 1883, p. 9). He recorded that: 'The natives had burnt all the grass at Gippsland late in the Summer. Heavy rains must have fallen before we reached there, in the month of March (Autumn)' (1883, p. 24). Further, many observations of fire were not recorded with any suggestion of cause. In March 1837, Governor Richard Bourke described the grassy plains between Geelong and Melbourne as 'much burnt'; the grass, he wrote, had been 'burned to the roots' (Jones 1981, pp. 104–10). Bourke did not attribute the firing of the grasslands to either anthropogenic (Aboriginal or non-Aboriginal) or natural causes.

The variation in recorded personal accounts of Aboriginal burning practices also prompts caution. Mary Thomas recollected her experiences in South Australia, describing fires lit by Aboriginal people as 'quite awful', suggesting that 'Adelaide has been nearly ambushed with fire' 'that may truly be said to burn like wildfire' (Thomas and Thomas 1925, p. 28). Conversely, another colonial observer from the same district, Simpson Newland, described how Aboriginal people took care to avoid major conflagrations: 'In my long experience [over 70 years] I have never known any serious bushfires caused by the blacks' (Newland 1921, p. 24).

A close reading of the historical record seeks to address these challenges attendant to the historical record. For instance, Hateley (2010) and Cahir *et al.* (2016) have argued that the majority of historical accounts of Aboriginal burning practices in south-eastern Australian ecosystems indicate that the application of fire was managed, was frequent and over generally small areas of grassland plains, partially owing to rainfall patterns and soil types. This observation of the ecological spread of fire usage is not overtly stated in the journals of early explorers and writers in south-east Australia during the 19th century. They do not explicitly state that low intensity fires were a frequent occurrence in the open forests and the grassy plains. However, scrutiny of work from contemporary writers such as JC Byrne indicates the observed variations in the landscape and how it featured fire. Byrne stated that while Melbourne had a 'park-like appearance', being either 'lightly timbered' or 'completely open and devoid of wood', land to its north and east was more heavily vegetated and mountainous, suggesting that different biomes were subjected to distinct management regimes (Byrne 1848, pp. 302–3). A comparable scenario was depicted in the observations made by Edward J Eyre, during his exploration of land north of Adelaide during 1839 to 1841. He noted the patches of timbered stands commingled with open plains and concluded that the landscape: 'may probably have been occasioned by fires, purposefully or accidentally lighted by the natives in their wanderings' (in Clarke 2005a, p. 429).

Similarly, observations made by Bennett, during her overland journey to Gippsland from Melbourne in 1844, provide some useful detail about the diversity of landscapes and the presumed evidence of previous burning. On 29 April, she describes travelling 'over a great deal of barren, healthy ground. Reached the Hurdee Creek in time to get our camp fixed before dark. A most miserable place, surrounded on every side with burnt scrubb [sic] and old gum trees' (Brodribb and Bennett 1883, p. 179). The next day says that their passage was impeded by 'thick scrub' and steep creeks (1883, p. 179). By 1 May, they were in land that was noted for being 'beautiful' and grassy (1883, p. 180).

Customary burning

In recent decades, researchers (Jones 1969; Whitehead *et al.* 2003; Clarke 2005a) have argued convincingly that the reasons for Aboriginal burning practices were extensive. Jones, for example, has argued that Aboriginal people burnt the bush to signal, to clear the ground, to hunt, to regenerate plant food and extend the human habitat, and for fun. In the context of south-eastern Australia, scholars (for example, Cahir *et al.* 2016; Gammage 2011) have extended this line of thought, noting that fire played and continues to play a significant part in domestic, social and ritualistic aspects of Aboriginal peoples' lives. The myriad uses of fire include: ceremonial occasions (life, initiation and death); illumination; protection against bad spirits; cooking; provision of personal warmth; curative purposes; and the warding off and driving out of unwanted animals and insects.

Many of these customary burning practices were identified in some form by colonial observers in south-eastern Australia. Fire was recognised for figuring prominently in important rituals such as initiation and mortuary ceremonies. Robinson reported on his interaction with an Indigenous man, known as Mr King, who described a ceremony in which fire sticks were thrown about in a way that 'resembles white man's book'; here the solemnity of the Aboriginal ceremony was likened to that of the Bible (Clark 2000, p. 128). During the early 1840s it was observed in the Riverine region of New South Wales and Victoria that during important ceremonies fire was used throughout the event and also the use of pyrography to mark the occasion. Brough Smyth, in his expansive ethnographic study of Victorian Aboriginal people, described how fire was used for warding off malevolent spirit creatures such as a large snake called Myndie which caused illness: 'When Myndie is known to be in any district, all the blacks run for their lives … They set the bush on fire, and run as fast as they can' (Smyth 1878, p. 426). Robinson was also informed that fire was a principle element in initiation ceremonies called 'Ko.ro.bine' (Clark 2000, p. 176), and Bulmer likewise describes both the ceremonial role that fire played in the initiation process of boys (Bulmer 1855–1908 [1999], p. 4) and the integral part it played in a young girl's play, as she tended to her 'own' small fire, roasting possums in imitation of her mother (p. 2).

Fire was also seen to have a role in the creation of artworks by Victorian Aboriginal people, according to Albert Le Souëf:

> The natives generally have a considerable taste for carving and drawing. I have repeatedly seen the inside of their mia-mias [houses] covered with rude etchings of the kangaroo or emu, or anything else that might occur to them. The sheets of bark are first blackened in the fire, and the drawings are made with a piece of pointed stone or a nail, and some are really very well done' (cited in Smyth 1878, p. 299).

Ceremonial rites for the deceased were also seen to feature fire. Escaped convict, William Buckley, who lived with the Wadawurrung people near Geelong (south-west of Melbourne) between 1804 and 1835, described how they used fire in their 'customary ceremonies' relating to mortuary practices. In one instance the body of their deceased was ritually cremated until it 'was burnt to ashes' and on another occasion the body was interred in the ground and flanked with fires (Morgan 1852, pp. 46–7). Robinson also provided examples of fire featuring similarly in the mortuary ceremonies of the Djabwurrung people near Challicum in central Victoria and in Melbourne (Clark 2014b, p. 74).

The historical record also suggests that fire was an essential tool used by Aboriginal people to execute their customary obligation to care for country and manage their resources and heritage, such as totemic or memorial sites and faunal enclaves. William Thomas reported that despite animosity from local squatters in the Cape Schanck area, the Boonwurrung people were intent on executing burns to execute their responsibilities to care for country (Atkinson 2005). Thomas describes on one occasion his unsuccessful efforts to prevent an Aboriginal Elder from setting alight bush near their encampment. 'Black fellows', he was told by the Elder, 'would not know where they were if he did not make fire, and made one so [effectively] about him that I could not get aside him' (Stephens 2014, pp. 141–2).

Motivations for burning

In many instances, the colonists offered explanations for the motivations they supposed were inspiring Aboriginal people to use fire. These explanations were often limited to those such as the assistance it afforded in the hunting of kangaroos, Emus and other large game (see Nicholson 1981; Gammage 2011). Buckley is typical in his brief explanation that recounted the Wadawurrung conducting burns 'round a kind of circle, into which they force every kind of animal and reptile to be found; they then fire the boundary, and so kill them for food' (Morgan 1852, p. 94). In the east of Victoria, colonist Harry Witham recalled a corresponding use of fire by the Yaitmathang in the Omeo district grasslands:

Kangaroos and wallabies were stalked and speared but when these larger animals were scarce fires were lit for the purpose of obtaining small game. The area burnt at any one time was limited to one or two acres [less than 1 hectare]. As the grass burned, the blacks, who had taken up positions around the fire, pounced on any lizard or small animal unfortunate enough to come into view in its efforts to escape (Fawcett 1955, p. 82).

From south- eastern Australian waters, mariners such as Captain King made comparable observations:

In passing Cape Howe, we observed large fires burning on the hills, made by the natives for the double purpose of burning off the dry grass and of hunting the kangaroo which were thus forced to fly from the woods and thereby fall an easy prey to the pursuers (King 1826, pp. 6–7).

On his approach to Adelaide in February 1837, Pastor William Finlayson and his fellow passengers aboard the *John Renwick* were gripped by the sight of fires 'which seemed to spread ... with amazing speed' across Adelaide's hills (cited in Clarke 2005a, p. 428). The cause of this fire, Finlayson explained, was the Aboriginal people who sought to more easily 'obtain the animals and vermin on which a great part of their living depends' (cited in Clarke 2005a, p. 428). Similar reports were echoed by George McCrae regarding activity on the Mornington Peninsula (south of Melbourne):

The blacks set fire to the top of the mountain for the purpose of driving out the wallabies from the bushes and killing them. The fire gradually encircled the brow of the Mount like a diadem on the head of a monarch (McCrae 1934, p. 165).

Robinson noted the prevalence of Aboriginal fires on the river edges: 'On the verge of the river [Goulburn or Warring], numerous smokes of the natives' fires in all directions' (Clark 2000, p. 208). The record of Hume and Hovell's 'Journey of Discovery to Port Phillip' in 1824 is also punctuated with references to Aboriginal fires near watercourses:

The natives, from the appearance of their fires, seems numerous ... on the banks of the 'Ovens' [River] All the country in line of route to-day, had been burned, and a little to the Westward of this line, the grass was still blazing to a considerable height. At noon having travelled seven miles, rest near some water holes, on a small plot of good grass, which had most fortunately escaped the ravages of the flames ... The natives hereabouts [near present day Tatong] are evidently numerous, conclude[d] from their fires, the smoke of which is observed in every direction (Bland 1965, pp. 17, 22, 35, 42, 44, 48, 52).

Squatters such as John Dunmore Lang who travelled through the Western District of Victoria in the 1840s, often observed that summer (late January), was the season when 'considerable tracts' of the long grass plains country were burnt 'either accidentally, as is sometimes the case, or designedly by the black natives, that the young grass may shoot up after the next rain fresh and sweet for their cattle, the kangaroos; and wild turkeys …' (Lang 1847, p. 145). Ross echoed this view of mosaic-like burning off writing:

> Even before we white-folk made our appearance in this land [central Victoria], and for some time afterwards, the aborigines were in the habit of burning off the old grass on portions of their hunting grounds so as to have good fresh grass for their kangaroos, wallabies, emus, turkeys, and other smaller granivorous birds and animals which formed the chief portion of their food (Ross 1915, p. 4).

The Reverend Bulmer likewise described fire being used to hunt for multiple food sources and the judicious use of fire over a wide area. He wrote of how the women in Gippsland used fire to catch fish at night and in summer the men set fire to large areas of country so as to spear animals escaping flames, or take advantage of animals roasted by the flames (Bulmer 1855–1908 [1999]).

Smyth offers a general explanation for Aboriginal burning in Australia: 'It was their custom to burn off the old grass and leaves and fallen branches in the forest, so as to allow of a free growth of young grass for the mammals that feed on grass' (Smyth 1878, p. xxxiii). He also added there were other reasons which augmented what he perceived to be the primary one, and that is: 'they were at least careful to see that harm was not done to vegetables that yielded food' (Smyth 1878, p. xxxiii). Researchers Gott (1983), Nicholson (1981) and Cahir (2014) have argued that Robinson's account of winter burning by Djabwurrung people in central Victoria before the harvesting of *murnong* (Yam Daisy, *Microseris lanceolata*) (as opposed to dry season burning) supports the argument that Aboriginal people in Victoria deliberately used fire as an agent of greater yield change for tuberous food plant ecosystems. Robinson described Aboriginal (presumably Djabwurrung) women firing the plains and then harvesting *murnong* (a tuberous plant which was the staple food source of Aboriginal people across much of south-eastern Australia) during his journey through the western plains of Victoria in July 1841 (Oates 1977).

> Today the native women were spread over the plain as far as I could see them, collecting pannin, murnong, a privilege they would not be permitted except under my protection. I inspected their bags and baskets on their return and each had a load as much as she could carry. They burn the grass, the better to see these roots but this burning is a fault charged against them by the squatters (Clark 2014, p. 401).

The accounts from explorers and colonists also acknowledged that Aboriginal fire practices reflected a knowledge of fire's repellent properties. After Captain Woodriff 'Observed a great smoke to the northw'd' of Port Phillip Bay on 19 October 1803, a month later he wrote to Evan Nepean at the Admiralty, describing the poor quality of timber available around the Bay. He attributed this to sandy soil, lack of water and Aboriginal fire that was used, he postulated, 'to destroy the numerous snakes that the country abounds with' (Woodriff 1986, p. 10). Buckley, too, noted fire's use as a repellent agent, observing how the Wadawurrung people avoided 'suffering much inconvenience from the myriads of mosquitoes, and of a very large sort of horse fly by carrying their lighted fire-sticks, holding them to wind-ward' (Morgan 1852, p. 142). McDonald in Gippsland, similarly stated that the fire-stick was used as a 'disinfectant' and that it 'was thrust amongst the thatch when the huts were abandoned' (McDonald 1887, p. 78).

Other colonists, such as James Dawson, emphasised how important fire was for communication:

> Sometimes, instead of dispatching men to give notice of a meeting, a signal smoke is raised by setting fire to a wide circle of long grass in a dry swamp. This causes the smoke to ascend in a remarkable spiral form, which is seen from a great distance. The summons thus given is strictly attended to. Or, if there is not a suitable swamp, a hollow tree is stuffed with dry bark and leaves, and set on fire. Or, a fire is made on a hill top (Dawson 1881, p. 72).

Many explorers and early colonists also became aware that fire was repeatedly and skilfully being used to communicate the presence of the invading colonists quickly and at times over long distances. Major Thomas Mitchell journeyed from New South Wales to Portland Bay in Victoria. On 3 June 1836, he camped at the junction of the Murray and Darling Rivers and described his fear of attack from Aboriginal people, having seen their smoke 'signal columns [that] arose in the air' (Mitchell 1839, p. 115). In March 1840, Robinson reported that: 'Large smokes made by the Port Phillip natives, signal to the natives in the country to come in' (Clark 2014b, p. 127). Others, such as James Darlot, recalled how, 'as soon as we camped for the day a fire would be lighted about a half mile off and a couple of blacks would be seen evidently watching us' (Darlot 1834). This habit, which caused considerable anxiety for European explorers and squatters, suggests a use of fire by Aboriginal people designed to resist colonial incursion (Reynolds 1987). Thomas Learmonth, a contemporary of Darlot's, experienced a near identical scenario. Learmonth wrote to Governor La Trobe, recalling that in September 1837 he was one of a group of six returning from an excursion to Lake Corangamite in western Victoria when they surprised a large group of Wadawurrung at the mouth of the Pirron Yallock River: 'We came upon them so suddenly that they had time only to set fire to their miamias as a signal of danger to the other tribes ... we saw by the smoke rising in different quarters that the signal had been observed and answered' (Bride 1898, p. 40).

Third Lieutenant Nicholas Pateshall, of the HMS *Calcutta*, made a similar observation from the convict settlement at Port Phillip Bay in 1803. He became convinced that fire was being used to thwart and intimidate the new arrivals. He wrote on 20 November: 'We soon perceived the natives to be greatly alarmed, for the country in a short time was in a perfect blaze' (Pateshall 1803, n.p). Pateshall made a similar observation of native fires being a response to the British incursion on the *Calcutta*'s return trip to Sydney as 'the natives appeared to be much alarmed as we run along shore, by their innumerable large fires, kept up day and night' (Pateshall 1803, n.p.). A similar pattern of reportage of what were possibly signal fires was described by officers on the colonial cutter *Integrity* on a 'Voyage of Survey' at Western Port in November and December 1804: 'Back Mountains well stocked with trees sevl [several] Native Fires were seen ... Strong winds and cloudy during our excursion on this island saw two or three Native Fires ... A large Native Fire has been kept up abreast of the vessel on the Main since we anchord [sic]' (Integrity *et al.* 1804–1805, n.p).

Many travellers also remarked on the ease of travelling along paths created by Aboriginal peoples' use of fire. The explorer, Charles Sturt, described his journey on the Ovens River in April 1838 in which they travelled 'through narrow lanes, or openings which the natives had burnt, the reeds forming an arch over our heads and growing to the height of eighteen or twenty feet [5.5 to 6 m]' (Sturt 1838[1990], p. 34). He added that they could not have 'pushed' through this country 'but for the narrow lanes made in them by the Natives' (p. 35). Sturt observed on 23 May 1838 that the country near the junction of the Goulburn and the Hume Rivers had 'just been fired by the Natives, the trees were [scorched] to their very summits and the trunks of those which had fallen were smoking on the ground' (p. 35).

However, not all observers made explicit connections between the presence of pathways and Aboriginal people. At Sandy Bay (Western Port) in November 1826 Dumont d'Urville registered: 'fine stands of trees easy to get through … vast grass-covered clearings, with well defined paths linked by other tracks so regular and well marked'. Telling of his own knowledge on the topic, he reflected: 'it is hard to conceive how these could have happened without the hand of man' (Dumont d'Urville and Rosenman 1988, p. 59).

Knowledge of Aboriginal fire

Cahir *et al.* (2016) argued that many colonists were able to distinguish Aboriginal-managed fires from other fires. Such descriptions by colonists provide useful anthropological material about the form of fires lit by Aboriginal people. Byrne described a simple method of identifying Aboriginal fires, based on size, suggesting 'the whites are not so saving of the wood as the Aborigines, and the track of their camping place is often visible for days afterwards, by the yet unexpired fires; a large dry tree having perhaps been ignited' (Byrne 1848, p. 391). Murray offered a similarly basic classification based on the nature of the fires: 'Aboriginal controlled fires were common and docile during the summer' (Murray 1843, p. 201). Robinson's journal is peppered with references to 'native fires' and 'smokes'. In a period spanning 1839–1850, during his sojourns across south-eastern Australia, Robinson repeatedly noted the appearance of what, for him, were unmistakeably Aboriginal fires – particularly in the late summer and autumn periods and in 'grassy country'. For instance:

> [in Mameloid Hills in central Victoria] 'Saw large smokes of native fires on the Lodden [sic]. Saw a smoke of native fires among the hills of the Pyrenees' … [Mt Cole and central Victoria] The natives had been here and made a fire … Saw smoke at the N.end of the Pyrenees, smoke of native fires' … 'As I rode along saw large smoke for 30 miles along the Mur.ne.yong range ie Clark's Hill, also to the SE and SW … Saw from Koratanger large smokes of the native fires, N some NE … Campaspe plains, grassy country, north; Large smoke, native fire northeast. Large smoke, native fire … Saw to the NE. 25 miles distant, large smokes of native fires … Ascended the mountain called by Natives, Choorite Saw a smoke of Native fire 50 or 60 miles north and by east. I suppose Lake Hindmarsh (Clark 2014b, pp. 115, 120, 125, 651).

Forester Ronald Hateley has reasonably questioned how Robinson could have known with such certainty that these were 'native fires' from a distance of 25 or 60 miles [40 to 96 km]. Cahir *et al.* (2016) have reasoned that Robinson, after many years of mentoring by *waygeries* (Aboriginal messengers or ambassadors) with whom he regularly travelled, both in Tasmania and Victoria, had acquired a great deal of skill in this area. They further argue that the numerous references to fires that Robinson *did not* distinguish as 'native fires' underlines the importance of those thus described.

Moreover, in describing his coming to wisdom on the question of how Indigenous and non-Indigenous fires could be distinguished from each other, one colonist provided further insight into the perceived way that Aboriginal people used fire:

> Observing, one fine summer's evening, the well-known column of smoke peculiar to native fires rising and spreading up the steep sides of our sugarloaf hill, upon a green bank of which was a small spring of sweet water … An occasional halt was ordered, to ascertain if possible the character of the distant party. Stanmore soon decided that the smoke proceeded from a whiteman's camp; 'for the obvious reason,' remarked he, 'that, wherever a tribe of Aborigines locate themselves, each family kindles its separate fire at fourteen to twenty yards [13 to 18 m] apart. Consequently, on a calm evening, the smoke ascends in so many distinct columns, and thus reveals to the observant Bushman of what nature are the tenants of the wild and pathless forests' (Lloyd 1862, p. 57).

From south-west Victoria, Edward Henty likewise attributed many of the fires he witnessed in the summer or early autumn period, to Aboriginal people. In January 1835 Henty recorded seeing 'Many fires in the bush', suggesting that light showers in the previous week indicated Aboriginal causation (Peel 1996, p. 45). Similarly, John Webster, an overlander near the border of Victoria and South Australia noted in December 1840 how, 'from what I could judge, the firing of the grass had been quite recent' (Webster 1908, p. 144), and came to the view that the grassy river flats of the Murrumbidgee and the Murray were being deliberately managed by Aboriginal fire. In his journal entry for 17 December 1840 he noted:

> All the trees here … bear the evidence of frequent fires, some standing trees had arched so that one could drive a team of bullocks through, caused by many years conflagrations of grass. Here and there lay monarchs of the forest which had fallen from this cause (Webster 1908, p. 144).

Less than a fortnight later, Webster, near the banks of the Hume River, witnessed that the

> blacks have set fire to the grass … we saw that as far as we could see from the camp the country was burned black … The fire had consumed most of the grass, but had left sufficient for the cattle, and even where the fire had gone the dry portions of grass only were consumed (Webster 1908, p. 158).

James Kirby, a drover, made similar observations about fires lit by Aboriginal people in Swan Hill (northern Victoria) in 1840: 'In the distance where the blacks had not burnt the reeds, it looked like large fields of ripe wheat; and nearer where they had burnt them, it had the appearance of a splendid crop just before it comes into ear' (Kirby 1895, p. 28).

Historical records suggest, however, that colonists did not always accurately identify the cause of particular fires. On 1 March 1840, Thomas described Aboriginal people being alarmed by a large bush fire, which they maintained 'was not made by them' (Stephens 2014, p. 141). The following day Thomas wrote that: 'they return'd with plenty of roots & one opossum & a few eels, but in several places had again set bush fire they endeavour'd to make excuses by saying it was the former fire but it was in different directions. I scold them.' This reveals an important interplay between Aboriginal burning practices and colonial interests. Thomas describes the animosity and the confusion which existed between the commercial interests of the local squatter, Samuel Rawson, and the traditional food gathering methods of the Aboriginal people of that area (Boonwurrung):

> Mr R very angry wishes Blacks gone. Says that damages near 300 [pounds sterling] is loss of flesh in Cattle & stated that the Blacks have set fire to Cape Schanck station as to frighten near the whole of Mr Willoby's cattle. The Blacks hurge [urge] as a plea that they have always done it for to turn out opossums & wombats & c. Blacks state the necessity of setting fire to Bush to hunt & Drive opossums &c (Stephens 2014, p. 142).

In the interplay of Aboriginal customs and colonial interests, fire practices could be seen affecting colonial efforts to re-educate Aboriginal children. George Langhorne was a Government-appointed missionary to the Aboriginal people in the Port Phillip District who established the first school in Victoria in present-day Prahran, Melbourne. In a report to the Governor dated 31 January 1838, Langhorne explained that his attempts to establish a school for the local clans in the region had been thwarted due to the Aboriginal parents 'removing their children from the mission in the early part of the month'. Langhorne cited one of the reasons given by the Aboriginal parents was: 'that they required the attendance in their annual custom of burning the grass to catch the young kangaroo – the boys have not yet returned' (Langhorne 1838, n.p).

Fire as an offensive weapon

Overt uses of fire would also have played a role in relations between colonists and Aboriginal people. Aboriginal people are known to have fired the land to dissuade Europeans from venturing into their estates and thereby avoiding contact (Thomas 1839; Koenig 1935; Cannon 1991; Clark 2003; Collins 2006; Gammage 2011; Cahir *et al.* 2016). Thus, within the newly imposed colonial power structure, fire wielded by Aboriginal people served as an important weapon against Europeans who, at least initially, had little defence. The deliberate burning of the landscape by Aboriginal people was at times used to bewilder, confuse and ward off unwanted trespassers on their lands – and also to enable them to avoid contact. John McKinlay, an explorer travelling from Adelaide to the northern tip of Australia in 1861, frequently recorded in his diary the 'natives burning [the land] in all directions, but do not approach us' (McKinlay 1863, p. 28). McKinlay, an experienced bushman, was clearly confused about Aboriginal fire management regimes in general – but was fully cognisant of when fire was being used to repel him. In mid-May 1862 he wrote:

Natives burning grass close upon our right, on the way here to windward at a furious rate. What their particular object can be in burning so much of the country I cannot understand. No natives as yet have voluntarily shown themselves. I met the same lubra and child again near the same place that I before met her, but she did not attempt this time to fire the grass around me. A short way on further I met, or rather overtook, another lubra with two children; she at first tried to conceal herself, but when she saw she was observed she immediately set to work to burn the grass round us in all directions … Saw no natives since, but look where you may, except north, and you will see fires raging (McKinlay 1863, p. 27).

As Joseph Hawdon and Charles Bonney travelled along the Murray River, they made the first overland journey with cattle from Sydney to Adelaide. On 17 February 1838, below the junction with the Murrumbidgee, they encountered large numbers of Aboriginal people, whom they said at dusk:

set fire to the small patch of grass which I had selected for the stock to feed on … When they landed on the opposite bank they raised a shout, or rather a yell of defiance and set fire to the few reeds growing along the margin of the water (Hawdon 1952, p. 33).

For two days, Hawdon noted: 'the stock had scarcely anything to eat, the. Natives having set fire to everything in the shape of food that would burn.'

Mitchell, too, was very aware of Aboriginal tactics that were designed to hinder his progress. Near the Lachlan River, he recorded near that:

The first line of trees we crossed enclosed only a shallow channel, overgrown with polygonum; and we in vain sought the natives although we saw where portions of fire had been recently dropped. Three miles further we perceived a more promising line of trees and smoke arising from them also. There we found the yarra trees growing on a flat with a reedy channel meandering amongst them. The fire arose from some burning trees and grass; and there were huts of natives but no inhabitants … Abreast of him, but much more to the right, two of the old men, who had reached a fallen tree near the tents, were busy setting fire to the withering branches. Those who were further back seemed equally alert in setting fire to the bush and, the wind coming from that quarter, we were likely soon to be enveloped in smoke … (Mitchell 1839, pp. 50–1).

Webster described the use of fire as a combative tool, relating how Aboriginal firing of the land was deliberately used to try and destroy the overlanding party and noting how upon reporting to his leader that he had seen in the distance great columns of smoke the response was: 'My God, the &c., &c., &c., blacks have set fire to the grass' (Webster 1908, p. 155).

Among colonists attempting to establish stations, Aboriginal fire caused significant anxiety. In a letter Richard Bunbury wrote from western Victoria to his father in December

1841, he claimed that Aboriginal people in the surrounding area lit fires 'not unfrequently for the purpose of burning out a station; last year they made several most determined attempts to burn the huts of two of my neighbours' (cited in Clark 2017, p. 35). By the summer of 1848 there remained deep-seated concerns among squatters that Aboriginal people would destroy their property with fire whether intentionally (by malice) or unintentionally (by continuing their customary practices). AC Cameron, resident at Terinallum Station, near present-day Derrinallum in western Victoria, wrote to his neighbours stating: 'owing to the grass being so very dry I have to be constantly on the watch for fear the Blacks should set fire to it. Mr McPherson has had ~1000 sheep burned; the whole flock of nearly 1700 are more or less injured' (Brown 1952, pp. 270–1).

Such fears were not new. Indeed this fear of the Aboriginal use of fire on the frontier of colonial incursions was arguably present from the earliest points of contact. Historian Majorie Tipping maintains that the short-lived (October 1803–January 1804) convict settlement at Sorrento (south of present day Melbourne) was abandoned, in part, owing to implications of Aboriginal burning practices in the area (Tipping 1987). The commandant of the settlement, David Collins, held grave fears that the frequent fires in the area were 'danger signs' of impending trouble with the Aboriginal people, such as he had experienced earlier at Port Jackson in New South Wales (Tipping 1987). In another instance, a squatter reportedly 'had left his country station for the Yarra because he was fearful that the natives would set fire to the place in revenge for what had befallen some of their party' (Waterfield 1914, p. 108). In a similar manner William Godfrey wrote how the 'natives were attempting to burn them out' (Godfrey in Cannon and MacFarlane 1991). In particular reference to frontier violence on the Yambuck Station in the Western District, historian Jan Critchett likewise noted that arson was mentioned on the list of recorded incidents of Aboriginal violence between March 1845 and April 1847 (Critchett 1990). These examples indicate the way that colonists perceived Aboriginal use of fire to have been used as a purposeful tool in the development of frontier relationships.

In response, some colonists suppressed the Aboriginal traditional custom of burning off by forbidding or discouraging land management practices and usurping them of their more permanent camps. In the context of often hostile relationships with colonisers, the use of fire by Aboriginal people also appears to have been motivated by a need to avoid, retaliate, attack or defend themselves against the newcomers. Squatters were suspicious of Aboriginal peoples' motives for burning the landscape, particularly at shearing time or near their huts. The Henty family in south-west Victoria noted how, at shearing time in January 1838, they had had frequent visits from 'friendly natives', including a group of ten who 'came with each a new set of Spears'. Several days later Edward Henty complained that there 'were fires all round'. Subsequently, Aboriginal people were prevented from frequenting the huts of the sheep station. Much to Edward's chagrin a 'large tribe … fired the hills when they left' (Henty in Peel 1996, p. 205). This action was repeated a week later (and again in December of the same year). Edward Henty wrote that they were 'obliged to leave off [the shearing] in consequence of two natives setting fire to the grass all round us within a few hundred yards of the Hut. John came over in the afternoon, had great work to put the fire out' (Henty in Peel 1996, p. 206). The Hentys sought to protect their property with guns, shooting at Aboriginal people, but the burning of the grasslands continued. A week later and with shearing season in full swing the Gunditjmara returned. Edward Henty's frustration boiled over:

> Natives burning the grass by the River. On approaching them they put it out but when we turned they commenced again with double vigour, fired a ball over them which only frightened them a little, for one fellow returned making a circle and lighting a fire as he went, rode after him and frightened him away … rode over to the new station. On my return I found that the Natives had been burning close to us, which spoiled our days shearing in consequence of being obliged to put the fires out (Henty in Peel 1996, p. 207).

Katherine Kirkland's reminiscences of the Trawalla district in central Victoria exemplify this belief of colonists that they were under attack by fire. Kirkland wrote with some alarm in 1838 of how 'The fires in the bush are often the work of the natives, to frighten away the white men' (Kirkland 1845, p. 4). This belief is known to have continued even into the 1850s. One French colonist wrote: 'The woods are on fire every night. The blacks [probably Djabwurrung] set them alight in retaliation for being driven away' (De Chabrillan 1998, p. 125). In the Riverine region Webster noted several times on an over-landing journey in 1840 that they were convinced the landscape was being deliberately torched to drive them away (Webster 1908). Webster also intimated that the extensive 'firing of the grass by natives' was to deprive the large number of cattle of feed – and thus effectively directed his party on a route and at a pace dictated by local clans (Webster 1908, p. 144). Fire could also aid in the avoidance of the pale-skinned explorers by Aboriginal people, many of whom probably saw the newcomers as spirit people. GF McMillan's experience in Gippsland was common. Encroaching on Gunaikurnai land in December 1839, Robinson reported McMillan saying that: 'when the natives saw them coming they burnt their camps and ran to the reeds' (Clark 2000, p. 92). Mackay in his recount of the events of European exploration of the Gippsland area also includes references to the prevalence of campfires and the setting alight of bush by people fleeing on sight of white intruders (Mackay 1916, p. 80).

Many others made similar complaints. Captain Hepburn claimed that: 'the blacks had set fire to the grass closer to my home-station than I liked' (Bride 1898, p. 60). In the Western District of Victoria, surveyor Charles Tyers relates being near Hamilton, where he camped at Henry Wedge's station on 19 February 1840. Wedge's overseer, Patrick Codd, told him that: 'the natives had been very troublesome here lately, endeavouring to burn them out and steal the sheep. They succeeded in burning the country for some miles round the station and one of the huts of the out-station with everything in it' (Bride 1898, p. 190). In April 1840, GA Robinson recorded how the Jameson brothers (squatters at 'Tallarook' and 'Tarcomb') believed the local clan (possibly Daungwurrung) had set fire to their grasslands after 'provoking their resentment' by physically ejecting them from the station huts (Clark 2014b, p. 138). However, in a letter to Governor La Trobe, Blair's account attempted to moderate the intensity of claims against Aboriginal people, by stating that:

> Messrs. Whyte Brothers were the only settlers I heard of being annoyed by the aborigines [sic] as early as 1840, but they tracked those gentlemen on their

route from Melbourne, and harassed them in every way—setting fire to the grass round them, throwing spears at their shepherds, and stealing their sheep (in Bride 1898, p. 164).

Dissatisfaction with burning practices was not expressed by colonists alone. The historical record indicates that Aboriginal people expressed disdain and resistance to the way that colonists used fire. Gilmore recorded how an Elder spoke bitterly to her father 'of the white's carelessness in lighting bushfires, saying that they lit them and let them run like a child that loved destruction …' (Gilmore 1934, p. 219).

Recognising fire skills

It is interesting to note, however, the nuanced understandings and expectations of Aboriginal fire skills that colonists developed. It is the case that the skill with which Aboriginal people in south-east Australia utilised and managed fire was immediately recognised by some explorers and colonists. For example, on 4 January 1802, Captain Murray noticed that the Aboriginal people in Western Port Bay had 'retired back into the woods, and ~6 p.m. doused their fire at once, although it must have covered near an acre [0.4 ha] of ground'. Nevertheless, it was not until later, that colonists came to appreciate the value of the fire skills possessed by Aboriginal people, particular in the face of bushfires. Historian Madeleine Say (2005) has argued that William Strutt's historical painting *Black Thursday, February 6th 1851* is a curious mixture of the representational and the contrived. Strutt claimed, however, that the scene depicted in the painting was based on accounts of observers. Of particular interest here are the two Aboriginal men that feature in the foreground of the picture. One man is shown fleeing the cataclysmic fire on foot, with a non-Indigenous child in his arms. The second Aboriginal man is sitting atop a horse, wielding a stock whip, urging the stampeding stock to escape the threatening fiery maelstrom (Strutt 1864). The Macdonald family in the Portland Bay District recalls a similar experience, describing how the Black Thursday:

> fires swept through 'Retreat' [the family's sheep station], but the men were able to save most of the stock by putting them in the dry river bed. Mary was very ill and the [presumably Gunditjmara] aboriginal couple saved her and the children by taking them to a water hole and then returned to save the house (Meckel 1991, p. 92).

The memoir of Captain Hepburn at Smeaton in central Victoria similarly recalls how 'it was the local Aboriginal people who saved the Hepburn family from the inferno descending upon them on Black Thursday by directing them to a safe spot near a creek' (Quinlan 1967, p. 155). Local histories such as McGivern's 'History of Rutherglen' (located in northern Victoria) also highlight Aboriginal people's prowess in identifying an impending bushfire, and having the skills to fight it:

> Aboriginals had warned local land holders of a bad bushfire approaching from the north. All night the wind screamed and raged furnace-hot. About fifteen tribesmen arrived at the homestead the next morning … Carl discussed the fire threat with them. It was decided to make a firebreak at once on their advice.

Carl and the Aborigines, with the Chinese to help, started the firebreak, getting it well down in a triangle; the Victoria swamp tribe was brought up and went into the cart and buggy as the wind grew stronger ... Suddenly the fire jumped the river, and flying, flaming gum tops lighted all the trees round about. Fleeing figures sought shelter in cow-bails; the fires burned out all the paddocks ... the Chinese [and Germans] had quickly buried their plant, bedding, furniture and food on the south side of the sand-dunes, and thus saved all of it, even as the timely warning by the natives had saved the homestead, outhouses, and not least, the owners' lives (McGivern 1983, pp. 102– 3).

The seeking of Aboriginal people's counsel in the event of bush fires was echoed in New South Wales. Mary Gilmore (1934) wrote extensively of her pioneering experiences in New South Wales and vividly recalled how local Aboriginal people would educate them in how to fight fire by: 'running for bushes, put them into the immigrants hands, and show how to beat back the flame as it licked up the grass'. Gilmore further noted how white people in the bush considered Aboriginal knowledge and skills in extinguishing bush fires to be indispensable, noting how: 'Send for the blacks!' was the first cry on every settlement when a fire started' (Gilmore 1934, p. 219). Gilmore and her contemporaries she wrote were in 'constant wonder' how easily the Aboriginal teachers would: check a fire before it grew too big for close handling, or start a return fire when and where it was safest' (Gilmore 1934, p. 220). Gilmore continues to describe the variation in fire management techniques practised by Aboriginal people and the colonists:

There was a difference between the blacks' method and the white's. The white man used large bushes and tired himself out with their weight and by heavy blows; the blacks took small bushes and used little and light action. The whites expended the energy of panic; the blacks acted in familiarity, as knowing how and what to do. They used arm action only, where the white man used his whole body. Where, as a last resort, the white man lit a roaring and continuous fire-break, the aboriginal set the lubras to make tiny flares, each separate, each put out in turn, and all lit roughly in line. The beaters they used were so small that they hunkered to do the lighting and beating.

The aboriginals said that not only must fire be met by fire, but that it could only be fought while still not too hot to be handled closely; that when it became so hot that it burnt and exhausted men it had to be met from a distance. They also said that a big fire as a fire break was as dangerous as a big fire itself ... I have seen a whole station in panic – men, women and children nearly killing themselves with frantic and wasteful effort; and then a handful of blacks and lubras under their chief come and have the fire contained and checked in no time (Gilmore 1934, pp. 219–220).

Supporting evidence for Gilmore's assertion that the colonists held Aboriginal firefighters in high esteem is contained within some western Victorian station journals. Large and destructive bushfires that swept through much of western Victoria in 1854 were, according to one squatter, 'fought with as many of his Black troop as he could muster' and again in 1858 'The blacks are busy fighting fires for me'. On yet another occasion in the same year the diarist noted: 'I have had a lot of blacks with me at the fire' (Brown 1952, pp. 215, 499).

Newspaper accounts such as reports in the *Dunolly & Bet Bet Shire Express* attest to the accolades, and longstanding appreciation, showered upon Aboriginal people's firefighting expertise.

> EDDINGTON BRIDGE FIRE - [Constable Weekes] in hastening to the bridge, saw his deputy, black King Tommy, and the two, with great energy and exertion, and fifty buckets of water- obtained at a distance of forty yards by Tommy- soon extinguished the fire ... otherwise the bridge must have been entirely consumed ... (Anonymous 1876)

Earlier, in 1841, colonists in Victoria had noted how useful Aboriginal people proved to be in fighting fires for them. A letter to the editor of the *Port Phillip Gazette*, 13 March 1841, refuted an account that had been published the previous week stating that: 'the blacks had set fire to her property: in fact, they were helping her to put the fire out'. Byrt has identified several similar occurrences of Aboriginal people being incorrectly accused of starting fires that later were proved to be not their doing (Byrt 2011, p. 75).

Thus within the often conflictual relationship between Aboriginal and colonial uses of fire, there could also be found a recognition of the fire skills and knowledge Aboriginal people possessed. South Australian colonist, Simpson Newland summarised such a view by asserting that: 'trees that were always consumed by bushfires were all flourishing in the perfection of beauty and health when the whiteman arrived' (Newland 1921, p. 24). Newland noted his opinion was not a solitary one and that there existed sufficient proof that Aboriginal people were very proficient in managing the landscape with fire, in order to avoid major conflagrations:

> I have discussed this matter with many bushmen, and although in many cases they had not thought about it, they all agree that they have never noticed evidences of bushfires in country inhabited by the blacks ... the only explanation I can offer is that the blacks realised that if they allowed bushfires the birds and animals upon which they lived would be destroyed (Newland 1921, pp. 113–14).

Discussion

The historical record indicates that in south-eastern Australia, fire and its use by Aboriginal people was extensive and of great interest to European explorers and colonists. This chapter has presented an indication of the observations made by Europeans, illustrating their fascination with Aboriginal customary burning practices. Some of this appeal was due to fire usage in aspects that were considered novel, such as creation mythology, and mortuary and initiation ceremonies. It is evident that Aboriginal fires were seen as distinctive, and commonly purposeful. The Europeans frequently noted the impact that the fires had on the landscape.

In noticing fire practices, many explorers and colonists in south-eastern Australia also observed the significance of fire to Aboriginal people. Aboriginal burning practices and demonstration of their Biocultural Knowledge in this domain, although undoubtedly modified, continued to feature strongly in interactions that occurred during the colonisation of the area by Europeans. To the newcomers, the fire knowledge and practices employed by Aboriginal people, was source for both concern and comfort. The colonists readily perceived Aboriginal people to be using fire maliciously, both as a means to attack and deter the intruders. They also recognised, however, that their own ability to manage and suppress fire was insufficient, and so came to appreciate the skills displayed by Aboriginal people. In these ways, fire, and the Biocultural Knowledge of fire held by Aboriginal people, played an influential role in the development of relationships on the colonial frontier.

Chapter 8
Watercraft

Fred Cahir

Introduction

Aboriginal watercraft in Australia ranged from single log floats and floating platforms of buoyant weed, bark or lightweight woody material for rafting across creeks and lagoons, to more bulky and robust seagoing outriggers (Davidson 1932; Edwards 1972). While the use of single-log floats was recorded across Australia, the pre-European distribution of watercraft such as rafts and canoes was incomplete, even in some coastal regions. There were none recorded along the coast from North West Cape in Western Australia, south to Perth and then east to Yorke Peninsula in South Australia (Davidson 1932). Small rafts made from reeds and dry Grass Tree flower stems, and bark canoes, were used to travel across the relatively quiet inland waters of mainland south-east Australia (Clarke 2012). The antiquity of bark canoes in south-eastern Australia has been discussed for over a century (Thomas 1905; Mathews 1905; Davidson 1932; Edwards 1972). Gregory (1904, p. 121) pointed to the reported discovery of an undated Aboriginal canoe that was said to have been found in a deep lead under the basalts of Mount Tabletop in Dargo. The report from a member of the Geological Survey in 1904 noted that: 'this canoe was an ironstone concretion; its shape was that of the bow of an English canoe (of the Rob Roy type), and did not resemble the aboriginal Victorian canoe.'

Creation stories and cultural significance

The cultural significance of canoes is borne out by the many references to them in Aboriginal law and lore. Early colonists such as James Kirby (1895) wrote of the canoe's intrinsic economic and legal value for Aboriginal people in the Murray–Darling region: 'The canoes were made from the bark of the gum trees; and any gum tree that had a bend in its trunk of the proper shape for making a canoe of its bark was highly prized; and if any of the tribe were to injure that tree, they would if found out, be punished by death' (p. 34). Peter Beveridge (1861–1864, p. 22), a squatter in this district reiterated this theme when he noted: 'The natives inhabiting districts where large rivers or lakes abound, hold their canoes in higher estimation than they do any other of their possessions.'

The written accounts also reveal the canoe's pre-eminence in Aboriginal creation stories such as the following from the Gippsland region in eastern Victoria (Howitt 1904, p. 486): 'There is a legend that the first Kurnai man marched across the country from the north-west, bearing on his head a bark canoe in which was his wife Tuk, that is the Musk duck, he being Borun, the Pelican … There was a great flood which covered the land, and drowned the people, excepting a man and two women. Bunjil Borun, the Pelican came by in his canoe, and took the man across to the mainland, then one woman, leaving the better looking one to the last.' Similarly, the Kulin people in central Victoria held a comparable belief about the origins of their people. The Djadjawurrung people's legend (Gregory 1904, p. 121), recorded at Lalgambook (Mt Franklin), stated: 'that their ancestors entered Australia in a canoe, and that they travelled into Victoria from the west'. Clarke (2012) has documented canoes placed by Ancestors in the Skyworld and the use of rafts to transport human souls to the 'Land to the West' after death. Stone (1911, p. 461) chronicled the canoe's presence in a creation story at Lake Boga (north-west Victoria):

> At one time, long years ago, there was a very large redgum tree growing in the lake, and its branches supported a tremendously large nest, the property of an immensely large pair of wedge-tailed eagles ('Nurrayil'). One fine day a young mother wandered, carrying her baby, a long way round the lake, and far from the camp, when, feeling tired, she sat down and amusedly watched her baby playing in the warm sand, when suddenly, and without warning, the larger of the two eagles swooped down and, seizing the baby, carried it away over the water to its eyrie in the redgum tree. The poor mother, seeing her baby suddenly lost to her for ever, commenced a mournful wailing, which the curlews or stone plovers ('Will') in sympathy took up, and have continued ever since. The disconsolate young mother then hurried back to the camp and reported the occurrence, upon which the doctor or medicineman ('Barngnull') directed that every person with a canoe was to proceed to the tree, and after cutting it down to tear it into little pieces, and to boat it all away to the river, where it was to be thrown upon the water to be carried away. The doctor then decreed that no more trees should grow in Lake Boga.

Other writers were informed about the prominent role of canoes in the ceremonies (Angus 1847; Smyth 1878; Berndt 1940a) and spiritual world of the peoples living near the Murray River (Smith 1930 in Edwards 1972; Tindale and Maegraith 1931; Berndt 1940a; Clarke 2012), and to the north of Lake Boga. EJ Eyre (1845, p. 216), an explorer and colonial administrator wrote: 'they believe in the existence of an individual called in the Murrumbidgee Biam, or the Murray Biam-baitch-y, who has the form and figure of a black, but is deformed in the lower extremities, and is always either sitting cross-legged on the ground, or ferrying about in a canoe.' AW Howitt (in Smyth 1878, p. 407) recorded how he had been tutored about the canoe's importance in the spiritual world when being led by the Aboriginal guide Toolabar in the 1860s: 'Speaking from memory, this canoe was about ten feet long [3 m], and carried Toolabar, myself, and our saddles and effects over the 'Snowy [River]' but there was not much to spare between the edge of the canoe and the water. At the other side Toolabar pulled it up on the bank, and said, half seriously, 'Leave him here, I believe mraat (dead blackfellow - ghost) might want him.' The canoe also features prominently in Aboriginal artwork – in both the pre-colonial and contact periods. A bark painting from

Lake Tyrrell depicts a canoe and its pilot as do many drawings and paintings by Tommy McRae (Sayers 1994), a 19th century Aboriginal artist who produced a large body of artwork about traditional and colonial Aboriginal life in northern Victoria.

Distribution

Many writers (Cary 1904; Massola 1971; Gerritsen 2001) have discussed canoe distribution and on what type of waters they were used. Bark canoes were made in wetter forested regions of mainland Australia where suitable large trees grew – down the east coast and south to the Murray–Darling basin (Clarke 2012). Some of the early European visitors to Port Phillip Bay explicitly noted their absence. Flinders (1814) after exploring the bay for seven days in May 1802 commented: 'I am not certain whether they have canoes, but none were seen.' Similarly Tuckey, who was with the first official settlement at Sorrento in 1801–1802, commented on the absence of canoes: 'No appearance of Canoes or water conveyances of any kind was seen', although he did subsequently comment that canoes had been seen on the Yarra (Tuckey 1805, p. 121). Flemming exploring at the junction of two rivers (probably the Yarra and the Salt Water River, now known as the Maribyrnong), on 4 February 1803, wrote: 'saw a canoe and two native huts' (Cary 1904, p. 32). Early British exploratory parties in Western Port Bay found a canoe in 1801. The Captain of the *Lady Nelson* James Grant wrote that a crew member 'returned with part of a canoe which he had found sunk near the mouth [of the Bass River], together with the two paddles belonging to it, and some line used in fishing. This canoe differed from any before seen, as it was framed with timber, and instead of being tied together at the ends was left open, the space afterwards filled with grass worked up with strong clay' (Grant 1803, pp. 138–9). The general consensus is that canoes were not used on the open seas and that in central and western Victoria they may have been largely restricted to calmer riverine and lacustrine waters.

Gaughwin and Fullagar's work on canoe distribution has highlighted how several researchers have reached similar conclusions, and concludes that:

> Canoes became less common moving west around the Victorian coast and they disappear altogether west of Port Phillip. Coinciding with this absence, there are no accounts of travel offshore on the west coast of Victoria. The overall conclusion based on this data is that in the early nineteenth century islands were infrequently visited, there was little specialisation directed toward island use either in terms of the development of appropriate technology or in fixing their role as an integral part of the economy (Gaughwin and Fullagar 1995, p. 50).

It is clear that the distribution of canoes was dependent upon the climate as well. In the 19th century Aboriginal memories of significant fluctuations in the physical landscape, such as the drying up of lake systems demonstrates how dramatically canoe distribution changed. One traveller in the Mildura region (north-west Victoria) described how 'the plain or dry lake just passed was described by the blacks as having at no very distant period contained a very deep body of water upon which the blacks used canoes …' Later in his travelogue the same writer, when describing Lake Albacutya and the Wirrengren Plains: 'from the appearance of the gumtrees on the margin of the dry lakes and swamps from which bark canoes have been cut, and from the statement of those of the natives about thirty years of age, who remember in their youth that they were like Lake Hindmarsh [full of water]' (White 1851 in Kenyon 1914, pp. 54, 58), reinforced there had been changes to the landscape.

Traditional uses

The size and complexity of the canoes was often remarked on. Robinson for instance, while travelling in the King River region (a perennial river of the North-East Murray catchment of the Murray–Darling basin, located in the alpine and Hume regions of Victoria and New South Wales) commented upon the wide variety of canoe sizes – and their usefulness. Some canoes were built and designed for one person while others were made for family groups. Robinson wrote in November 1842 that he 'Crossed the King River over a tree and a lagoon in a small bark canoe sturdy just sufficient to sustain my weight' and a little later in his journey, on the east bank of King River he saw 'nine natives crossing in a bark canoe' (Clark 2014b, p. 479).

Robinson was also a keen observer of the wide applications for Aboriginal canoes, the various modes of manufacture and customs surrounding them. In his travels across south-eastern Australia he encountered different types of canoes, and like many writers in the colonial period he often compared one style with another. His journal entries provide illustrations and brief notes about canoes that underscore the wide variety of purposes they were used for and the varying levels of craftsmanship embedded in their construction. For instance, in May 1840 he 'saw numerous indications of natives and on the bank large pieces of bark taken off the bend of the gum tree for canoes to cross the river when up … Some small corong, ie. canoes, were here in which they make the yeerip korr, the sweet beverage … Some large canoes were near the camp and covered up with boughs, stowed away for further use' (Clark 2014b, p. 184). While at Cape Liptrap he wrote of how he saw an 'abundance of swans and trees barked when the natives of Gipps land had made their canoes. These natives and Western Port natives cut the bark short across and take it of[f] a straight tree thus sew the ends and gum it like Sydney blacks. The Murray Blacks make their canoes like Waverong [Woiwurrung] and Dargoongerong [Daungwurrung] … The Gippsland blacks had been getting swan with canoes … This morning crossed the river in a frail canoe. The natives had made four canoes and this was the only one that could be used and very bad. Natives made a large and fine canoe this morning … Natives made a good sound canoe and was launched …'(Clark 2014b, p. 548). At Lake Wellington a similar picture presented itself. Robinson noted: 'Swans geese ducks and other birds are innumerable on these lakes. Bark canoes with natives in great numbers float on their waters … Canoes are different from others I had seen; are sewed at the ends …' (Clark 2014b, p. 562). At Mallacoota, Robinson again made comparative analyses of the various types of canoes he had witnessed elsewhere, noting how 'the native canoes are made same as Gipps Land; the two I saw on lake were so: viz. bark taken off tree one sheet square at ends and folded or puckered at ends with two ribs and two cross beams are safer and better than Melbourne canoes …' (Clark 2014b, p. 595). Robinson also drew illustrations of the canoes he saw and took note of their seaworthiness, and briefly noted how adept they were at handling their craft: 'Black fishing off coast in canoe, their canoes with both their hands alternately thus and bale their canoes with their hands from behind their backs thus …' (Clark 2014b, p. 607). At Lake Alexandrina in South Australia during June or July 1846 Robinson noted that: 'the natives live chiefly on fish, the men may be seen in their canoes spearing fish and the women diving for mussels. For this purpose they have a singular custom, they construct a raft made of reeds, sufficiently large for six or seven with a rail round fire to warm themselves, once there, they push boldly out and to the mussel grounds, eight or nine miles [13 to 14.5 km] from the shore. They are covered with sea weed, on them the natives make fire to warm themselves. They cook their fish and seen in the distance and have regularly on account of the smoke been mistaken by some, as I was, and have thought them small steam boats' (Clark 2014b, p. 732). A report in the *Geelong Advertiser* in May 1844 affirms Robinson's observation that in eastern Victorian waters canoes were used for coastal travel.

> Toby the king, native name Budyimbree, the king of the Nullika blacks, whose country extends from the south side of Twofold lay towards Cape Howe. The tribe are a fine, bold and manly race, and in a great measure, from living on the coast their habits are aquatic, and in their bark canoes they may be seen in the most boisterous weather buffeting seas where a whale boat can scarcely live (Anonymous 1844b).

Haydon was informed that Aboriginal people in Gippsland used their canoes to travel to islands close to the coast. He wrote of how an unspecified group of Aboriginal people in Gippsland had travelled to Rabbit Island, which was 'two miles distant from the mainland' in their canoes and 'insisted on the occupants of the island [whalers] leaving' and then burnt the whaler's huts (Haydon 1846, p. 51). Other writers (Gaughwin and Fullagar 1995) also recounted how, in a later account (Howitt 1904), annual trips were made to Rabbit Island to collect muttonbirds and their eggs. On this journey the long route was taken from Port Welshpool passing Snake Island, across Corner Inlet and then along the Wilsons Promontory coast, a trip of some 20 kilometres. Other accounts of canoes being used for seasonally travelling between the mainland and islands are found in William Thomas (February 1840 in Stephens 2014, pp. 131–2) who noted that: 'the Coast tribe however used in former times to make very large canoes and go over to French Island at a certain season after eggs'. Thomas also recorded a conversation in which he asked a Boonwurrung man, Pinterginner, whether they travelled to French Island (situated ~5 km distant from the Mornington Peninsula, south-east of Melbourne). Pinterginner told him the story of 11 named men – himself, Billy Lonsdale, Nerenuner, Munmanginer, Buller Bullup, Mr King, Lively, Devilliers, Burenun, Bobinerring and Mr Hill who once, only once, a long time ago before the white men came, made a bark canoe that apparently fell to pieces: 'Cut plenty bark and made boat went over only once plenty tumble to pieces; plenty of swans, kangaroo rats, but no Kangaroos or Oppossum. They were he said plenty frightened …' (Thomas, February 1840 in Stephens 2014, pp. 131–2).

Transport and communication

Upon approaching the Hume (now called the Murray) River in northern Victoria, Robinson recorded how canoes were used as a means by which Aboriginal people conveyed messages to one another. Robinson wrote: 'Natives got two of my cards to send to the Ovens and Goulburn Blacks who they were anxious to come. A messenger was cano[e]ing down river for this purpose, very swift and a fire in front, go in canoe rugs flying about …' (Clark 2014b, p. 629). Colonists in south-eastern Australia, such as Hugh Jamieson, also witnessed how canoes were a very ready source of transporting people and goods. Jamieson noted that in 'cases of sickness, much kindness and watchful attention is shown to male relatives. I have never seen a case in which they were neglected. When seriously unwell, they frequently express a wish to be removed from one place to another; the wish is complied with at all times, and they are removed either by means of a canoe or by a rude litter made for the occasion' (Bride 1898, p. 273). A report in the *Illustrated Sydney News* highlighted its use in in Aboriginal mortuary practices. In the article titled 'Native Funeral on the Murray' the writer states how 'The corpse is laid on a sheet of bark, the body is conveyed to a canoe formed out of a large sheet of bark, and the other members of the tribe, in similar crafts, convey it to the place of burial, chanting as they proceed … Most of the canoes have fires lighted in them, the blacks believing that in this manner they drive away evil spirits' (Anonymous 1864, p. 9).

Night fishing

Many travellers in colonial south-eastern Australia were particularly taken by the romantic beauty of Aboriginal canoes, particularly when viewed at night. The explorer EJ Eyre wrote eloquently of how:

> It has a singular and powerful effect upon the imagination, to witness at midnight a fleet of these canoes, gliding about in the distance like so many balls of fire, imparting a still deeper shade to the gloom of darkness which surrounds the spectator, and throwing an air of romance on the whole scene. Occasionally in travelling at night, and coming suddenly upon the river from the scrub behind, I have been dazzled and enchanted with the fairy sight that has burst upon me. The waters have been alive with brilliant fires, moving to and fro in every direction, like meteors from a marsh, and like those too, rapidly and inexplicably disappearing when the footsteps of strangers are heard approaching (Eyre 1845, p. 159).

GA Robinson was equally effusive in his appreciation of them. In May 1846 near the junction of the Murray and Darling he wrote:

> Their canoes were filled with small pieces [of] wood and a piece of bark covered with clay in the bow for the fireplace. They were preparing for fishing at night, said when sun went down spear fish in water from canoe. The native sports are exciting and various, their nocturnal spear fishing is interesting, by day a fleet of canoes is pretty, by night beautiful, canoes filled with small billets of wood are seen moored along the banks. Wet clay covering a piece of bark on the bow the fireplace, at night the fishermen launch into the stream and the noiseless

sport begins. From some points of view the brilliant fire lights, like floating meteors have a singular appearance. The scene at times fairy-like and enchanting (Clark 2014b, pp. 698).

Krefft (1862–65) writing principally about northern Victoria and New South Wales and George Haydon (1846) on the Gippsland lakes were similarly lavish in their descriptions of the panoramic night setting – and the dexterous manner in which (crayfish and) fish were hunted via canoes.

> Some of the canoes made by these people … are capable of carrying from six to ten persons; as they subsist chiefly on fish; the traveller may often notice them engaged in the peaceful occupation of catching the several species found in these waters. The night is the favourite season when their fragile canoes having a fire in them placed on mud or stones may be observed creeping along the shores in great numbers. Torches are lit and held high above the head of the fisherman, who waits patiently, scarcely moving except to wave his light and so throw out a stronger glare upon the waters. Presently a fish is discovered at a great depth … But look! he has speared his fish - it is a monstrous bream; and now all his art is necessary to preserve an equilibrium; a trifle either one side or the other will capsize his canoe and oblige him and his wife to swim on shore … At length after playing his fish for some time and still keeping hold of his spear, he proceeds to draw it in gently towards the bow of the canoe … (Haydon 1846, p. 44).

Customs and ceremony

Colonists were occasionally informed by Aboriginal people of how canoe making in south-eastern Australia was instilled with law and lore. In October 1844 Robinson was informed of a small detail regarding Aboriginal people's customs while travelling near 'Black Jack Creek, Reed's Ovens River' in northern Victoria, specifically regarding the manufacture of canoes: 'The Natives yesterday got three canoes, one good one, three men … appointed to work them. A law among them or custom: all cannot work [on] canoe …' (Clark 2014b, p. 630). Peter Beveridge also elicited information about canoes from Aboriginals of the Murray, Murrumbidgee and Darling River areas. Beveridge wrote of the care taken in the construction of canoes and the great value imbued in the proprietorship of canoes by Aboriginal society in general:

> They make canoes from the bark of the red gum; they generally select a tree with a bend in it for this purpose, as that saves them a great many hours work in the manufacture of their tiny craft; because if they use the bark of a straight stem they have to give it the necessary curve at each end by means of fire. They use these frail vessels very dexterously in the pursuit of fish, which they spear with the paddle, which has hooked grains at one end made of kangaroo leg bones. They also chase swans and other aquatic birds during the moulting

season, and capture canoe loads of them. Each male adult is the proprietor of a canoe, and he values it more than any other thing he possesses (Beveridge 1861–1864, p. 22).

Fison and Howitt, prominent anthropologists in the late 19th century and early 20th century discussed how canoes in the Gippsland region also figured in the giving of names:

> Another of the Briakolung was Bunjil Dauangiin. He was renowned for making canoes much turned up in front (Dauangun = to turn up) … As an illustration of the way in which such names are acquired, I note the following: —The Mitchell River flows for some thirty miles [48 km] through a gorge-like and inaccessible valley. In order to examine this, I caused two Kurnai to make canoes at the upper end, and therein we floated down together. The gorge was unknown to them, and the navigation was regarded as a great feat in consequence of the numerous rapids. For some time it was spoken of among the Brabroling, and the name of Bunjil Guyurgun (rapids) was given me (Fison and Howitt 1880, p. 211).

While Howitt was made privy to Aboriginal songs about canoes, including one about an attempt to cross the Snowy River in a leaky canoe during a flood sung by the Kurnai, he also related some detail about mortuary practices related to canoes. He wrote: 'in one case a [deceased man's] canoe was cut into pieces so that it could be put in the grave' and further noted how in some instances the body 'was placed in a canoe cut to its proper length … The canoe with the corpse was carried on the heads of the two natives to the grave …' (Howitt 1904, p. 463).

How to make a canoe

The colonial records are replete with descriptions of canoes and how Aboriginal people in south-eastern Australia made their canoes so quickly and seemingly with little effort. According to Thomas, bark canoes in south-eastern Australia are a single sheet of bark-type of canoe and that the single sheet type are divisible into 'four species distinguished by the four ways in which they are finished off: these are: by tying the ends; by lacing or sewing them; by skewering them sometimes with a lashing to make sure; and by blocking the ends with clay' (Thomas 1905, p. 56). Squatters overlanding stock from New South Wales to Victoria in the 1830s also often noted that the first evidence they had of Aboriginal people in the district they were travelling through was the presence of scarred trees denoting canoes or the Aboriginals' canoes themselves. Captain John Hepburn, an ex-sea Captain and squatter who took Djadjawurrung land in December 1836 described how upon his reaching the Goulburn River, 'we saw a specimen of their knowledge of naval architecture, in the shape of a canoe ~12 feet [3.6 m] in length, cut rudely from a half round of a box tree, which was bent, and gave the canoe a good spring at both ends. This sheet of bark would carry four men easily' (Bride 1898, p. 47). Robinson is typical of many colonists fascinated with the speed that Aboriginal people could manufacture their canoes. He wrote in October 1839:

'Saw the natives strip the bark off a bent tree and make a canoe and went on the water. The whole was not more than the work of an hour.' Robinson, and others, were clearly impressed by Aboriginal ingenuity and skill, and wrote frequently with illustrations in this vein: 'Saw several bark canoes along the banks of the Ovens, low down towards the junction, one of which was at least 12 feet long, thus: made by the bark from gum trees by taking it off whole … This is the most ready expedient I know of. A few minutes if urged by necessity will suffice to prepare this machine. They soon denude it from the tree, then dry it by placing the hollow part over a fire, made with boughs and dried grass' (Clark 2014b, p. 162).

EM Curr recounted his experiences managing his father's properties in northern Victoria 40 years earlier, including his interactions with the local Aboriginal people. He provides a vivid description of the awe with which many non-Aboriginal people on the frontier viewed the canoe-making expertise of the Aboriginal people. The settlers were greatly impressed at the local people's ability to manufacture canoes of a considerable size with great rapidity. Curr's views on Aboriginal canoes are also an exemplar of how many writers' opinions were interspersed with notions of white racial superiority. Curr wrote:

> how the Black makes his boat serves to illustrate not only the ready way in which he obtains from nature all that he requires, but also the skill with which he avails himself of the few implements of which he is possessed … and one is particularly led to contrast the ways of the savage and the civilized man. The first arriving at a stream, with the aid of a stone tomahawk provides himself in half an hour with a boat – frail and perishable, no doubt, but sufficient for the occasion, and passes over whilst the white man, checked for the time, sits down deliberately, and after a long delay produces an article of wood or iron which may serve him for years (Curr 1883, pp. 50–1).

Curr described further how the canoe he witnessed being manufactured 'was ~18 feet long [5.4 m] by 2 feet 6 inches [70 cm] wide, floating eight inches [20 cm] clear of the water, and would carry five persons' (Curr 1883, p. 51). Smyth wrote of several types of canoes and the various ways in which they were manufactured. In general he wrote of how the canoes used in south-eastern Australia are 'usually made of the bark of some species of gum-tree. The bark of the real gum-tree (Eucalyptus rostrata) is generally preferred; but in many districts the bark of other trees is taken, not because it is the best, but because it is easily obtainable of the sizes required' (Smyth 1878, p. 407).

Smyth also noted that there was much consultation among the community before a canoe was made and that it was a prestigious occupation only undertaken by the Elders.

> The Koor-ron or canoe is not made unless there be immediate occasion for its use. When it is necessary to cross a stream, a lake, or an arm of the sea, the natives assemble near the point of departure, and earnestly discuss questions relating to the means of transport. Some may be able to swim well and swiftly, and these would take to the water at once, if it were not for the goods they must carry - their shields, their weapons, and their cloaks. When it is finally settled

Illustrations of two different types of canoes. The top drawing is of a canoe from the south coast which is made from bark that does not allow for the ends being tied together. The water is kept out by walls of clay at each end. The bottom illustration depicts a bark canoe from Lake Tyers which has its ends tied together. Source: Smyth RB (1878) *The Aborigines of Victoria*. Ferres, Melbourne, p. 408.

> that the water must be crossed, the oldest and wisest of the tribe have devolved on them the duty of making a suitable canoe. If the numbers be large, the canoe must be large - so as to carry as many as possible at one time (Smyth 1878, p. 407).

Other writers too (see Krefft 1862–65) were intensely interested in regional variations in the construction of the canoe. Smyth drew on correspondence from non-Aboriginal people who had direct experience of being tutored by Aboriginal people in the skilled trade of making a canoe.

> I [Howitt] am acquainted with two kinds of bark canoe. One kind which is folded together, and tied up at the ends to form the stem and stern, and another kind, which is not tied at the ends, but is usually completed by a lump of mud at one or other end, as may be required by the shape of the canoe. The first kind of canoe is used, I think, alone by the Gippsland blacks (Smyth 1878, p. 410).

A little later in his book Smyth relates more detail about the type of trees used – information derived directly from Toolabar, a skilled Kurnai canoe maker.

1. Mountain ash [Eucalyptus regnans], a variety of ironbark, not turned inside out, but tied.
2. Stringybark [Eucalyptus agglomerata], turned inside out and tied.
3. Red-gum [Eucalyptus camaldulensis], generally from a bent tree; may be tied, but not turned inside out.
4. A variety of blue-gum [Eucalyptus globulus] (Ballook), turned and tied.
5. White-gum [possibly Eucalyptus elata] of river valleys, turned and tied; likewise the Snowy River mahogany [possibly Eucalyptus botryoides] (Binnack).
6. Peppermint; [Eucalyptus nicholii] 'no good;' according to Toolabar; as also a thin yellow barked stringy bark (Yert-chuck), the good kind being Yan-goura. (Smyth 1878, p. 411).

Colonists often corresponded about regional variations in the manufacture of canoes. Typical of this commentary is a letter by a gold miner Frances Vautier, at Warrandyte north-east of Melbourne, who wrote in 1886:

> Had a ride in a Native Bark Canoe last week. At first I was dubious, but soon I was enjoying it. Boonuka a Wurundjeri Native made the canoe from a bark of a Red Gum down river from Warrandyte. It was about twelve feet in length and two and a half feet wide [5.4 m x 70 cm], with both ends tied with a rope made from hair and stringy bark. Three yellow box sticks held the canoe in shape by placing them in the centre across the canoe and about a quarter of way along from each end (Woiwod 2012, p. 110).

Making use of Aboriginal canoes

Early colonial writers such as Kirby typically offered descriptions of the canoes that were so critical to the opening up of the Australian interior in a dichotomous manner. These writers noted that they were rudely built but also emphasised how well suited canoe and the maker were to the untrammelled bush – a bush not yet tamed by Western civilisation. Kirby further noted that he had 'seen, and been in these canoes, that would carry eight persons across the Murray', adding how adeptly Aboriginal people ferried white people's highly valued commodities: 'the blacks had tied light lines to the horses, and swam them over, holding the other end in the canoe' (Kirby 1895, p. 45). Most colonial commentators on the Aboriginal and his canoes (Jenkins 1850; Angas cited in Mathews 1907b) merely noted its simplicity – and almost grudgingly its efficacy:

> Rude canoes, fourteen feet long [4.2 m], and three feet [90 cm] wide, are made by the natives from the bark of the gum tree. For this purpose the tree is girdled, and a piece of bark, of the proper size and dimensions, is stripped off; this is folded in at either end, and fastened together with cords made of the fibres of the bark, or wooden pins. The canoe is then completed, and though not very strong, answers their purpose in coasting along the shores within the surf, or ferrying across the creeks and rivers. It is customary among them, as with the Fuegians, to build fires in the bottom of their canoes, on layers of earth or clay (Jenkins 1850, p. 251).

Haydon, too, noted the extraordinary usefulness that the Aboriginal canoes were to exploratory parties on the Tarwin River in the Gippsland region.

> Our canoe about ten feet [3 m] in length, with a beam of two feet six inches [70 cm], being completed and launched a native called Mumbo took charge of it, and by crossing thirty times managed to get all the lighter baggage and the individuals composing the party across in safety, but not without many narrow escapes from getting swamped, in consequence of the number of snags in the river (Haydon 1846, p. 48).

Later in the journey Haydon emphasised how the Aboriginal canoes were also used to transport very large and heavy items such as their dray across the Tarwin River. Haydon explained how the ends of their 'heavy conveyance' were fastened to bullock chains and the 'ends taken across [the creek] in the canoe' (Haydon 1846, p. 48). In earlier times explorers in south-eastern Australia such as Major Mitchell found that Aboriginal people used their canoes to escape from danger. Mitchell wrote of how on one occasion near Swan Hill in northern Victoria he witnessed a large group in 24 canoes flee from his advancing party (Mitchell 1839).

A study of colonial texts reveals a popular representation of inland canoes and Aboriginal people as: simple, crude, uncivilised and barbaric. However, they are also represented in a positive light – as witnessed by Major Mitchell's description of Aboriginal people as a race, before being sullied by contact with whites, was quickly followed by a summary of their skills in canoe construction and piloting:

> My experience enables me to speak in the most favourable terms of the Aborigines. The quickness of comprehension of those in the interior was most remarkable ...They are never awkward; on the contrary, in manners and general intelligence they appear superior to any class of white rustics that I have seen ... The men can strip from a tree in a very short time, a sheet of bark large enough to form a canoe, and they can propel this light craft through the water with astonishing ease and swiftness (Mitchell in Pridden 1843, p. 76).

There was a discernable need filled by Aboriginal canoe makers and pilots on the colonial frontier as evidenced by an account of an unidentified Wadawurrung man in Victoria rescuing a white woman from being lost in the bush that reveals their pre-eminent position as saviours to whites in peril.

> I wandered about for some time, not knowing which way to turn, then I was attracted by a fire ... As I neared the fire I was surrounded by a number of aboriginals each holding a tomahawk in his hand ... We were soon on the banks of the Barwon, where the native with his tomahawk cut a large piece of bark from a tree, and, in less time that it takes me to tell, placed it on the water, placed me on it, and plunged into the river beside me. I was conscious of being slowly paddled across the stream. All the time, I could feel his hot naked body touching my face ... Soon I was lifted up on the other side, and, in the same manner, almost dragged on until we reached Kardinia. The Dr [Thompson] rewarded the native by giving him food to take back to the camp (Cary 1904, p. 36).

Reports of Aboriginal perceptions of oafish white people, especially in water, are not unknown (Edwards 1972). Newspaper reports acknowledged – albeit in condescending terms – that *only* Aboriginal people had the requisite skills to handle canoes safely. A *Bendigo Advertiser* correspondent duly reported that female settlers on the Murray River routinely dealt with their stations being 'seventeen miles [27 km] under water in consequence of the Murray overflowing by boldly entering into a canoe, in the charge of a blackfellow'. The

THE LADIES WANT TO CROSS A RIVER

IN MID-STREAM

This illustration accompanied a news story about the perils of travelling on the inland waters of south-eastern Australia. Source: Unknown artist, *The Ladies want to cross a stream; In mid-stream*, 1883, illustration taken from *The Graphic*, wood engraving on paper. Art Gallery of Ballarat, Purchased with funds from public donation, 2014.

account continued: 'she steered along courageously and crossed the Edward, the Gulpa, and other deep water. She landed, and met the governess, who inspired by the example of her employer, ventured herself in the frail canoe, and traversed the seventeen miles of water to her new home in the bush' (Anonymous 1863, p. 1). In a similar vein one illustrated newspaper report (*The Graphic*) noted how on the 'back cattle stations of the interior a visit from ladies is a rare occurrence' and continued:

> When such an event takes place all the stockmen and native assistants gather together and stare their utmost. In the third illustration the ladies wish to cross the river, but they are a little uneasy at the sight of the boat, and of the ebony-skinned mermen who are acting as attendants. But the truth is that these blackfellows are indispensable. The boat is only a frail bark canoe, and to keep this from upsetting the balance must be very exactly maintained. Few Europeans possess the delicacy of touch and sight requisite for this, so the safest way especially for ladies who cannot swim, is to be piloted by a couple of blackfellows, one swimming on either side of the canoe (Anonymous 1883, n.p.).

Reliance upon Aboriginal knowledge of canoes

Perhaps one of the most undervalued contributions Aboriginal people made to the new colonial economy was that of guiding people and stock across the river systems of Australia. Explorers and drovers utilised Aboriginal ferrying expertise on a constant basis as it afforded them the most efficient and safest mode of river pilotage. Explorers such as EJ Eyre were not circumspect in their admiration of Aboriginal guides, their canoe technology and skills. Eyre noted: 'In travelling about from one place to another, I have always made it a point, if possible, to be accompanied by one or more natives, and I have often found great advantage from it … They are useful also in cutting bark canoes to cross a river, should such impede the progress of the party …' (Eyre 1845, p. 132). A decade later, during the gold rush period, the same admiring voices are still to be heard. Short references in miners' diaries relating the great value of encountering Aboriginal guides on inland waterways are not uncommon. Thomas Blyth, having kept a diary on the goldfields of Bendigo and Ballarat in 1852, devoted two small fragments relating to his contact with Aboriginal people: 'Proceeded ~3 miles [5 km] and camped near a gentleman with two blacks … Crossed the Campaspe taking the horses and cart through the R[iver] and paying a native with a canoe to cross our goods' (Blyth c.1850s, n.p.). Frederick Burchett wrote of how they now used 'the blacks canoes' to ford swollen rivers and that during floods 'we had to carry rations to outstations in a bark canoe … manufactured by the blacks in a very few minutes' (Burchett 1840s, p. 82). The occurrence of floods stymied many travellers and made travelling a misery. Diaries by early colonists such as Jane McCartney are a testament to how dependent many colonists were on Aboriginal skills in the early period of colonisation – not just for their labour, but for transport. McCartney's diary entry for 27 January 1859 tells how she and her family, in order to survive dangerous flood waters, were 'obliged to cross the creeks one by one in a boat made of a small piece of bark and only holding the native beside, who paddled it' (McCartney 1859, n.p).

Hubert De Castella's description of Aboriginal people taking on the risky task of guiding large numbers of people, cattle and supplies across the Murray River in the 1850s was a common one: 'Crossing the Murray, which is half a kilometer wide at that spot [junction of the Murray and Darling], was a large number of savages, [who] were camped on the river banks and had boats ready to help the travelers cross' (De Castella and Thornton-Smith 1987, p. 128). Other contemporary observers noted how traversing the Moorabool River in southern Victoria was a hazardous affair without highly skilled Aboriginal guides to ferry their goods across. Charles Sievwright wrote in June 1840 of how: 'The Surveyors who were proceeding to Portland Bay, can bear testimony to the skill and safety with which their provisions and equipment were transported across the Nar-ra-hil [Moorabool River], in a bark canoe, when without such assistance they must have remained some weeks upon its banks ere the river subsided' (Lakic and Wrench 1994, p. 129). Many travellers such as Alfred Howitt, who conducted geological research in Gippsland in 1875, also depended on their Aboriginal guides to construct and pilot vessels for ferrying them across rivers and entrusted them to deliver vital stores and provisions to forward positions.

> I wanted to examine a long portion of the Mitchell River which runs through horizontal strata and which are almost unknown, I therefore sent up two blackfellows 'Long Harry' and 'Charley Boy' under the care of a trustworthy man to Tabberaberra station at the head of the Gorges. Here they made two bark canoes by the time I arrived from Crooked River and the following morning we started on our voyage … Long Harry [sat] behind with a piece of green wattle bark in each hand about 6 in. by 12 in. [15 by 30 cm] which he used as a paddle … The other canoe contained Charley and the provisions for three days (Attwood 1994, 139).

At Murray River crossings into Victoria numerous miners overlanding from New South Wales and Adelaide shared the same experience as JH Trevena. In his 'Reminiscences of a Journey to the Victorian Diggings' he relayed how a party of miner families were paddled across by Aboriginal people in canoes and their bullock drays pulled across by Aboriginal people on the opposite bank (Faull 1983, p. 52). George Sugden, a roving bush worker in south-eastern Australia during the mid-19th century likewise recorded: 'The men would get over [swollen rivers] in blacks dugouts … dray ferried over by blacks' (Sugden c.1840–1870, n.p). Similarly, on a journey from Adelaide to the Victorian diggings, JH Walker's party in the 1850s employed a group of 'blackfellows' to 'drag a horse out of a morass' and take their womenfolk, their dray 'and all your things' across several rivers by 'canoes which are very useful, made out of a log hollowed out' (Walker 1993, p. 227). One cattle drover recalled how in 1872, '18 blacks came to his help with bark canoes' to get 3500 cattle across the Murray River. He wrote: 'When we drove the cattle into the water the blacks came in canoes behind them and formed a moving fence on the downstream side, which stopped the cattle ringing. By this means we soon got the remainder of our mob over' (Conrick 1923, p. 4).

Many parties were circumspect about the 'primitive constructions' to which they were entrusting their lives but there appears to be no record of them capsizing – while being piloted by an Aboriginal guide (Cahir 2012). EM Curr, as noted earlier, described the dimensions of the canoe he witnessed being manufactured as '... ~18 feet [5.4 m] long by 2 feet 6 inches [70 cm] wide, floating eight inches [20 cm] clear of the water, and would carry five persons' (Curr in Smyth 1878). A further description follows which notes the canoe's remarkable functionality – and cross-cultural purposes it was put to – on the early colonial frontier: 'Having thus got a canoe ... and with the assistance of my men and the three Blacks, I got one of my flocks transported across the Goulburn. The operation was performed by placing the sheep, six at a time, in the canoe with their legs tied, when a Black punted them over' (Smyth 1878, vol. 2, p. 334).

Judging by some accounts, the assistance of Aboriginal guides when fording rivers was not merely time saving or allowed for an 'uncommon neatness' – as Thomas Woolner described his observations of two Aboriginal people crossing the Ovens River (Woolner c.1855, n.p). Some gold seekers attested that it was only with the assistance of an Aboriginal guide that their party survived the Murray River crossing.

> We had some difficulty in fording the back-water course of the river, which we were compelled to do in consequence of the accident to the bridge; and unless we had had the assistance of a native, who directed us which way we should incline when we were in the river, we might have failed in safely getting over (Mossman and Banister 1853, p. 134).

Quips in some 19th century sheep station journals reveal the reliance on Aboriginal canoe technology. FR Godfrey at Boort in north-central Victoria was struck by the usefulness and utilitarian nature of Aboriginal canoes, especially noting in his journal several times the debt owed to the Aboriginal water carriers. These men rescued 'two tons of trussed hay in a fine canoe made by the blacks' on one occasion in September 1852 and more generally: 'The Aboriginals were often sent across by canoe for urgently needed goods – flour, tea, sugar, tobacco and the like, which were loaded onto waiting drays' (Fernihurst District History Committee 1992, p. 21). GA Robinson likewise noted how the colonial mail was at times dependent upon the use of Aboriginal canoes. In April 1844 he recorded how the 'Wordonger [Wodonga] under 3 1/2 miles [5.5 km] water. Three weeks mail white man came in canoe. Town of Albury, mail could be conveyed [only by] canoe' (Clark 2014b, p. 631). And likewise a news report three years later merely noted: 'the mail man had to take to a bark canoe for upwards of one mile and a half to get the bags safely through' (Anonymous 1847d).

Saving white fellas

Russell Ward's research on the 'Australian Legend' highlighted how 'stringy bark became a symbol of the outback' but, remarkably – and despite acknowledging that 'many a lost white man, and child, owed his life to the charity of the dark people' – glossed over how Aboriginal people saved countless people using stringy-bark canoes (Ward 1958, p. 201). Soerjohardjo has memorialised the heroic deeds of Yarrie, Jackey and two other unidentified Aboriginal people who rescued 49 people from drowning in the infamous Gundagai flood of 1852. Yarrie, a Wiradjuri bush worker on the Nangus sheep station that belonged to the

Gormly family, is reputed by the newspapers and white oral accounts of the period to have rescued approximately one-fifth of the population from drowning in what is recognised as the biggest flood disaster in Australian history (Soerjohardjo 2012). A report in the *Argus* (1867) similarly reminded its readers of the courage and readiness displayed by Aboriginal people ('Billy Wood and his gin') by building a canoe and rescuing a 'Mr McLachlan who could not swim from the swirling flood waters of the Macalister River in Gippsland' (Anonymous 1867c). Others, such as gold miner James Dannock, also praised their indispensability when crossing the Murray. Dannock, suffering badly from dysentery, and not responding to Aboriginal medicines, entrusted his life to some Aboriginal people who got him across the Murray.

> I took bad with the dysentery and the black lubras kindly got me wattle gum and when I did not get better they said 2 days that fellow go bung [dead] so I thought I had better clear out and got the blacks to put me over the river in a canoe (Dannock c.1855, p. 63).

Economic dependence on canoes

Squatters such as Frederic Godfrey reported how floodwaters would force them to be dependent on Aboriginal canoes to get them out of perilous situations. Godfrey described how rivers such as the Loddon, in central Victoria, were prone to flooding and one year [1851]: 'All the country on both sides of the Loddon was flooded, and the wagons could get no nearer than four miles [6.5 km] from the homestead, so supplies had to be brought in by bark canoe' (Stevens and Boort Historical Society 1969, p. 28). Similarly, Thomas Brown (aka Rolf Boldrewood) who penned several books about his experiences as a squatter in Victoria explained how dependent he became upon Aboriginal resourcefulness:

> I succeeded in inducing two blackfellows, by promise of unlimited shirts and tobacco to return with me. We walked to the edge of our plain, when they cut out a bark canoe, and after a miserable journey, raining in torrents all the time, I got home again. We now do all our work in bark canoes. (Here follows a sketch of Fred standing up in light array, with a large pipe in his mouth, navigating a bark canoe) (Boldrewood 1889, p. 52).

Likewise, experienced bushmen and squatters such as Albert Le Souëf, who had grown up in colonial Victoria and having been tutored by Aboriginal people in making and piloting canoes, did not fail to devote a sizeable portion in his reminiscences to memorialising the adaptation of canoes by colonisers for transporting sheep and people and to lauding the extraordinary skills of the Aboriginal canoe pilots who had mentored him (Smyth 1878). The use of bark canoes by early colonists was not limited to the transport of people, supplies and livestock. One observer wrote of the difficulties experienced by woolgrowers at the Mildura sheep station in northern Victoria. Allen wrote: 'The process of carriage is very inconvenient. Their station is on the Victoria side of the river {Murray], from which they have to convey it across the stream bale by bale, in native canoes, which though so very fragile in appearance, are capable of conveying a bale of wool across, weighing 400 pounds' (Allen 1853, p. 29).

Colonisers using canoes – badly

A substantial number of death by drowning reports in the colonial newspapers and contemporary books (Edwards 1972; Allen 1853) attests to the fact that colonisers were acculturating Aboriginal canoes and were, more often than not, ill-equipped with the requisite skills to either build canoes (Edwards 1972) or to safely pilot Aboriginal canoes. There are also many reports (including from Hovell cited in Andrews 1981) of ineptitude, like that of explorer and squatter Angas McMillan, who discovered much to their chagrin that – while they wished to emulate Aboriginal sagacity and skills in making a canoe – they could not. McMillan wrote of how in 1840 he attempted to ford the Macalister River.

> After travelling about ten miles [16 km] we encamped in the evening on a large stream, which I named the Macalister. This river appears deep and rapid, and is about 40 yards [36 m] wide … Started early in the morning, and tried to cross the river, but could not succeed, and followed the River Macalister down to its junction with another very large river called the La Trobe, which river is bounded on both sides by large morasses … We were busy all the evening endeavouring to cut a bark canoe, but did not succeed (quoted in Bride 1898, p. 258).

Discussion

Reminiscences about the fabled Aboriginal canoe continued into the 20th century. However, it is interesting to note how the spotlight of nostalgia shines brightly on the stoic white pioneers. An obituary in the *Mornington Standard* in 1917 is a vivid example of how Aboriginal people's ingenuity and consummate skills have been hijacked in the 20th and arguably by default, the 21st century. The obituary of a 'District Pioneer' relates the perilous journey of a family who were the first to overland a mob of cattle from Namoi in New South Wales and how 'hemmed in by flood on one spot for three months' called on 'those qualities of initiative and resource characteristic of the Australian, and making the 'ANZAC' of a soldier, the admiration of the world.' The obituary goes on to describe in vivid detail how 'the family proudly points to large articles of furniture ferried across the rivers Lachlan, Murrumbidgee and Murray in aboriginal canoes of bark'. Moreover, the white pioneering family, not the Aboriginal builders and pilots, even after this admission, is lauded for its 'wildest resourcefulness and generalship of no mean order' (Anonymous 1917, p. 3).

This chapter has demonstrated that one of the most undervalued contributions Aboriginal people made to the 19th century Colonial monetarist economy was the one of guiding people and stock across the inland river systems of Australia using their canoes. It has presented evidence that supports Reynolds' notion that it was Aboriginal guides – and specifically guides on their canoes – that smoothed the moments of colonial crisis, often articulating an economic limit to non-Indigenous occupation and a philosophical/spiritual understanding of the land.

Chapter 9

Shelter: housing

Fred Cahir

Introduction

Memmott's (2007) research encompassing all of Australia reveals that a range of well-crafted and technologically designed shelters and residential camps existed and that these varied from temporary shelters to substantial houses for large kin and community groups. He further noted that the materials of construction varied across geographic regions depending on the availability and supply of materials and that the types of construction varied from 'dome frameworks through to clad arc-shaped structures, to tripod and triangular shelters, and to elongated, egg-shaped, stone-based structures with a timber frame, and pole and platform constructions'.

The importance of shelters from an Aboriginal perspective is exemplified by their inclusion in the Creation stories of south-eastern Australia. By way of example it was explained by the 'Aborigines of Encounter Bay Tribe, South Australia' to H Meyer in 1846 that 'The Milky Way is a row of huts, amongst which they point out the heaps of ashes and the smoke ascending' (Meyer 1846, p. 202). Similarly, the Ramindjeri people of the Lower Murray region associated star constellations with the houses and camps of Ancestor Spirits (Meyer 1846, p. 201). Memmott (2007) notes that drawings done on the inside of bark shelters to illustrate Dreaming stories prefigured the later production of bark paintings. The painting or etching of bark shelters of Victoria was also remarked upon by Phillip Chauncy (Smyth 1878, vol. 2, p. 258) who wrote that: 'the young men [on the Murray River] often amused themselves with carving, or drawing with charcoal, on the inside of the bark, various objects and scenes in illustrations of any events which they desired to record …' Robinson, too, saw 'excellent sketches by natives on inside of bark [shelters]' in Gippsland and elsewhere in Victoria.

Many colonial writers wrote descriptions about how Aboriginal people constructed their shelters and also of the wide variety they encountered across south-eastern Australia. The extensive Australian English borrowing of words, such as *willam, humpy, gunyah, mia-mia* and *wurley* from Aboriginal languages in the 19th century to mean an Indigenous form of housing is also evidence of how Indigenous housing formed a significant part of colonial notions about Aboriginal culture (Dixon *et al.* 1992; Cahir 2006). Frances Perry (Perry and de Quetteville 1984, p. 167), a visitor to Buangor (central Victoria) in April 1852, described the familiarity which non-Aboriginal people had with the name and structure of traditional Aboriginal housing:

> We took a walk amongst the wooded hills, and came upon the largest (deserted) native encampment we had ever seen. One of the Mia Mias (you know what that is by this time – the a is not sounded) was as large as an ordinary-sized circular summer-house, and actually had rude seats all round, which is quite unusual.

Types of shelters

Based on the ethnographic literature a classification of various shelters constructed in south-eastern Australia included:

1. Small individual huts designed as hides for hunting

Both Eyre and Robinson remarked upon the clever mode of hunting for animals by constructing small huts. Eyre described one such hide thus:

> A small hut of reeds is made near the springs, or water holes, in those districts, where water is scarce; and in this, or in the top of a tree, if there be one near, the native carefully conceals himself, and patiently waits until his game comes to drink, when he is almost sure to strike it with his spear, seldom quitting his lurking place without an ample remuneration for his confinement (Eyre 1845, p. 276).

Similarly Robinson (Clark 2014b, p. 364) described how in the Grampians region and Lake Bonney in South Australia's riverlands there 'are hiding places ['about two feet six high'], made in the trees, that are situated in the gullies and where the natives used to conceal themselves and watch for the emu passing when they would spear them from the tree. I have seen in my travels a great number of these little bush huts in trees, mostly cherry trees.'

2. Clusters of seasonal (temporary) windbreaks frequently called 'screens'

Seasonal windbreaks were made of boughs, sticks, bark and grass or stone often in the shape of a half-cupola. On some occasions it was observed that animal skins and/or turf were also utilised in the construction of these quickly made shelters. The earliest written observations of seasonal Aboriginal shelters in south-eastern Australia are sourced from maritime explorers such as Matthew Flinders. On his journeys along the south-east coast Flinders made frequent, but brief, references to Aboriginal huts he saw on the shorelines, often merely remarking that there were 'many natives huts'. In late April 1802 Flinders and several members of his crew went ashore at Indented Head (near Geelong) and 'found a hut with a fire in it ... In the evening we found two newly erected huts with fires in them, and utensils' (Flinders 1814, vol. 1, pp. 280, 284, 340). Unfortunately Flinders provided no description of the huts. Collins, in 1803 (cited in New South Wales 1896), wrote briefly about Aboriginal shelters, merely noting: 'Those living in the bush, at some distance from the coast, contented themselves with, for each, a sheet of bark, bent in the middle and placed on its two ends on the ground' (p. 360). Robinson writing almost 40 years later described how the seasonal huts in the Grampians region were called 'worns' and that they 'are constructed of sticks, some of them neatly in the form of a half cupola … and of a variety of forms, the sticks are the frame work over which bark is laid, and the whole covered with turf, the grass laid inwards'. Robinson distinguished between the seasonal *worns* of the Grampians region and of the Kilambete or Hopkins district by noting that in the latter district they 'consist of 3 skins sewed together. The fur side is worn inwards, the flesh side is ornamental with various desines [designs]'. By utilising Robinson's journals it is possible to reconstruct to a degree many of the types of bark shelters used in numerous regions of south-eastern Australia. Some representative examples of Robinson's keen observations on principally bark shelters include:

> Ovens River: Saw numerous indications of the natives huts - these were of rather singular character to what I had been accustomed. They were made by large sheets of the box bark [possibly Eucalyptus macrocarpa], some of them 4 feet wide and a length sufficient for the hut purposed to constructed … They were of different forms … Devils River: Saw a native camp which had been occupied not long ago. The huts were constructed of green bark placed against sticks. There were 12 huts of different construction … Cape Howe: Saw several native camps today, old ones, double hut. Huts made by one sheet of bark knotched in middle and bent like dog kennel above is double hut … [near Nelson] Passed several recently formed camp huts differently constructed to those at Twofold Bay — probably because no trees were large enough to afford large sheets of bark or season when it would not strip. Three sticks or more covered with strips of bark sometimes with grass … [Near Albury] Native house: build huts with cross beam thus … laid bark against it bent over when bark crooked lupe [loop] one over the other like tiles … [near Tarong] Native huts: at this camp were well constructed. They are in various forms, principally a half cupola. A strong and sometimes neat frame work of sticks are first made, then a covering of bark and turf laid over the whole with the grass side downwards. These are warm and durable and are sufficient to prevent a native weapon from penetrating … [Mt Tappoc] Passed the frame of a small native hut; made very substantial and neat and placed on the slope or declivity of a hill with an oven at the back … Saw several native worns or huts, one 10 feet [3 m] in diameter (Clark 2014b, pp. 159, 178, 593, 614, 631, 299, 324).

William Thomas, like Robinson, experienced firsthand the efficiency, efficacy and aesthetic beauty of bark shelter villages. Thomas (in Bride 1898, p. 93) wrote with some effusion that: 'in wet weather a few sheets of bark make a comfortable house … in one half hour I have seen one of the most beautiful, romantic, and stillest parts of the wilderness become a busy and clamorous town, and the beautiful forest marred for materials for their habitation, and as much bustle as though the spot had been located for generations.' However suitable and useful Robinson, Dawson, Thomas, Eyre and other correspondents thought the seasonal shelters were for their stated purpose, there was a wealth of reports by the colonists about Aboriginal shelter that was derisive. In general it was considered to be primitive and flimsy housing. For instance, non-Aboriginal respondents' answers to the question 'What are the kinds of habitations in use among the people?' the 20 respondents' answers (Victorian

Parliament Legislative Council, Select Committee on the Aborigines 1859, p. 68) were typically brief and included phrases such as: 'miserable', 'temporary and primitive', 'merely a break wind of boughs' and 'a few sheets of bark stuck up for the night'. Hermann Koeler, an Adelaide colonist writing in 1840 was equally as disparaging. Koeler wrote: 'Homes are unknown to the South Australian: he lives out in the open the whole year round, and when he makes camp here today and there tomorrow he has to do nothing more than break off a few branches, which he piles up into a sort of wall 3 - 4 feet [1–1.2 m] high …' (Koeler and Mühlhäusler 2006, p. 78).

3. Seasonal cupolas

The seasonal cupola-shaped shelters were described as similar to the above but typically covered with reeds, limbs, bark, grass and a variety of foundations according to available materials. These shelters were generally considered to be seasonal accommodation that would be used for only a few months.

One of the earliest and slightly more expansive written descriptions of seasonal cupolas was provided by Collins (1798–1802, p. 360), who reported in 1803 that he 'saw on the sea-coast huts formed of pieces of bark from several trees, put together in the form of an oven, with an entrance, and large enough to hold six or eight persons. Their fire was always at the mouth of the hut, rather within than without' (p. 360). Oxley (1820, p. 253) exploring in central-western New South Wales wrote about the ubiquitous presence of Aboriginal huts, writing that 'their guniahs (or bark-huts) are in every direction'. Other colonists, such as William Thomas, recorded unusually constructed shelters they encountered from time to time such as in June 1849, near Brighton, a suburb of Melbourne. Thomas (1849 in Stephens 2014, vol. 2, pp. 387–8) related how he had 'come upon a Native Miam … it is a curious make … found it went in like a cave a small entrance where had been a fire, I crept in it was at least 3 or 4 yards [2.6 to 3.5 m] deep and evidently at the end was where someone had lain.' The following quotation from the early ethnographic literature made by colonist Peter Manifold at the Leigh River in western Victoria, demonstrates that low stone walls were used as windbreaks in areas where timber was hard to come by:

> The natives there formed these break-winds of stones, placed on edge in a circular form, some of them very perfect, leaving the opening generally towards the east. These circles are common on the plains or eastern part of this property (Purrumbete), where branches of trees could not be procured for giving shelter. When we first occupied this country (1839) it was common for the natives to use these circles as camping places, always having the fires in the centre … The circles are generally formed of large stones set on their edges and bedded in the ground close together, without any other stones on the top, thus forming good protection from the wind as they lay around the fire … The circles were about the size of the ordinary mia-mias, that is from ten to twenty feet [3–6 metres)] in diameter (Memmott 2007, pp. 189–90).

Colonists often described how different types of housing structures were made not just according to the materials available, such as the basalt rock in western Victoria, but also with a variety of different materials according to the season. Edward Curr (1883), writing about the Bangerang in northern Victoria, noted that they 'built their huts of boughs in fine weather, and of bark in winter'. Moreover in some districts, such as near Mt Eccles in south-west Victoria, Robinson (19–20 March 1842 in Clark 2014b, p. 453) occasionally saw both 'plenty [of] huts of dirt and stone houses'. In the same region (near the Hopkins River) he recorded: 'we saw a large mound of earth at least four feet high and 10 feet long, five wide [1.2 x 3 x 1.5 m]. My native companion said it was a black man's house, a large one like what [white] man's house' (p. 301). Again Robinson's journals highlight the wide variety of materials and modes of construction used by Aboriginal people to construct this classification of shelters. An extensive series of journal entries from Robinson's journals illustrate the variety of types of shelters and materials used:

> [near Mt Arapiles] Saw a reed hut … [near Glenelg River] Two reed huts … Lake Bonney: five huts strongly built with timber and covered with grass and turf were in the neighbourhood … [near Keliambete] The natives since this rain have covered their huts with turf … Passed a deserted Elengermot native camp of nine huts of recent construction; each hut was large enough to contain seven or eight persons (adults). They were made in form of a cupola with bark and sods over them with a doorway … [Emu Creek] Found two large native huts … [Coorong] Saw a native hut covered with turf at Salt Creek … Saw a vast number of old native encampments and huts, called in the language of the country, worn (Tcharcote). The native huts were like those on the west coast, in form of a neich. But the most singular huts or shelters were the [ones] with a piece of thatch upon it … And small light wood and box tree overhanging branches, were converted into habitations by having thatch laid on the top of the branches or crown (Clark 2014b, pp. 652, 663, 722, 295, 375, 282, 283).

EJ Eyre described the ubiquity, layout and structure of Aboriginal shelters in the Murray River area of South Australia during the early period of colonisation in south-eastern Australia (1830s–1840s):

> The huts of the natives were numerous, and of a large and substantial description; in the evening, the huts of the different tribes are built as near to each other as practicable, each tribe locating itself in the direction from whence it came. The size and character of the huts, with the number of their occupants, vary according to the state of the weather, and the local circumstances of their position. In fine weather, one hut will contain from two to five families, in wet weather more, each family however having a separate fire (Eyre 1845, p. 192).

Similarly, GA Robinson, the Chief Protector of Aborigines in Victoria, in his travels across south-eastern Australia frequently made note of the wide variety of Aboriginal camps or villages he encountered. One of these near the Murrumbidgee River 'comprised thirteen grass covered huts' and at 'Tower Hill [near present day Port Fairy] is a native village, an assemblage of huts' (Clark 2014b, p. 308).

There are many accounts in the historical record which vividly describe large populations of Aboriginal people living for extended (unspecified) periods of time in what was invariably described as villages. By way of example colonial writer William Westgarth stated:

Gill's illustration depicts what is probably a seasonal family shelter which many writers, including William Thomas described as 'a comfortable house made with a few sheets of bark, within half an hour'. Source: ST Gill, The Australian Sketchbook, mia mia image Gold Museum (The Sovereign Hill Museums Association), Ballarat.

> There was a 'native township' as it was termed, on the banks of the Eumaralla [Eumeralla] Lake or swamp ... The Aborigines generally encamped there during a portion of the year, for the purpose of fishing, with occasional rambling over the neighbouring country. At the period above alluded to [June 1844], these Eumaralla blacks were stated to be about two hundred in number; but two years previously [1842], when this locality was first taken up for pasturage the 'township' was said to contain five hundred (Westgarth 1848, p. 8).

4. More permanent village huts

Aboriginal huts in many colonists' diaries and reminiscences were often described as 'substantial' and constructed of sheets of bark – or alternatively of timber and turf and frequently coated with clay on the interior. Some huts of this type were described as being made from sand or earth and in certain regions were observed to have been built using large purpose-built mounds of earth. In some coastal areas huts were also described as being constructed of whalebones.

A considerable number of colonial writers observed and reflected on what many considered to be permanent or semi-permanent Aboriginal habitations including Charles Sturt (1834), who found on his *Two Expeditions into the Interior of Southern Australia* 'a group of seventy huts, capable of holding from twelve to fifteen men each' on or near the banks of the Macquarie River (p. 89). Sturt wrote: 'They seemingly occupy permanent huts', and also observed: 'The huts were large and long, all facing the same point of the compass, and in every way resembling the huts occupied by the natives of the Darling' (p. 89). Other early explorers and squatters who trespassed on Aboriginal land also often remarked on what they perceived as 'permanent villages' that they had seen. Major Mitchell (1839, vol. i) has provided us with some descriptions of these villages. Mitchell wrote:

> There were also permanent huts on both banks [of the river Darling] the first of the kind I had seen; these were large enough certainly to contain fifteen persons, and in one there had recently been a fire; they were semicircular, and formed of branches of trees, well thatched with straw, and forming altogether a covering of about a foot in thickness: these afforded a ready and dry shelter for a whole family, in bad weather for instance … [on the Lower Darling] we found a native village, in which the huts were of a very strong and permanent construction. One group was in ruins, but the more modern had been recently thatched with dry grass. Each formed a semicircle; the huts facing inwards, or to the centre, the open side of the curve being towards the east (Mitchell 1839, pp. 194, 261).

Mitchell (1839, p. 261) witnessed a variety of architectural types as he travelled through a large number of Aboriginal territories. He often compared one region's housing and architecture with another – such as his journal entry when he was near Corrong, close to the Lachlan River, he 'found several large huts similar to those of the Darling' (Mitchell 1839, p. 69). While travelling in the Grampians near White Lake [Mt Arapiles] Mitchell noticed some of huts which were of:

> a very different construction from those of the Aborigines in general, being large, of a circular form, and made of straight poles meeting at an upright pole in the centre; the outside had been first covered with bark and grass, and then entirely coated over with clay. The fire appeared to have been made entirely in the centre; a hole in the top had been left as a chimney, and the place seemed to have been in use for years as a casual habitation (Mitchell 1839, p. 194).

Mitchell's deputy, Stapylton (1836 in Andrews 1986, p. 146), also noted on the same day: 'passed to day several Guneaks [huts] of very Large dimensions one capable of containing at least 40 persons and of very superior construction.' Gerritsen (2000) contends that Stapylton and Mitchell's accounts taken together appear to strongly imply that:

> Aboriginal people in the southern Wimmera built well executed dome, tepee or tent-shaped structures capable of holding 40 people. A single residence of these dimensions, if this conjecture is correct, is qualitatively different to any previously reported in the Western District, and a most remarkable structure by Australian standards (Gerritsen 2000, p. 2).

Mitchell (1839) also expressed his amazement at how large and 'commodious' the Aboriginal tomb huts were. On one occasion he wrote: 'On the side of the hill of Tombs, there was one unusually capacious hut, of a very substantial construction, and on a commodious plan … which might thus have easily contained twelve or fifteen persons' (Mitchell 1839, p. 262). In tracing the course of the Gwydir, Mitchell found:

> [Sometimes there were] three or four huts together fronting to one and the same fire. Each was semicircular or circular, the roof conical, and from one side a flat roof stood forward like a portico supported by two sticks. Most of them were close to the trunk of a tree and they were covered — not, as in other parts, by sheets of bark, but with a variety of materials such as reeds, grass, and boughs (Mitchell 1839, p. 77).

In the photographic records too there is evidence of substantial housing. Memmott's (2007, p. 152) insightful analysis of photographs at or near the Yarra River (Melbourne) and central Victoria dating from the mid or late 19th century reveals a complex picture about what he terms 'substantial cold-weather houses'. The textual records describing substantial bark shelters concur very closely with Memmott's critical appraisal of several different bark shelters depicted in photographs as:

> a triangular prism (or tentlike) form, comprising large sheets of rigid bark supported on a central ridge pole. Sheets of bark are also used as a ridge capping along the top. A domiciliary space surrounded by a windbreak and a main shelter, both of durable construction using bush timbers, sheets of bark and other foliage. The main shelter is a triangular prism form supported by two forked posts and a central ridge pole … The well-constructed shelter is built of poles and sheets of rigid bark [possibly a type of stringybark or similar), and is close to 2 metres high. It contains a substantial space, in which the residents can stand up. The bark cladding is restrained by external poles tied together at the ridge of the house … [there are] several interconnecting spaces, constructed of sheets of rigid bark and poles. The house is thus a fairly sedentary one, reflected by its large size and its height (Memmott 2007, pp. 152, 155).

In the pictorial record, too, is evidence of what appears to be semi-permanent styled housing in the Portland region that was corroborated in a report by Henry Gisborne (1839, n.p). It succinctly noted: 'the natives build for themselves much more substantial huts than those used by the Northern Tribes'. This description was mirrored and expanded upon by George French Angas's (1847b) description and illustration of a shelter at Portland Bay in Western Victoria in the 1840s. This shelter appears to have a high, long-spanning domed roof and was probably occupied during the wet, cold winters of this region. The illustration is titled 'Native Encampment at Portland Bay - "Cold Morning" and his family'. The description notes:

> In the southern portion of the Province, and about the Portland Bay district the natives built larger and warmer huts than those to the northward: the native

name among the Portland Bay tribe for these huts is *miam miam* ... which was constructed of boughs of banksia and eucalyptus, thatched with reeds and dry grass (p. xxxiv).

Dawson (1881), also writing of the huts in the Western District, considered that Aboriginal shelters had superior insulating qualities to the European cottages in use during the late 19th century:

> In some parts of the country where it is easier to get stones than wood and bark for dwellings, the walls are built of flat stones, and roofed with limbs and thatch. A stony point of land on the south side of a lake near Camperdown is called 'karm karm', which means 'building of stones', but no marks or remains are now to be seen indicating the former existence of a building there. These permanent residences being proof against all kinds of weather, from excessive heat in summer to frost in winter, suit the constitutions of the aborigines very much better than the wooden cottages used at the Government aboriginal stations (p. 12).

Gideon S Lang (1865), who took up squatting at Kentbrush in South Australia on the coast west of Portland in 1842, described not only the habitations in rare and significant detail, but supplies the provenance of the huts as well:

> The huts are generally about nine feet [2.7 m] in diameter, five feet [1.5 m] high, and in the shape resembling half an orange. They are built in the first place of ... dry stiff branches ... the lower row set in the ground, and the rest interlaced above in the manner of a bird's nest. Upon this they place branches of trees, reeds, or long grass; over this they again place grass, turf, and above all sand if they have it, the top being rendered around and smooth like the Esquimaux winter hut. There is one low opening or door at one side of the hut, and in the opening is placed a fire. The largest of these huts I ever saw was on the Koorong [The Coorong], an arm of the sea behind the coast sandhills [Younghusband Peninsula], between Adelaide and Portland; it was fourteen feet [4.2 m] in diameter and quite eight feet [2.4 m] in height inside, and rose perpendicularly at the sides, and could accommodate an unusually large number of people (p. 26).

Other accounts only afford glimpses into what were perhaps semi-permanent Aboriginal shelters such as that of Joseph Lingard (1846) at Mallacoota in 1842. Lingard met two Aboriginal men and, 'made bold to go into their retreat, which I found to be like a house inside; their implements of war were reared in one corner, the floor was strewed over with all kinds of shells and fishes' bones, they had two fires of wood' (Lingard 1846, p. 57). Similarly, a comment by William Westgarth (1848, p. 47) reported a particular habitation, 'belonging to the Wimmera blacks, of a fanciful structure' but provides no further detail. The latter description presumably refers to sod cladding on a dome frame, which was used in this region. A small reference by William Thomas in March 1844 (cited in Coutts 1981, p. 241) is similarly slim in detail but hints at a class of substantial housing, merely stating that: 'the huts were of the first rate, and they are said to have woven straw huts of their own manufactory.' Smyth (1878, p. 128) likewise, while not citing specific references, observed that some villages 'seen by the first explorers occupied a considerable space, and were intersected by neatly-made walks, running in graceful curves; others consisted of well constructed huts, thatched and secured with a net.' Krefft (1862–65, p. 366) noted, also, that in the Mildura region 'they lived in closed sort of huts, which had somewhat of a permanent character'.

Historical and archaeological research by Memmott (2007) and Williams (1988) has revealed that a surveyor, Mr Alex Ingram, found several circular stone walls c.1898 on the stony rises south of Lake Condah and around Mt Eccles. Ingram was told by an old Aboriginal at Condah Mission Station that they had been roofed over with boughs and bark like an ordinary hut. Another informant recollected his grandfather describing decaying bark and sapling roofs on the stone houses on his family property near Louth Swamp. Williams (1988) contends that the construction of what contemporary observers referred to as the 'beehive'-shaped hut type seems to have been restricted to the south-western region of Victoria and also posits that there are indications that sizeable earth mounds recorded by Robinson and Dawson's Aboriginal informants were used as ovens, general camping areas, and as formations for substantial huts.

Gerritsen (2000) concurs that the best-recorded example of the beehive-styled housing was one near Caramut in south-west Victoria c.1840. Gerritsen has calculated that the residences described by William Thomas' informants were 'in the order of 3–3.5 m in diameter, 2.5–2.8 m high, with a semi-oval (vertical orientation) opening ~1 m high, and a circular aperture at the apex (20–22. 5 cm in diameter).' Thomas (1858) described the houses thus:

> … by Mustons and the Scrubby Creek to the westward … first settlers found a regular aboriginal settlement. This settlement was about 50 miles NE of Port fairy. There was on the banks of the creek between 20 and 30 huts in the form of a beehive or sugar loaf some of them capable of holding a dozen people. These huts were about 6 feet [1.8 m] high or little more, about 10 feet [3 m] in diameter, an opening about 3 feet 6 inches [1 m] high for a door which they closed at night if they required with a sheet of bark, an aperture at the top 8 or 9 inches [20–22 cm] to let out the smoke which in wet weather they covered with a sod. These buildings were all made of a circular form, closely worked and then covered with mud, they would bear the weight of a man on them without injury. These blacks made various well-constructed dams in the creek which by certain heights acted as sluice gates in the flooding season … (n.p).

Squatters in other parts of south-eastern Australia likewise recalled large numbers of Aboriginal huts frequently congregated around water sources including: the Aboriginal Protector, GA Robinson (Clark 2014b, p. 93) travelling in the Mt Alexander district (central Victoria) in 1840. Robinson was advised by the resident squatter that close to the creeks 'the huts of the natives are still very numerous, like villages'; Captain Hepburn (Bride 1898, p.

47) in 1839 noted a 'very large encampment, ~70 mia-mias' near the Goulburn'; George Haydon (1846, p. 136) near the Tarwin River in Gippsland came upon 'upwards of a hundred huts'; Joseph Gellibrand (Bride 1898, p. 284) at an unspecified place near Port Phillip Bay in January 1836 reported that: 'about one hundred native huts were found near water'. Edward Henty (Bassett 1962, p. 303) at Portland Bay in 1834 described a cluster or village, noting that: 'we have seen today at least 50 native huts'. Occasionally the historical records reveal substantially more about these villages, such as an oft-quoted passage from George Robinson (Clark 2014b) who wrote on 10 May 1841 that he had encountered several camps at the Great Swamp at 'Konnung-i-yoke', (probably near Hamilton in western Victoria). He describes one village with 13 houses as follows:

> This place, previous to its occupation by white men, was a favourite resort and as this was the only permanent supply of water, a village had been formed. I counted 13 large huts built in form of a cupola. When seen at a distance they have the appearance of mounds of earth. They are built of large sticks closely packed together and covered with turf, grass side inwards. There are several variations. Those like a cupola are sometimes double and have two entrances; others again are like a niech. Then there are some made of boughs and grass. And last are the common screens. The permanent huts are those in form of a

This photograph, more probably taken in the 1860s, depicts a scene in Strathalbyn, South Australia, which closely resembles GA Robinson's description of a village in Hamilton (western Victoria) of 'fine large double huts', cupola shaped and some with two entrances. Source: Flour Mill and Grain Store (1842) [1860s] with an Aboriginal camp nearby. Strathalbyn Branch of National Trust of South Australia.

cupola. Three of these huts had been occupied a day or two previous to my visit … Saw a fine large double hut, 10 feet [3 m] diameter with two entrances and four feet [1.2 m] high in centre. I went in at one door and came out of the second (Clark 2014b, pp. 321, 326).

In a later letter to the House of Commons (Great Britain 1844), Robinson's descriptions of this Aboriginal village and other types of Aboriginal shelters provide us with further valuable ethnographic information about Aboriginal architecture, spatial arrangements, the types of materials used and modes of construction.

Sometimes each married man will have a hut for himself, his wives, and family, including perhaps occasionally his mother, or some other near relative. At other times, large long huts are constructed, in which, from five to ten families reside, each having their own separate fire. Young unmarried men frequently unite in parties of six or eight, and make a hut for themselves. The materials of which the huts are composed, are generally small branches or boughs of trees, covered in wet weather with grass, or other similar material. At other times, and especially if large, or made in wet weather, they are formed of thick solid logs of wood, piled and arranged much in the same way as the lighter material, but presenting an appearance of durability that the others do not possess. In this case they are generally well covered over with grass, creeping plants, or whatever else may appear likely to render them waterproof (p. 240).

Robinson's journals again are a storehouse of information about this housing class across south-eastern Australia (Clark 2014b). Some representative examples of his journal entries in respect to this class of architecture include:

[near Coorong] Passed a Native hut, made like those of Tapoc, with logs and turf … [near Murray River] Saw a fine hut of logs made like Portland natives … Near Mount Schank: The huts of the Aborigines are strongly built with logs in the form of half a beehive and turfed over … [at Mt. Burr, to the north of Mt. Gambier] Past a Native hut, made like those of Tapoc [Mt. Napier], with logs and turf … [Encounter Bay SA] Visited the native camp and huts about the fishery [whaling station] … Rude huts, logs and small farm of beehives huts covered loosely with boughs and dirt … [Wellington] Five or six large huts at police station, made of wood and reeds about 8 feet [2.4 m] diameter and 6 feet [1.8 m] high … [Encounter Bay] some of the huts there I observed were constructed of whale bone … at Wellington there were two of these huts of the lake natives made of reeds high enough to stand upright and large enough to contain two or three families or more and this was the case at Encounter Bay (Clark 2014b, pp. 661, 718, 720, 731).

Several reports of whalebones being used for the structures of coastal houses exist. For example, a painting from 1842 by WA Cawthorne (Memmott 2007) depicts two whalebones used as a structural frame in an Aboriginal shelter. George French Angas (Angas 1847a), an explorer and also a painter, copied some of Cawthorne's paintings and his subjects and the following is the caption accompanying one of Angas's copies:

> Natives of Encounter Bay. The view here given represents a part of the Shores of Encounter Bay, with a native hut formed of the ribs of a whale. Numerous carcasses of whales being cast upon the shores adjoining the fisheries, most of the native huts are constructed with a framework of bones, the interstices being filled in with boughs and dried grass. The present group consist of a man called Ginginnana, and his two wives, Kundarkey and Wuddugar (cited in Memmott 2007, p. 206).

Robinson's and Thomas's testimonies to Aboriginal villages in south-eastern Australia, and particularly in south-west Victoria, have been instrumental in the quest for Aboriginal and archaeological evidence of semi-permanent and permanent villages. For instance, in 1839 William Thomas, wrote an account of what would appear to be a seasonal fishing village '-2 miles further from Arthur's seat were 12 Miams, Most substantially built which I no doubt not in their periodical fishing excursions they return to, and from appearance they return to and inhabit their own individual huts' ((Stephens 2014, vol. 1, p. 47). William Thomas, again in early 1844 (cited in Gerritsen 2000, p. 49), recorded in his journal seeing: 'Settled Villages ... on the Banks of Thorn's River ... [of] well built slab huts some capable of holding 15 or 20 persons' and that one village '[is] supposed to contain 400 Inhabitants'. Robinson, in one of his reports on his visit to western Victoria (Clark 2014b) recorded how he had:

> Led our horses into the stony rises ... plenty huts of dirt and others built of stones. At the native camp they had oven baking roots ... Stone houses, stone weirs (p. 453).

Aboriginal and archaeological research by many scholars including Coutts *et al.* (1978), Williams (1988), Gerritsen (2000), Builth (1998) and Memmott (2007) have argued that in south-west Victoria the local groups of the Gunditjmara owned different estates, each estate containing a set of eel traps and associated structural complexes, including a village; and that they were passed on to their descendants. Gerritsen (2000, p. 41) has argued the grounds for identifying embedded mobility are very limited but insists 'it would be unsafe to eliminate this as a possibility', and highlights a comment made by Robinson in his journal dated 30 April 1841 indicating one particular group did undertake residential moves. Gerritsen (2000) writes:

> Near Tarrone, on the Moyne River, north of Port Fairy, Robinson met an unnamed man who, 'took me to several spots where he had resided and had worns or huts'. This man was a member of several families that resided in the area who, it appeared, owned a 'very fine and large weir' for catching eels. The clear implication of this was that this particular group had huts, probably of a

permanent or semi-permanent character, in different locations that they shifted location with unknown frequency but they were 'tethered' to their weir (p. 42).

5. Permanent stone house villages

A select few writers in the colonial period noted houses and/or villages built with low stone walls and regularly roofed with branches and bark. Archaeologists have extensively researched the south-west region of Victoria in particular for evidence of intensification – including stone houses. Builth (2000, 2002, 2014) has analysed the edge of one area of the lava flow and found what she believes to be 103 stone dwelling and storage remains, the latter occurring in clusters, with some having shared walls. Further archaeological and historical research combined with Aboriginal oral testimonies led Builth to believe that there existed higher walled structures as well as low windbreaks which were roofed with branches and bark. In addition to the archaeological research there is an intriguing description of Aboriginal people living in permanent houses, not caves, made of stone. Aboriginal informants told William Thomas (1858), the Guardian of Aborigines in Victoria, that there were Aboriginal people:

> in the Australian Alps, who continually live in stone Houses, Great Wise Blacks who make blacks Dream or appear to them & shew [show] them Dances, I have a Drawing from an Old Black Representing one of these druids coming forth from the Rocky habitation, & 6 Druids Dancing. In Report 1843 one of the Devils River Blks [Blacks] gave me the following History of the Gaggip. He stated that there are in the Australian Alps a race of Blks who live in Stone Houses made by themselves (not caves) and that some of these blacks give them, that these Blacks are *very* good like our Sunday, that they teach Omeo, Devils River, Broken River & other Blks dances & singing, that other Blks go to these blacks & learn & when one tribe has gaggip with another from that time they are friends (n.p).

6. Complex hut structures

Intricate shelters usually described as 'sugarloaf', 'double cupolas' or 'tent/tiwi' which were thatched with reeds or local timbers and generally had a central timber support structure were also noted in the historical records. Some of these structures were observed to have multiple 'apartments', were made of sods of earth or of bark – and a door made of skins or bark. One of the few accounts received directly from Aboriginal people of this class of housing in south-eastern Australia was obtained from Kaawirn Kuunawarn, a Kirrae Wurrung Elder. King David (his assumed Anglo name) was one of the chief informants for James Dawson's (and his daughter, Isabella Park Taylor) *Australian Aborigines: The Languages and Customs of Several Tribes of Aborigines in the Western District of Victoria, Australia* (1881). Kuunawarn's (in Dawson 1881) and the 'united testimony of several [other] very intelligent aborigines' provides this more general description of west Victorian habitations – including multi-apartment dwellings:

Habitations - *wuurns* are of various kinds, and are constructed to suit the seasons. The principal one is the permanent family dwelling, which is made of strong limbs of trees stuck up in dome-shape, high enough to allow a tall man to stand upright underneath them. Small limbs fill up the intermediate spaces, and these are covered with sheets of bark, thatch, sods and earth till the roof and sides are proof against wind and rain. The doorway is low, and generally faces the morning sun or a sheltering rock (p. 10).

Charles Griffith (1845), a squatter in the Port Phillip District (now Victoria) was generally disparaging of Aboriginal shelter but considered that the Aboriginal shelters in the south and west of Victoria to be superior. Griffith wrote:

towards Port Fairy and Portland Bay they construct a kind of hut for the winter season, which is of a more durable character. This they do by heaping sods and clay on the top, of the original mi-mi; they add a new piece to it at every shift of wind, so as still to make the entrance from the lee side, and by this means, when they remain in one place for any length of time, these earths reach to a considerable size: I have seen one fully fifteen feet [4.5 m] long, and high enough for a man to stand upright in (p. 152).

Similarly, the Assistant Protector Sievwright, in his report to GA Robinson on 1 June 1840 (Lakic and Wrench 1994), was taken aback by the size and structural integrity of the Aboriginal residences he encountered in south-west Victoria:

I came upon what may be called an Aboriginal village. There were five huts, within sight of each other, where the bush was thick, two of them were large, and capable of containing fifty people each, they were compactly and solidly built, - in the form of a cupola - the beams were curiously though rudely interwoven with each other, forming a perfect done, about seven feet [2 m] high, there was a central prop, to sustain the roof which was made of a mixture of mud and grass, laid over beams and was of such strength and solidity that the boy rode his horse upon the top of one of these without it having the slightest effect of the building. There was an entrance to each, in the opposite direction of the prevailing wind, they were curious and wonderful specimens of primeval architecture, and must endure for years (p. 130).

Dawson (1881, p. 10), in discussing traditional habitations of the same region as Sievwright, Griffiths and Robinson, wrote 40 years later in 1881 that, in 'what appears to be one dwelling, fifty or more persons can be accommodated'. Gerrittsen (2000) explains that Dawson had received this information in a letter from Rev. J Francis, manager of the Lake Condah Mission in 1868. This advised Dawson that people at the Mission, who had come from all over the western part of the Western District, had told Francis they had formerly lived in, 'communities of 30–40 and even more, occupying one Mia mia'. William Stanbridge and George Angas noted that in South Australia there also existed very commodious dome-shaped houses:

In the neighbourhood of Mount Gambier some of the lodges or oolahs are very pretty, being dome-shaped, and constructed of boughs closely interwoven, with a small arched opening for the entrance, and a little aperture in the crown for the escape of the smoke (Stanbridge 1861, p. 300).

[Sand hills of the Coorong] The people inhabiting the margin of the lake, build for themselves winter huts, resembling beehives, to protect them in these exposed situations from the cold south and west winds that prevail during that season. These huts are composed of turf and mud, over a framework of sticks, and have a small entrance on the leeward side. Along the shores of the Coorong they cover these huts with sand and shells, so as to form a hollow mound, impervious to the wind, beneath which they creep in stormy weather (Angas 1847, p. 64).

Use and reliance by colonisers on Aboriginal shelter

Judging by some accounts the expertise of Aboriginal builders when constructing shelters was not merely time saving or allowed for an 'overcoming of mechanical difficulties', as GS Lang's (1865) anecdote about Aboriginal hut building by colonists demonstrates. Lang, an experienced bushman and others, attested that Aboriginal proficiency in building huts was truly the technical preserve of Aboriginal people – a fact that caused much mirth among Aboriginal people.

> Every native can build one of these huts with the greatest ease, but I never knew a white man who could do so. One of our overseers, a very ingenious man, singularly skilful in overcoming mechanical difficulties, I saw over and over again attempt the construction of a hut, native fashion, under the direction of the blacks, and with a blackfellow beside him building up another, as an example; but he never got his edifice to stand the weight of the turf, and it generally fell before he had the framework completed, of course to the intense amusement of the natives (Lang 1865, p. 26).

Other squatters openly adopted the local Aboriginal architectural style as evidenced by this testimony cited in Smyth (1878):

> A squatter — who was one of the earliest settlers in the Wannon district — says that the natives had comfortable huts at the time he first occupied the country. They were dome-shaped, made of branches of trees, and covered with grass and clay. The opening, protected by a porch, was always towards the north-west, whence came only gentle breezes occasionally never strong winds or storms. Observing this peculiarity — and having ascertained that a house presenting such a front was protected from gales — he built his own bush residence with its doors and windows towards the same quarter (p. 126).

Many travellers and government officials, such as William Thomas who travelled frequently accompanied by Aboriginal people, also depended on their Aboriginal confidants to construct shelters for them and entrusted them to also build shelters for their children, wife and children. Thomas' (Stephens 2014, vol. 2) journal frequently contains entries such as this one in 1844: 'got the blacks to make me miam which they do and a good one' and adds with great pleasure that he was not left waiting for very long as 'it took two of them 1 h' (p. 58). Joseph Parker, also an Assistant Protector of the Aborigines, similarly wrote of how in 1839 the Djadjawurrung (central Victoria) had travelled with his two children ahead of Parker and his wife, and that to his amazement they had very quickly constructed shelters so as to care for his children. Parker (in Clark and Cahir 2016, p. 14) wrote of how he found 'the blacks had formed a mia-mia, and the two little chaps [Parker's children] were on a rug, each with a cooked leg of an opossum in his hands'. Parker, like many of his contemporaries, was an avid recipient of Aboriginal fashion, lore, bush craft – and hospitality while being guided through the bush of Aboriginal Australia. He wrote of Aboriginal dwellings (in Clark and Cahir 2016) thus:

> The natives generally have a considerable taste for carving and drawing. I have repeatedly seen the inside of their mia-mias covered with rude etchings of the kangaroo or emu, or anything else that might occur to them. The sheets of bark are first blackened in the fire, and the drawings are made with a piece of pointed stone or a nail, and some are really very well done.

> Let me tell you once for all our proceedings in forming our camps, and making ourselves as comfortable as possible for the night - First find a clear patch as near as possible to a good log to form the back of our fire - Then pull down sufficient dead Fern fronds to make a bed a foot thick - Then cut a couple of forked sticks and a long sapling these to support the back of our Mia Mia formed of the long green frond of the Tree Fern leaned against it at an angle of 45° forming a tight little place - The bed being covered with a 6 foot square of waterproof Calico - a good fire in front, enrolled in a pair of blankets we were snug for the night.

> The police as part of their equipment, had two small calico tents. This provided for the Government Officers. The bushmen assisted by the Natives, soon rigged up capital Mia Mias of Tee tree, screening them on the windward side, and a good fire in front made things comfortable for the night (p. 242).

Most colonial commentators on Aboriginal shelters merely noted their simplicity – and, almost grudgingly, their efficacy. Early squatters such as WT Mollison (Bride 1898, p. 183) on the Coliban River in central Victoria reflected on how in 1837: 'We lived in reed mia-mias and tents comfortably enough for some time'. Others noted the admixture of traditional shelter structure with the colonists' shelter materials such as Henry Godfrey (1926, p. 91) at Boort in central Victoria who 'put up a tarpaulin mia-mia for the shepherds'. Likewise GA Robinson (Clark 2014b, p. 702), in his extensive travels throughout south-eastern Australia in the 1840s, frequently remarked on an eclectic assemblage of materials and structures used by shepherds to build their huts such as '… a rude tent hut covered with dry reeds their only habitation … this was their dining room and their dormitory.' Another example of the medley of Aboriginal and non-Aboriginal styled housing commonly used by non-Aboriginal people on the colonial frontier was recorded by Patrick Costello (1841, p. 16). He noted on his arrival at a sheep station in central Victoria that a 'bark hut 'Gunyah' as it is termed was fixed up for us to live in'. Sheila MacDonald (c.1850s, n.p) at Heathcote, north of Melbourne, similarly wrote: 'We lived in a bark gunyah with three bunks in it – a bunk being simply a sheet of bark laid on four posts with a possum rug for blankets.'

Pascoe (2014), too, has highlighted how 'quips' in some 19th century explorer's journals reveal the understated appreciation of both Aboriginal housing for shelter and its frequent use by colonists. For instance, Major Mitchell documented this style of accommodation in his expedition through south-eastern Australia. He noted that, on their journey along the shore of Portland Bay in Western Victoria, he and his party had 'encamped on the rich grassy land just beyond and I occupied for the night a snug old hut of the natives' (Mitchell 1839, vol. 2, p. 247). Later in his journey, near present-day Mt Napier, he took up residence in an Aboriginal hut – and further extolled the twin virtues of tasteful Aboriginal planning and careful construction. Mitchell (1839) wrote:

> Two very substantial huts showed that even the natives had been attracted by the beauty of the spot and, as the day was showery, I wished to return if possible to pass the night there, for I began to learn that such huts with a good fire before them made very comfortable quarters in bad weather (p. 247).

Early squatter's journals also reveal the occasional use of Aboriginal shelters, such as this small reference by Edward Henty (Peel 1996, p. 96), at Portland in November 1835, who wrote how he 'stopped the night in some Native Huts'. In colonial south-eastern Australia, Aboriginal people were frequently employed by squatters and miners and others to supply them with bark slabs for huts (Cahir 2001, 2013). Diaries by early colonists such as Charles Hall (Halls Gap and Grampians Historical Society 2006, p. 3) in the Grampians-Gariwerd region are a reminder of how dependent many colonists were on Aboriginal skills in the early period of colonisation – not just for their labour or providing the raw materials (bark and timber) for colonial-styled architecture, but for Aboriginal-styled architecture. Hall, like many of his contemporaries, built himself a 'reed mia mia' by the banks of the Mt William Creek in 1840.

A decade later, during the gold rush period, the same admiring voices are still to be heard. Short references in miners' diaries relating the great value of Aboriginal shelters are not uncommon. Henry Gray (c.1850s, n.p.), like many miners, wrote to relatives describing the 'blacks mia mias' in some detail and others such as Police Magistrate, Eveleigh Johns (c.1870s, n.p.) repeatedly refers to 'his mia mi' in his correspondence. Goldfields artist and miner William Strutt (c.1850s, n.p) also noted how extraordinarily useful constructing *mia mias* was to mining parties as they journeyed from one mining area to another. Strutt reflected in his autobiography how 'we erected a mia mia for shelter' while on his journey to Ballarat. This was a theme repeated not just in reminiscences and letters but also in maps produced in the gold mining period. This was evidenced by several references, such as 'women's mia mia' and 'our mia' (relating to non-Aboriginal miners), appearing on a 'Map Depicting the Discovery of Bendigo Goldfields' (Anonymous 1851).

Many researchers including Reynolds (1990), Clarke (2008a) and Cahir (2010, 2014) have explored the extensive transmission of information from Aboriginal people to the colonists and how non-Aboriginal families, travellers, miners, squatters, entrepreneurs and government officials were provided with tutelage in not simply surviving in the bush – but thriving. This guidance in bush living often involved Aboriginal people demonstrating to the colonists how to quickly and cheaply construct a shelter in an environment which colonists often found hazardous. In a very similar vein to how Aboriginal canoes were viewed by early colonists, writers such as Walter Woodbury (1854), who was surveying in the Buninyong district (central Victoria), typically offered descriptions of Aboriginal shelters in a dichotomous manner. These shelters that were so critical to the opening up of the Australian interior were often noted as being very 'primitive' but there was also emphasis on how well suited an Aboriginal shelter was to the climate and the immediate needs of the 'new chum' colonists.

> We have had a tribe of the native Blacks camped near us for the last week so that we have had an excellent opportunity of seeing how they live … they construct what they call miamias, consisting of two forked sticks placed in the ground with one stick running across the top of them, they then rest large pieces of bark or branches of trees on these which gives them a shelter from the wind (Woodbury 1854, n.p.).

Alfred Selwyn (1859), a Victorian Government geologist, like many of his contemporaries also used the term *mia mia* with a great deal of familiarity, but unlike Woodbury was very appreciative of Aboriginal bush craft and had learnt how to use it: 'Mr. Daintree and I have had many a hard day's work in penetrating it [the Dandenong Ranges 35 km east of Melbourne], whilst at night like blackfellows, we built a mia mia, and rolled ourselves in our opossum rugs' (p. 4). Cahir (2012) has chronicled how there were many miners who concurred with Selwyn and Daintree about the suitability of constructing *mia mias* for short-term mining camps and that they drew on Aboriginal Biocultural Knowledge to achieve this. By way of example the Reverend Arthur Polehampton (1862), a visitor to the Victorian goldfields, explained the economic and functional rationale of emulating Aboriginal shelters at the Ballarat diggings in central Victoria:

> We had determined to remain in Ballarat for a few days as we could not afford speculating in deep sinking. We could not afford to remain longer looking for employment in a place where all necessities were so terribly dear. We lived in a sort of hut built of branches and bark, not unlike the mia mia of the blacks. The weather being warm and dry it was quite a sufficient shelter (p. 226).

German miner and artist, Eugene von Guérard was, like many of the 300 000 miners who flocked to Ballarat from around the world, trying his luck there in the 1850s. His 'Journal of an Australian Gold Digger' is somewhat typical of many educated miners' writings on his experiences as a 'new chum' to colonial Australia, and in particular, the diggings of Victoria. Von Guérard journeyed in January 1853 from Geelong to the Eureka diggings at Ballarat, along the way noting and painting a picture of temporary Wathawurrung shelters. Though von Guérard (11 January 1853) had been in the Antipodes less than a fortnight, he was conversant with the Aboriginal name for temporary camp shelters, writing in his journal near the village of Batesford that he passed:

> three or four mia –mias, the abode of some eight or ten aborigines. In front of each burned a little fire, and some spears lay at hand. The mia-mias are made of the branches of trees in the form of half an open umbrella of large dimensions. Some were covered with the skins of animals (Von Guérard in Rich 1990, p. 27).

A further journal entry dated 24 September 1853 reveals that his memory of the Wathawurrung's 'mia-mias' had not grown dim as he described how a great number of miners at Golden Point (Ballarat) had appropriated Aboriginal modes of shelter, similar to the ones he had seen earlier that year: 'Besides a vast number of tents, many diggers are contenting themselves with a kind of mia-mia, simply made of green branches, to sleep under at night.' Miners such as JG Smith (2002, p. 23), hard up on their luck at the Ballarat district diggings in the early 1850s, with only a few shillings left to their names, opted to 'live in a Mia Mia for several weeks'. Others, such as travel writer William Howitt (1855), more explicitly noted that the gold miners had mimicked Aboriginal methods of building shelters. At Spring Creek in central Victoria he wrote that there were 'huts of mingled boughs and sheets of bark; and here and there simple mimies, in imitation of the mimi of the natives, that is, just a few boughs leaned against a pole, supported by a couple of forked sticks, and a quantity of gum-tree leaves for a bed' (Howitt 1855, p. 139).

Destruction of shelters

Many scholars including Critchett (1990) and Clark (2003, p. 222) have highlighted that colonists often 'wished to destroy the Aboriginal pattern of land use' and that the destruction of capital sites also extended to what colonial observers described as villages. Gerritsen's (2000) study of the traditional settlement pattern in south-west Victoria similarly has revealed that capital sites, such as a village of large Aboriginal huts, were deliberately burnt by colonists in a bid to spatially dislocate Aboriginal people from districts which the squatters coveted. George Robinson (Clark 2003) wrote with some despair about how colonists regularly destroyed Aboriginal people's capital sites across Victoria: 'it frequently happens that runs change hands and often their best and kindest friends are succeeded by those quite the reverse, an instance lately occurred where the natives had their huts knocked down, their fires kicked out, and they had to leave their camping ground' (p. 213).

Decades later, after the bloody frontier decades of physical violence had largely subsided, officials writing reports (Victorian Parliament 1859) and giving evidence to the Victoria Royal Commission into the condition of Aborigines (1877) noted that Aboriginal communities, who were by now coerced or forced onto Government reserves, stoically maintained a great predilection for their traditional styled dwellings.

There are about a dozen good huts, some of which are clean and neat, with even some attempts at ornament. There are also four or five bark huts, which some of the older people prefer … Some of them prefer bark huts to the slab houses, and accordingly there are yet three of such huts, which are tolerably roomy and well closed in. The manager is endeavouring to get rid of such structures altogether, but finds it difficult at once to overcome the prejudices of the old people (Questions 1555–59).

Some government officials were acutely aware of how the environmental ravages of colonisation during the preceding 30 years had impacted the viability of Aboriginal people being able to sustain their traditional ways of building shelter. Charles Gray (Board for the Protection of the Aborigines in the Colony of Victoria 1872), at Nareeb-Nareeb (Western Victoria), in May 1872, reported: 'If your Board has not already provided at Purnam several small huts for the use of the Aborigines, I would recommend its doing so, as the blacks complain of the distance they have now to carry materials for making mia-mias, great part of the trees having been long since stript' (p. 14). Other government officials and writers acknowledged, too, that the imposed European-styled architecture was a poor cousin to the traditional shelters – and that severe illnesses and deaths were a consequence of forcing Aboriginal people to give up their traditional ways of constructing shelter. Woiwod (2012) has chronicled how European-styled housing arrangements at Aboriginal reserves such as Coranderrk (north-east of Melbourne) led to the untimely deaths of some Aboriginal people and that visitors to Coranderrk such as Rev. Robert Hamilton had claimed 'that the huts they occupy now are not as good as their previous thick-set branches of mia-mia that did not expose them so much to draughts [as the prefabricated huts provided by the Government]' (p. 54).

Discussion

The historical record indicates that in south-eastern Australia, the types of shelter used by Aboriginal people was extensive and of great interest to British explorers and squatters – and in particular gold miners. This chapter has presented an indication of the observations made by colonists, illustrating their fascination with Aboriginal shelters. Appeal was held by examples of complex shelters in aspects that were considered novel, such as their design, the rapidity with which they were constructed, the materials they were constructed of and mortuary ceremonies. It has been shown that in the colonial period Aboriginal shelters were seen as distinctive, and commonly purposeful evidenced by the pervasive borrowing of Aboriginal words for shelter into Australian English and also the widespread use of Aboriginal labour to build shelter and the adoption of Aboriginal architectural style and materials.

Aboriginal shelter construction practices and demonstration of their Biocultural Knowledge in this domain, although undoubtedly modified, continued to feature strongly in interactions that occurred during the colonisation of the area by Europeans. To the newcomers, the presence of large villages in some regions and the semi-permanent or permanent appearance of their cold-weather shelters, was a source for both wonder and concern. Reports of Aboriginal villages or clusters of huts and in particular of substantial huts were commonly seen by squatters as evidence of a recognisable sedentary-styled occupation of land by Aboriginal people – land which the colonists wished to usurp. While the colonists generally regarded Aboriginal people from a Eurocentric perspective to be on a lower scale of civilisation, there was often an acknowledgement that Aboriginal people were masters in bushcraft. It was frequently remarked how their skills and knowledge in selecting materials and building temporary shelters with extraordinary speed was unsurpassed and the colonists came to appreciate the skills displayed by Aboriginal people. In these ways, shelter and the Biocultural Knowledge of shelter held by Aboriginal people played an influential role in the development of relationships on the colonial frontier.

Chapter 10
Clothing

Fred Cahir

Introduction

Anthropological and ethnographical studies of the use of clothing among Aboriginal people of south-eastern Australia indicate that garments for thermal protection were often not worn in the warmer months of the year. The increased use of Aboriginal clothing in the colder months of the year point to it being a behavioural adaptation to exposure to cold (Gilligan 2008). Colonial writers often remarked on Aboriginal people's nudity or being scantily clad, which was really a Eurocentric social construct rather than an actual state of nakedness as Aboriginal people were known to make significant behavioural adaptations to cold weather in order to keep warm, including feathers, ochres and animal fat (Stephens 2014; Massola 1971; Smyth 1878). Dawson (1881) described some of the methods that Aboriginal people used to protect themselves from the cold that did not require 'clothing':

> The aborigines are very fond of anointing their bodies and their hair with the fat of animals, and toasting themselves before the fire till their skin absorbs it. In order to protect their bodies from the cold, they mix red clay with the oily fat of emus, — which is considered the best, — or with that of water fowls, opossums, grubs, or toasted eel skins, and rub themselves all over with the mixture. Owing to this custom very little clothing is necessary. During all seasons of the year both sexes walk about very scantily clothed. In warm weather the men wear no covering during the day time except a short apron (p. 8).

Animal skin and fur clothing

Historians and anthropologists have remarked that the principal form of clothing worn in south-eastern Australia were capes or cloaks (Clarke 2003b; Howitt 1904; Stokes 1986; Smyth 1878; Massola 1971). The journals of colonists confirm that they were worn by both sexes (Angas 1847b; Beveridge 1889). The garments made from the skin or fur of animals worn by Aboriginal people were principally manufactured from marsupial skins, mainly kangaroo or wallaby hides, or, especially in colder areas, from several possum (*Phalanger*) skins sewn together. The following observation about the distribution of possum skin rugs was made by one colonist: 'On the east coast the opossum rug comes into use in Latitude 26°, in the central country in about latitude 28°, and from that prevails throughout to the south coast' (Curr 1883, p. 93). Later studies confirm this (Stokes 1986; Kamminga 1982). There are also many accounts of what was commonly called a fringe, girdle or apron being worn. This was described by William Stanbridge (1861), a colonist writing principally about central Victoria, as 'suspended strings, twisted from bark, which is always worn by females until maturity' (p. 289). In reference to the High Plains of Victoria and New South Wales Massola (1962) wrote:

> The only clothing worn was a fringe of narrow strips of hide, suspended back and front from a belt around the waist. This belt was made of a string of twisted possum fur, twelve to fifteen feet [3.6 to 4.5 m] long. To put it on they fastened one end to a tree, and holding the other end to their waist would turn around and around until completely wound. Both sexes wore this fringe. During cold weather both sexes also wore a possum fur cloak, or a mat of kangaroo skins. When not used as a garment this latter would be used by the men as a carry-all (p. 321).

The terms 'cloak' and 'rug' appear to be interchangeable terms for the same object and a similar situation seems to exist with the terms 'fringe', 'girdle' and 'belt' (Wright 1979; Stokes 1986). In south-eastern Australia, the records describing how skin rugs and cloaks were worn are rich in ethnographic detail. Among the Aboriginal people of the Murray, Murrumbidgee and Darling areas it was stated that the possum skin rug was the only covering of both men and women, and 'in both it is worn in exactly the same manner - that is somewhat after the fashion of the Roman Toga, across the shoulder, with one arm free' (Beveridge 1889, pp. 28–9). A similar observation was recorded in western Victoria: 'The original clothing, both of men and women, seemed to be two mats, made of skins joined together, the one hanging before, the other behind' (Bride 1898, p. 278). In contrast, the Aboriginal men and women of the Lower Murray in South Australia were reported as wearing cloaks in different ways:

> The cloak is worn with the fur side outwards, and is thrown over the back and left shoulder, and pinned on in front with a little wooden peg; the open pa[r]t is opposite the right side, so as to leave the right arm and shoulder quite unconfined, in the male; the female throws it over the back and left shoulder, and brings it round under the right arm-pit, and when tied in front by a string passing round the cloak and the back, a pouch is formed behind, in which the child is always carried. In either if the skin be a handsome one, the dress is very pretty and becoming (Eyre 1845, vol. 2, p. 210).

One observer in central Victoria also reported that: 'Opossum skins, sewed together with kangaroo or emu sinews, form the rugs with which both sexes are usually clothed' and likewise noted the different mode of wearing their rugs. He observed that the men 'wear the rug secured over the shoulder by a peg, which leaves the right arm at liberty. Women wear it as a cloak, the string of the net or basket worn at the back keeps it in its place, and there can be formed when required a hood for carrying an infant' (Stanbridge 1861, p. 287). Aboriginal people across south-eastern Australia were observed wearing cloaks with the fur next to the skin (Curr 1886–87, p. 48), *and* with the fur to the outside (Beveridge 1889; Eyre 1845). It was explained that:

In wet weather, the rug is invariably worn with the fur to the weather. Worn in this manner, they are almost impervious to rain, whereas when the flesh side is exposed to the wet the cloak becomes saturated and consequently unpleasant in a very short space of time (Beveridge 1889, p. 139).

Occasionally a parallel was made by colonists between the dress of Aboriginal people and the Scots. Writing about the clothing of Aboriginal people in western Victoria, Dawson (1881) drew some comparisons between the Scottish sporran and kilt and the coverings worn by Aboriginal people.

In warm weather the men wear no covering during the day time except a short apron, not unlike the sporran of the Scotch Highlanders, formed of strips of opossum skins with the fur on, hanging from a skin belt in two bunches, one in front and the other behind. In winter they add a large kangaroo skin, fur side inwards, which hangs over the shoulders and down the back like a mantle or short cloak. This skin is fastened round the neck by the hind legs, and is fixed with a pin made of the small bone of the hind leg of a kangaroo, ground to a fine point. Sometimes a small rug made of a dozen skins of the opossum or young kangaroo is worn in the same way. Women use the opossum rug at all times, by day as a covering for the back and shoulders, and in cold nights as a blanket. When they are obliged to go out of doors in wet weather, a kangaroo skin is substituted for the rug. A girdle or short kilt of the neck feathers of the emu, tied in little bunches to a skin cord, is fastened round the loins (p. 8).

James Dredge (1845a) also noted the prevalence of the wearing of possum skin cloaks among Aboriginal people (writing of the Daungwurrung people in the Goulburn district) and also drew similar comparisons between that of the possum skin cloak and the Scot's 'vestment'.

The dress of the men consists of a large cloak formed of the skins of the opossum and kangaroo, neatly sewn together with the sinews of the animals; this is fastened together so as to slip over the head upon the shoulders, whence it hangs loosely to the knees or the middle of the legs. This single article serves as their dress suit by day, and in it they wrap themselves, for bed and bedding, beside their fire at night. There seems to be some analogy between the form and uses of this rude but not uncomfortable vestment, and those of the Arab and the Highlander (p. 6).

There are also a large number of colonists' accounts testifying that possum fur was an extremely versatile source of clothing material – not just for cloaks and rugs. Some of the uses included the fur being spun into cord or yarn (Curr 1886–87; Smyth 1878) and the possum fur cord being used to make headbands (Smyth 1878) and necklaces (Curr 1886–87). Beveridge (1889) wrote that the men 'wear a belt round the loins under the cloak, whilst the women wear a band round the same portion of the person, said band having a thick fringe all round it of about a foot in depth. The fringe is made of innumerable strips of opossum or wallaby skin' and added that: 'Both sexes wear armlets made of opossum skin on the upper portion of both arms, and a netted band of around an inch and a half [3–4 cm] wide around the brow' (p. 29). Anthropologists (Mathews 1905; Howitt 1904) and writers such as ALP Cameron (1885), who wrote principally about the Barababaraba in New South Wales, noted how in initiation ceremonies a covering of possum skin was worn and that: 'Each youth is invested with a belt made from the twisted fur of opossums, and a fringe made from strips of skin of the same animal hangs down in front' (p. 358). William Haygarth (1861), a colonist who lived for eight years predominantly in the Monaro district, also wrote of the variety of clothing uses the skin and furs of marsupials were put to:

> The skin and sinews [of the kangaroo] are articles of great importance to the native, with the one he provides himself covering, and the other serves the purposes of thread in sewing the different skins together, the needle being made of a piece of sharpened bone, by which holes are bored and the sinews are afterwards introduced and drawn tight although common English needles can be procured in most parts of the settled districts, the natives still adhere to the old method … Their only dress consists of rugs made either from the opossum or kangaroo skin, a small bandage round the head, and a quantity of string made from opossum hair twisted, which is wound around the neck in a great number of folds. The women wear a belt of emu feathers to hide the person, and the men a wallaby skin cut into a number of narrow slips for the same purpose (p. 97).

Unfortunately the possum species from which the skins were obtained were rarely identified (Stokes 1986). Possum skins, particularly those of the Common Brushtail (*Trichosurus vulpecula*) were likely to have been the species used to make rugs and cloaks in south-east Australia (Wright 1979, p. 54). The Mountain Brushtail or Short-eared Possum (*Trichosurus caninus*) were almost certainly being taken for pelts, and, to a much lesser extent, the Greater Glider (*Petauroides volans*), and *Petaurus australis* or *P. norfolcensis* (Kamminga 1982; Stokes 1986). The dominance of Brushtail Possum skins is substantiated by Stokes' (1986) research, noting that:

> with a few exceptions all rug-making accounts refer to the use of the skins of the 'opossum' or the 'common possum', which is known to be the Common Brushtail, although the skin of the Common Ring tail (*Pseudocheirus peregrinus*) is sometimes mentioned as being used (p. 52).

Dawson (1881) is one of the few writers to consider the value of various marsupial skins:

> The ring-tailed opossums were more plentiful than the common kind but the skins were less esteemed. Rugs were also made of the skins of the wallaby and of the brush kangaroo, which are likewise inferior to the common opossum (p. 9).

Manufacturing possum skin garments

Colonists from across south-eastern Australia have also left us written records of their observations of Aboriginal people preparing possum skin rugs for use as both cloaks and doubling as blankets at night (Krefft 1862–65). As recorded by many of his contemporaries,

John Elder, a Scottish colonist writing in 1850 reported how he too had frequently noticed gum trees that showed signs of Aboriginal garment industry as he travelled through the bush near Geelong:

> On many of these can be seen oblong pieces of the bark cut out about 2 feet [60 cm] in length by 18 inches [45 cm] wide, these having been cut by the natives to dress their opossum skins upon by nailing them down on the pieces of bark by a number of small wooden pegs until dried and kept to the original size (Elder and Taylor 2013, p. 54).

Other writers proffered that the manufacturing process was a simple one.

> The Kurnai in their primitive state usually went about without any covering. But they made what are now called 'opossum rugs'. These were made of the dried pelts of the opossum sewed together with sinew. They did not dress the skins but merely tied them, and to make them more pliable cut markings on the skin side by means of mussel shells (nanduwung). These marking[s] are called waribruk, and each man had his own (Howitt 1904, pp. 741–42).

A select number of colonists such as William Buckley, a convict escapee who lived with the Wadawurrung in the Geelong region for over 30 years (1802–1835), had been tutored by Wadawurrung in the making of possum skin garments. Buckley made some direct comparisons between the English and the Aboriginal methods used to prepare skins.

> They taught me to skin the kangaroo and opossums with muscle-shells, in the same way sheep are dressed with the knife; to stretch and dry them in the sun; to prepare the sinews for sewing them together for rugs; and to trim them with pieces of flint … their only covering being skin rugs, sown together with sinew using as needles fine bones of the kangaroo. These rugs serve them also to lay upon (Morgan 1852, p. 40).

Others such as Beveridge (1889) observed an incremental approach to the making of a possum skin cloak which may suggest that, in the areas (Murray, Darling and Murrumbidgee) where the possum skins were more difficult to acquire, the making of possum skin cloaks was somewhat more *ad hoc* than in the southern regions where possum skins were known to be easier to procure.

> When a native has a rug to make he does not wait until he has acquired a sufficient number of skins to complete it, for as soon as he has two or three skins he sews them together and wears them mantilla fashion across his shoulders going on day by day adding thereto as he procures the skins, but wearing it all the time, until it becomes a finished cloak (Beveridge 1889, p. 122).

Early colonial records contain many references to squatters and their workers being struck with how finely made the Aboriginal possum skin cloaks and rugs were. Often these accounts do not describe the tools and apparatus required to manufacture these garments.

Fortunately some writers were privy to the intricacies of Aboriginal manufacturing processes and have provided some evidence about preparing a possum fur garment. Isaac Batey (1916) who had grown up in Melbourne in the early 1840s wrote a detailed first-hand description and was at pains to emphasise the exacting and demanding nature of the task. Batey was an appreciative and long-term observer of Aboriginal culture:

> In preparing the skins … Blackfellows used a square of bark taken from a gumtree, at least I saw a native use it. It was a smooth even-surfaced sheet. As a beginning he had stretched the pelt much the same as we do, the only difference was that … little waste material. He had a bag of short pegs, all of uniform length, beautifully round and bevelled to a point the same as lead pencils. Beginning at the bottom corner w[h]ere a wooden nail was already driven, grasping the hide with forefinger and thumb he drew it very slightly beyond the first inserted spike and drove in another peg. He kept on repeating, I did not see him finish his slow task - but it is distinctly remembered that the pegs touched each other throughout. From what is minded of the sheets of bark the pelt had been drawn to an exact square and when removed would show perforations such as we see in postage stamps. On next visiting him the bark with pegged skin was seen placed just near enough to the fire to dry gradually. The Aboriginals are very fastidious in their choice of skins for they only used those taken from the largest and best furred opossums (Batey 1916, n.p).

Similar reports were recorded in central Victoria.

> One woman presented me with an opossum skin. Pegged out on sheet of bark with the fur inwards …, for which purpose they carry about with them between 2 and 300 small pegs. One possum skin that I counted and which had been so prepared had 200 holes, hence requiring 200 pegs. The skin is cut square and then pegged out. It measured 14 inches by 21 inches [35 x 53 cm]. These skins are then sewed together and made into rugs (Robinson May 1839 in Clark 2014b, p. 34).

In different regions there were different traditions and practices associated with the manufacture of possum skin garments. One colonist writing principally about north-east Victoria wrote that it was men's work to hunt, prepare the skins and manufacture the garments.

> When the nights began to get cold in autumn, and after the first fall of rain made the forests remote from the river - in which opossums abounded - accessible, it was usual for the men of the Bangerang to set about making new opossum rugs; their old ones, in due time, being given up to the women, children, or aged persons. Whilst engaged in this business, the men hunted opossums during the day, skinned them on their return to camp, and, after feasting on their flesh, pegged out the skins, each on a small sheet of bark,

which were then placed in front of the fire, so as to dry gradually. This done, the skins were scored with a mussel-shell in various ornamental patterns, and were then fit for use. When enough had been collected, they were sewn together, a sharp bone used as an awl to pierce them with; the sinews of the animal itself, or of a kangaroo, serving very well for thread (Curr 1883, p. 131).

Regional differences in the sewing materials and in the methods to adorn themselves using possum skins can also be discerned in the historical literature. For instance, Taplin (1874) in South Australia and Haygarth in the Monaro district note that the thread used to sew the skins together was made not from sinew but was in fact plant based – and that a fastener made from bone was used.

Those of the party that wore cloaks made of the skins of the opossum, about forty of which, stitched together with a strong thread made from the 'stringy bark' tree, formed a kind of wrapper reaching nearly to the ground, and fastened at the neck by a crooked piece of bone (Haygarth 1861, p. 105).

In the main kangaroo or possum bones were fashioned into sewing needles but the spines from the Echidna (*Tachyglossidae*) were also used (Massola 1971; Smyth 1878). It was observed in northern Victoria that the sewing tools were used with great expertise:

a bodkin made of bone was the instrument by which the punctures to receive the sinews are made, and the quickness and dexterity displayed in the manipulation of these rude instruments and appliances is truly astonishing (Beveridge 1889, p. 139).

Smyth (1878), while not indicating which region his information was sourced from, noted the sewing technique and the type of thread utilised in the making of a garment from possum skins.

In making an opossum rug some skill and knowledge are employed. In the first place, it is necessary to select good, sound, well-clothed skins. These, as they are obtained, are stretched on a piece of bark, and fastened down by wooden or bone pegs, and kept there until they are dry. They are then well scraped with a mussel shell or a chip of basalt, dressed into proper shape, and sewn together. In sewing them the natives worked from the left to the right - not as Europeans do - and the holes were made with the bone awl or needs, and instead of thread they used the sinews of some animal - most often the sinews of the tail of the kangaroo (p. 273).

Possum rug decoration

The artistic designs which Aboriginal people commonly used to decorate the flesh side of the pelts and completed rugs were frequently commented upon appreciatively by the colonists. One writer not only extolled the accuracy and the dexterity with which the garments were made – but also waxed lyrical about the artistic and utilitarian skills of the garment maker as displaying more than a 'modicum of taste approaching the artistic' (Beveridge 1889, p. 138). Others such as Batey (n.d) observed that the designs were often coloured with ochre or some other substance and Smyth (1878) added: 'They inscribed the lines on the skins and darkened them with powdered charcoal and fat, or with other colours' (vol. 1, p. 288). Blacklock (2002) and Stokes (1986) concluded that judging from some of the accounts (Howitt 1904; Fraser 1892), it would seem that at least some of the decorations were the personal marks of the owners or makers i.e. their 'signatures' or 'trademarks'. It was observed that the scraping of the pelts was done by shell or stone and in the post-colonisation era by a sharpened spoon to achieve pliancy – often in a manner resulting in patterns and designs. Beveridge (1889) remarked that this was so well done that when the separate pelts were joined together patterns matched accurately:

> When an opossum skin has been thoroughly dried and all the fat removed it requires to be scraped to ensure pliancy; this part of the operation is performed by means of sharp-edged mussel-shells, and scraping so performed is generally done so as to represent a pattern of some kind. They succeed in doing this so well that the various portions scraped upon the separate skins join together most accurately when sewn into the rug. Should a rugmaker be entirely lacking in artistic taste he merely scrapes the skin diagonally. When the whole skin has been thus scraped he turns the skin round and again scrapes diagonally, this time, of course, the scraping crosses the lines first made, thus forming a lozenge-shaped design. This latter method, as regards utility, is unequally by any other, for thus treated, if the pattern be a small one, the skins become pliant and soft as well-prepared doe-skin (p. 138).

Straight lines, herringbone, chevrons and diamonds were the usual designs (Smyth 1878; Massola 1971) in Victoria, while Worsnop (1897) reported that South Australian Aboriginal people decorated their rugs with either a zigzag or diamond pattern. Sometimes animals and men were represented (Smyth 1878). However, it is uncertain whether this is a post-contact invention as both Beveridge (1889) and Dawson (1881) suggest that the use of figures in rug decorating was due to innovative artistic changes that Aboriginal people brought to their garments with the advent of new materials and was not done before colonisation.

> Before the advent of Europeans these cloak patterns usually took a scrolly shape, and striking objects in nature, such as flowers, foliage, or animals, were never copied. Since then, however, we have seen the great glaring design common to cheap druggists very successfully reproduced, even to the colours. These colours are made by mixing pigments of different shades with fish oil, and laying the shades on the respective portions of the designs requiring them, thus produce an exact counterpart of the copy (Beveridge 1889, p. 138).

Beveridge's critique, principally written about the Murray–Darling and Murrumbidgee regions, concerning the appearance of the new designs being a product of colonisation, is mirrored by Dawson (1881) whose informants were from the Western District of Victoria.

Previous to sewing the skins together, diagonal lines, about half-an-inch apart, are scratched across the flesh side of each with sharpened mussel shells. This is done to make them soft and pliable. The only addition to this kind of ornamentation is occasionally the figure of an emu in the centre skin of the rug. It may be stated that, although many of the opossum rugs of the aborigines are now ornamented with a variety of designs, some of which are coloured, nothing but the simple pattern previously described, with the occasional figure of an emu, was used before the arrival of the white man. The figures of human beings, animals, and things, now drawn by the natives, and represented in works on the aborigines of the colony of Victoria as original, were unknown to the tribes treated of, and are considered by them as of recent introduction by Europeans (p. 9).

Garments other than skin, feather or fur

While most colonial commentators spoke only of animal fur and hides, especially the possum and kangaroo, as the principal dress or covering of Aboriginal people in south-eastern Australia, there are solitary accounts of other animal furs being used as well. One of these was Frenchman Charles Brout, who was astounded by 'coats made from platypus (*Ornithorhynchus anatinus*) skins'. There are also single accounts of slippers fashioned from skins (Akerman 2005). There are, however, numerous accounts of garments being manufactured from a range of materials other than animal furs including seaweed, reeds, tussocks of woven grass and bark. In the first 'Annual report of the Aborigines Department for the year ending 30th September 1843' by the protector of Aborigines in South Australia, there is reference to Aboriginal clothing from several different sources.

> Clothing. - The clothing in their natural state is exceedingly simple. In those districts where the gum and peppermint trees are to be found, the skins of the opossums are made into coverings; in other districts - as on the sea-coast, borders of salt water rivers and lakes, where fish forms their principal food - seaweed and rushes are manufactured into coverings (Foster 1990, p. 39).

EJ Eyre (1845, vol. 2), writing principally about Aboriginal people of the inland Murray River section, wrote more glowingly about the 'ingenuity' of their dress when habituating in areas where animal skins were difficult to procure.

> On the sea coast, where the country is barren, and the skins of animals cannot readily be procured, sea-weed or rushes are manufactured into garments, with considerable ingenuity. In all cases the garments worn by day constitute the only covering at night, as the luxury of variety in dress is not known to, or appreciated by, the Aborigines (p. 128).

In a similar vein, George Angas (1847b), writing about Aboriginal people on the Lower Murray in South Australia, observed how they 'manufacture round mats of grass or reeds, which they fasten upon their backs, tying them in front, so that they almost resemble the

This illustration depicts an Aboriginal man wearing a cloak made from seagrass (possibly *Ruppia tuberosa*) which was commonly used in coastal regions where animal skins were more difficult to procure. Source: 'Illustrations of seagrass cloaks from the Murray Mouth area', illustrated by George French Angas in South Australia Illustrated (1847) p. 85.

shell of a tortoise. In the loose portion of these circular coverings the mothers carry their children astride round the shoulders' (p. 85). GA Robinson in his extensive travels across the plains of Western Victoria both deplored the paucity of covering afforded to the Aboriginal people (which he attributed to being driven from their homelands by white colonists) and, like Eyre, applauded their ingenuity in clothing manufacture.

> A gentleman, Mr Ryree, told me that to the westward on the numerous plains country of Geelong he said natives naked when the cold and rain was so severe that he could not keep himself warm with thick heavy covering, said the natives some of the time hang tusks of grass in front of their chest when the rain was in front and turned around when behind them (GA Robinson September 1840, in Clark 2014b, p. 209).

Headbands and bonnets

Colonists and ethnographers commented infrequently on the wearing of headbands and bonnets. On the other hand, Robinson, for example, during his lengthy sojourns through south-eastern Australia made much of the few times he saw bonnets being worn in the western district of Victoria. At first sighting, Robinson considered the indigenous bonnet as rudimentary and was disparaging about its utilitarian purpose.

> Saw an old woman with a bonnet or hood made of long tusicks [tussocks] of grass, which she wore on her head. They were tied together at the upper part as a shelter from the rain; primitive and wretched covering (GA Robinson September 1840, in Clark 2014b, p. 299).

Robinson further considered the bonnet and concluded that it was in fact well suited to its purpose. He also reflected that the bonnet in effect symbolised how the usurping of all their lands by the colonists had pauperised the Aboriginal people – and that the Indigenous bonnet was a clever adaptation to their situation.

> They were severely destitute of covering and, living as they do on these bleak plains, where in some parts there is not a tree to shelter them from the cold blast and pitiless storm which sweeps over the plains, their state was deplorable. Their camp evidenced complete destitution. A few branches of banksia was all they had to screen the wind and several pundarerer, bonnet or head covering of rushes, that lay on the banksia, too plainly indicated their state of destitution to which these people are subject. The head covering of rushes was the first I had seen. Indeed, on these extensive plains where there are but a few herbs and short grass thinly scattered some plan for such an invention, was well suited to their condition (Robinson April 1841 in Clark 2014b, p. 393).

Eyre (1845, vol. 2) wrote of similarly made bonnets on the Lower Murray, but for a different purpose, noting that in general no covering is worn upon the head, although in extreme seasons of heat, and 'when they are travelling, they sometimes gather a few green bunches or wet weeds and place upon their heads; but this does not frequently occur' (p. 129). Smyth (1878) was advised that they [the Lower Murray tribes] tied a band round the head 'extending from the occiput over the parietal bones to the place of the frontal suture' and is called 'Mar-rung-nul'. Further discussion about the intricate headbands worn by the people of the Lower Murray was provided:

> This ornament is closely woven, and to the eye resembles a thick coarse cloth, but it is really soft and pleasant to touch. It is made of the fibrous root of the wild clematis (Mo-u-ee). It is exceedingly strong. The length of the band is twelve inches [30 cm], and the breadth one inch and a quarter [2.5 cm]. Dr. Gummow says that these bands are usually made by the women. Wing feathers of the cockatoo are stuck in the band, one on each side of the head. The feathers are called Wyrr-tin-nay. This band is worn by males only … The band of

network Dr. Gummow says is named Moolong-nyeerd. It is worn across the forehead, with the kangaroo teeth as pendants, which, when lashed together, are known as Leangerra. When stretched, as it would be when on the head, the broader part of the network is nearly twelve inches [30 cm] in length and three inches [7.5 cm] in breadth. The open network on each side up to the knot is four inches [10 cm] in length. The material is the fibre of some aquatic plant, twisted and formed into a fine, hard, durable twine. The teeth are fastened neatly with the tail sinews of the kangaroo (Wirr-ran-nee). It would not be easy to find anywhere a more highly-finished piece of work of its kind than this. The wider part is beautifully knitted. This band was worn both by males and females (Smyth 1878, vol. 2, p. 276).

In the Western District of Victoria headbands were similarly observed to be in popular use.

A band of plaited bark surrounds the head, and pointed pins, made of wood or of the small bones of the hind foot of the kangaroo, are stuck upright at each side of the brow, to keep up the hair, which is divided in front and laid over them (Dawson 1881, p. 31).

Colonist use of Indigenous skins and furs, and of Aboriginal garments

During the 19th century Aboriginal possum fur cloaks or rugs were clearly sought after by the colonists who greatly admired the ease with which Aboriginal people procured the possum skins, as well as the aesthetic nobility the possum skin cloaks afforded the wearer. Hermann Koeler (Koeler and Mühlhäusler 2006, p. 70), having recently arrived in Adelaide in 1840, wrote: 'The women have large pelts with which the women cover themselves', while William Thomas writing in January 1839 was more effusive in his admiration: 'The [Aboriginal] woman was dressed comfortably and had a good opossum rug on, close wrapped around her ... Saw one particularly fine young man as I was going to Melbourne ... In his clean [possum skin] blanket down to his legs he had a majestic appearance. His countenance was dignified and frank' (Cannon 1982, p. 726). Similarly, Beveridge (1889, p. 29) wrote: 'On the very old men and the young women it is an exceedingly graceful garment ... infinitely more becoming to them than the conventional garb of civilised life is to those belonging to that higher order of humanity'.

Moreover, many white people rapidly developed a keen appreciation of the usefulness of possum skins. Official reports and personal correspondence describe the colonists using possum skins for a range of different purposes, most of them mimicking the traditional uses (Cahir 2005). Edward Curr, a young squatter at Port Phillip in 1841, wrote of a typical overseer's hut having an 'opossum-rug' spread over the bed (Curr 1883, p. 18). On one occasion a colonist's friends 'buried him, having wrapped him up in an opposum rug' (Cannon 1982, p. 465), while Katherine Kirkland (1845, p. 16), one of the first white women in the Ballarat district, described how she hung her baby at her side in a possum skin basket

Mary Phillips, a Djabwurrung woman from central Victoria stands wearing a possum skin cloak which as William Thomas and others noted was worn 'closely wrapped around her'. In the photograph Mary is wearing the possum skin cloak over Western dress. She holds a basket, boomerang, digging stick and spear, at her feet are another 2 baskets and a boomerang. Source: F Kruger, 'Queen Mary' (Ballarat Historical Society) Ballarat Historical Society, c.1870s.

as she had seen the local Wadawurrung women do. On some occasions the colonists made innovative adaptations of the possum skins traded to them. A number fashioned fur-lined caps and jackets for themselves (Cahir 2005). Andrew Russell, a squatter like Kirkland, saw the advantages of purchasing Aboriginal garments and also noted their wider applications and popularity in colonial society. Indeed Russell (1840) documents other garment sources being used for Aboriginal dress besides the pervasive possum skin.

> The[ir] covering, when used, is a rug or mat, made of kangaroo skins. These are sewed together with the sinews of the same animal or the emu, using a small pointed bone for perforating the skins in sewing. The skins of these animals [kangaroos and possums] are converted into many purposes, such as shoes, rugs, &c. The kangaroo is of so many species, that much taste is often shewn in

sewing these skins together for rugs; with the many different colours and sizes, they form some good designs. I got one such made up, consisting of many sorts, having the opossum and native cat skins as borders. This latter animal is not quite so plentiful as the others, and therefore more prized. It has various shades, but generally of a dark colour with white spots. Those of a superior kind are worn by ladies as tippets, boas, &c. The rugs are much used by travellers in the bush, suiting, from their warmth and softness, the purpose of mattress and cover at the same time, as I have experienced in the chill of night … Those [Aboriginal women] far up in bush spin from the rude distaff a cover from the wool or hair of the oppossum which is worn as a girdle (pp. 119–120).

The demand by pastoralists and their servants for possum skins and especially possum skin rugs was widespread. So popular were the latter that a small number of white entrepreneurs established a lucrative trade in the skins by merchants in Melbourne and the 'settled districts' and through many government officials, such as the Assistant Protector of Aborigines. One such, Thomas, reported that he had been canvassed by the Aboriginal people to find markets for them as they were 'bringing in some thousands of skins for sale' (Cahir 2005). Reports from all the Aboriginal Station Managers across Victoria in the following four decades document how Aboriginal people carried on their lucrative trade unabated.

Possum skin commerce in the gold rush era

Similarly, gold miners readily took up the Aboriginal skills of possum skin rug making across south-eastern Australia. Newspapers often reported how 'diggers were very fond' of hunting possums and 'making beautiful rugs of them, by sewing their skins together' (Cahir 2012). There was also an extensive monetary trade with Aboriginal people of possum skin furs for a range of non-Aboriginal purposes, including instances of non-Aboriginal stockmen who 'had on possum skin caps' and the making of a tobacco pouch from an 'opossum skin' – a tradition that had continued from the pastoral period (Cahir 2005; Woolner 1855).

It is worthy of note that miners and others more commonly accepted that Aboriginal people were more adept at the trade and thus engaged in what miners described as a 'good item of commerce'. Miners commended their manufacturing skills and noted: 'they are very adept in curing skins perfectly' which are 'taken into the townships for sale' (Leatherbee 1984). Some miners were taught by Aboriginal people how to make possum skin rugs. Accounts from colonists described in great detail the differing processes of rug manufacture and they considered the Aboriginal manufacture to be infinitely superior, describing how 'the aboriginals were fastidious in their choice of skins' (Batey no date). Others recognised it was a skilled trade which was extremely useful.

I learnt the art of curing skins from Gardner at least simply drying them after a fashion he had learnt from the blacks. Small wooden pegs are cut and the skin is stretched with them on the back of a tree and left a couple of days in the sun after which they are ready for use most frequently they are sewn together for rugs 50 or 60 making a covering more durable and much warmer than a blanket (Johnson 1855, n.p.).

Many miners were clearly enamoured with possum skin products sold to them by Aboriginal people. Artist and miner on the goldfields of central Victoria in the mid-1850s, George Rowe, wrote that the 'opossum fur is beautifully soft and makes a warm covering to sleep under and is what most diggers have as it is very light a good one costs four pounds' (Rowe 1854, n.p.). Furthermore there is clear evidence that Aboriginal people moved quickly to grasp the economic opportunities presented to them by the miners flooding to the gold diggings. JF Hughes, a Castlemaine (central Victoria) pioneer also in the mid-1850s, recalled that possum skin and kangaroo skin rugs were 'sold to settlers and lucky gold diggers at five pounds a-piece' (Castlemaine Association of Pioneers and Old Residents 1972, p. 224). Likewise miner James Arnot (1852, n.p.) bought a possum rug in Melbourne made of 72 skins sewn together with sinews, also for five pounds.

In the goldfields literature the ubiquity of references to the possum skin rug's superiority to European blankets, and its aesthetic allure, is testimony to its enduring importance. John Erskine's (1852) comment is atypical of miner's opinions about possum skin rugs' utility, noting: 'the frost was severe towards the morning, making a good covering of blankets and opossum rugs necessary' (Erskine 1852, n.p). Others recorded its practicality in treating cases of hypothermia while others were content to sing its praises, such as AA Le Souëf (1890), an experienced bushman in Victoria from 1840 until 1902, who noted: 'I know nothing more delightful than camping out … wrapped in an opossum cloak or blanket, with your feet to the campfire' or JS Prout (1852, n.p.), a writer and artist on the goldfields of Mt Alexander, who pronounced that 'an opossum rug leaves the comfort of a feather bed unwished for'. George Henry Wathen (1855, p. 131), a visitor on the Victorian goldfields, also enthusiastically praised the virtues of possessing a possum rug. Further, he acknowledged, perhaps grudgingly, that people on the goldfields considered the possum skin rug to be undoubtedly the most highly valued intercultural trade item in Victoria: 'I was soon asleep on the ground, by the fire, under an overbowering banksia, wrapped in the warm folds of my opossum rug. For a night bivouac, there is nothing comparable to the opossum-rug; and it is perhaps the only good thing the white man has borrowed from the blacks.' Others such as HW Wheelwright (1979[1861]) confirmed this opinion, noting: 'for of all the coverings in dry cold weather, an opossum-skin rug is the best, as I can well testify' and adding that he recommended: 'If any blacks are handy, it is best to get them to sew the skins, for a black's rug beats any other.'

Discussion

In the past few decades there has been a pronounced return to traditional practices in south-eastern Australia (Jones 2014). The current role of 'possum cloaks' as being seen as iconic south-eastern Aboriginal garments is evidenced by major projects which feature the revitalisation of this Aboriginal industry. The act of mining the museum and historical archives to create contemporary works which have featured in major exhibitions and events such as the Commonwealth Games was revolutionary, and since then 'we have witnessed recurring acts of institutional decolonisation with many artists breathing new life into the past by reclaiming and honouring their ancestors' work' (Jones 2014, p. 38). Contemporary Aboriginal artist, Vicki Couzens, who helped to start the revitalisation in making possum skin cloaks in 1999, noted 'there has been a real renaissance of cloaks being used back in the communities at ceremonies' (Keeler and Couzens 2010, p. 79). This theme is reflected in projects such as the 'Our Possum Skin Cloak' project at the Victorian Aboriginal Child Care Agency (VACCA) where Elders teach the traditional skill of making possum skin cloaks to young people in out-of-home care. Aunty Muriel Bamblett (2014, p. 134) emphasised that the making of possum skin cloaks is 'not simply the action of making the traditional cloaks but it is relational, a journey of people coming together, sharing knowledge, healing and growing in pride of their identity and culture. As well as teaching a skill, traditional stories are told and connection to community made.'

Chapter 11
Wellbeing

Ian D. Clark

Introduction

Indigenous Australians have their own traditions of healing founded upon a non-Western view of the body and its relationship to the spirit. Australian Indigenous healing is founded upon the intersection of both material and spiritual realms. In this ontology, 'the human body is matter which is animated by each individual's spirit – a spirit, which in turn, may be lost, taken or damaged by external spiritual forces' (Suggit 2008, p. 3). The Indigenous body is 'a sacred wholeness whose invasion or penetration is a matter for either deep spiritual concern or celebration. A body is entered for the purposes of harming or healing the human identity (soul), or to enable the transformation of that individual into a more sacred identity (ceremonial initiate or clever man/woman)' (Suggit 2008, p. 4).

Indigenous healers play significant roles in the religious, judicial and therapeutic foundations of Aboriginal community life. They are often described as clever men and women. The word 'clever' resonates with both a 'respect for the healer's extensive therapeutic knowledge and skill, as well as a degree of fear for their presumed mystical, supernatural and spiritual capabilities' (Suggit 2008, p. 3). Diane Bell has argued for a greater recognition of women as healers in Indigenous society and their role in health maintenance. She considers the fetishisation of the 'medicine man' has obscured the interdependence and complementarity between male and female practices in the domain of health (Bell 1982, p. 220 cited in Suggit 2008, pp. 20–1).

Medicine men and women

Aboriginal societies across south-east Australia had their own healers, who were commonly referred to in 19th century literature as 'doctors', 'medicine men', 'sorcerers', and 'sacred men'. The work of Indigenous clever men and women does not easily respond to fundamental Western dichotomies of mind and body, and the integrally related oppositional pair of matter and spirit (Suggit 2008, p. 40). Philip Clarke's (2008b, p. 9) assessment is that the closest equivalent in contemporary Western medicine would be a professional who is both a general practitioner and a psychiatrist. As Clarke explains, the role of the healer was to diagnose problems, advise on remedies, suggest and perform ritualised healing procedures, explore the impact of community social and cultural issues upon the illness, and to reassure their patients that they could be cured. This chapter will consider the 19th century sources across south-east Australia.

Christopher P. Hodgson (1846, p. 204) discussed the role and place of the doctor in Aboriginal society:

> The next officer in importance is the 'Tanjoor', or priest, or doctor, or lawyer, who assists at every ceremony, and is regarded with great reverence by the females, as well as the men, inasmuch as the sanction of their ultimate disposal is vested in their hands. Besides his official capacity, he is otherwise an important functionary. All those who are diseased, unhappy, wanting to marry, with cracked heads or broken sculls (sic), with a full set of teeth, or with the desire of losing one, make him their confidant; and thus he is intimately acquainted with the feelings, wishes, loves and dislikes of the rest, and with the sagacity of a knowing politician, ready to avail himself of his knowledge to his own advancement and interest.

Haydon (1846, pp. 108–9) explained that the 'The doctor of a tribe is generally speaking a personage of much importance, who by a slight knowledge of a few herbs useful in some diseases, and well versed in the natural superstition of savages, is looked up to with awe and respect'.

James Dawson noted that: 'Every tribe has its doctor, in whose skill great confidence is reposed; and not without reason, for he generally prescribes sensible remedies. When these fail, he has recourse to supernatural means and artifices of various kinds' (Dawson 1881, p. 57).

> If diseases will not yield to these ordinary remedies, the doctor invokes the aid of spirits. Visiting his patient in the evening, and finding that the case is beyond the reach of the ordinary remedies, the doctor goes up to the clouds after dark, and brings down the celebrated spirit, 'Wirtin Wirtin Jaawan,' who is said to be the mate of the 'good spirit, pringheeal.' When he is expected to arrive, the women and children are sent away from the camp, and the men sit in a circle of fifty yards [45 m] in diameter, with a banksia tree in the centre. The doctor and spirit alight on the top of the tree, and jump to the ground 'with a thud like a kangaroo.' The spirit gives his name; and, after the doctor has felt all over the body of his patient, they both go up to the clouds again. It is supposed that the patient must get well. Occasionally the doctor brings down with him the spirit of the sick man, in the form of a doll wrapped in an opossum rug. This doll produces a moaning noise. The sick person is placed sitting in the middle of a circle of friends, supported behind by one of them, and the doctor presses the rug containing the doll to the patient's chest for some minutes, and then departs.
>
> If the sick person is a chief or a chief's wife, or of superior rank, and the doctor, on visiting him at sunset, finds it beyond his power to remove the disease in the usual way, he goes up to the clouds after dark, and fetches down ten spirits. These he places at a distance of fifty yards from the sick person. He then has a conversation with his patient, and, after kneading him all over to ascertain the seat of the disease, he informs the spirits, and they tell him what to do. Having received his instructions, he warms his right hand at the fire and rubs it

over the affected spot. The spirits then depart, with a croaking noise 'like the cry of the heron.' The doctor repeats the rubbing for three nights, and then, telling the patient he will soon be well, he departs for his home, with his followers. If, at the first meeting thereafter, his patient is cured, the doctor receives presents of food, rugs, and weapons; but if he dies the doctor gets nothing.

The doctor pretends to cure pains of every description, and makes his patients believe — not unwillingly — that he extracts foreign substances from the body by sucking the sore places. He actually spits out bits of bone, which he had previously concealed in his mouth. He also, by rubbing, apparently makes stones jump out from the affected part (Dawson 1881, pp. 57–8).

Alfred William Howitt explained that he adopted the term 'medicine-men' as a convenient and comprehensive term for those men who are usually spoken of in Australia as 'Blackfellow doctors' – men who in the native tribes profess to have supernatural powers.

The term 'doctor' is not strictly correct, if by it is meant only a person who uses some means of curing disease. The powers which these men claim are not merely those of healing, or causing disease, but also such as may be spoken of as magical practices relating to, or in some manner affecting, the well-being of their friends and enemies. Again, the medicine-man is not always a 'doctor'; he may be a 'rain-maker,' 'seer,' or 'spirit-medium,' or may practise some special form of magic.

I may roughly define 'doctors' as men who profess to extract from the human body foreign substances which, according to aboriginal belief, have been placed in them by the evil magic of other medicine-men, or by supernatural beings, such as Brewin of the Kurnai, or the Ngarrang of the Wurunjerri. Ngarrang is described as being like a man with a big beard and hairy arms and hands, who lived in the large swellings which are to be seen at the butts of some of the gum-trees, such as the Red Gum, which grows on the river flats, in the Wurunjerri country. The Ngarrang came out at night in order to cast things of evil magic into incautious people passing by their haunts. The effect of their magic was to make people lame. As they were invisible to all but the medicine-men, it was to them that people had recourse when they thought that a Ngarrang had caught them. The medicine-man by his art extracted the magic in the form of quartz, bone, wood, or other things. Other medicine-men were bards who devoted their poetic faculties to the purposes of enchantment, such as the Bunjil-venjin of the Kurnai, whose peculiar branch of magic was composing and singing potent love charms (Howitt 1904, p. 404).

Knowledge of some particular clever men and women in colonial south-east Australia has survived. Dawson has memorialised the celebrated Tuurap Warneen, who was both a 'doctor' and 'clan-head' of the Kulurr gundidj clan belonging to Mt Kolor (Mt Rouse), near present-day Penshurst in south-west Victoria:

When the white men came to Victoria, there was one doctor of great celebrity in the Western District, Tuurap Warneen, chief of the Mount Kolor tribe. So celebrated was he for his supernatural powers, and for the cure of diseases, that people of various tribes came from great distances to consult him. He could speak many dialects. At korroborees and great meetings he was distinguished from the common people by having his face painted red, with white streaks under the eyes, and his brow-band adorned with a quill feather of the turkey bustard, or with the crest of a white cockatoo. Tuurap Warneen was unfortunately shot by the manager of a station near Mount Kolor; and his death caused much grief to all the tribes far and near.

On one occasion, when the tribe had a great meeting at a lake called 'Tarrea Yarr' to the north of Mount Kolor, doubts were expressed as to his power to summon spirits, and make them appear at mid-day. To show he could do this, he went up to the clouds and brought down a gnulla gnulla gneear, in form of an old woman, enveloped in an opossum rug, tied round her waist with a rope of rushes. In order to thoroughly frighten the people, he held her tethered with a grass rope like a wild beast, as though to prevent her chasing and hurting them. He did not allow her to go nearer to the wuurns than about fifty yards. After exhibiting her for half an hour, he led her off. Everyone was intensely terrified at the gnulla gnulla gneear, and the doctor found her a profitable invention, as he received numerous presents of weapons, rugs, and food to keep her away. When he was in want of a fresh supply, he could always command it by a threat of another visit from the gnulla gnulla gneear (Dawson 1881, pp. 58–9).

Ethel Shaw, whose father Joseph Shaw managed Coranderrk from 1886 to 1908, recalled a Djadjawurrung doctor known as 'Sergeant Major' or 'Old Major':

Old Major was, or claimed to be, an Aboriginal doctor. He was tall and gaunt, and looked as if he had been in many wars. His face was scarred, with one eye missing, the socket of which he kept always covered with a rather discoloured handkerchief. On visiting a sick man one day, Mrs Shaw found Major busily engaged in 'making 'im better'. Three or four Aborigines were watching the process. The sick man looked rather apprehensively at Mrs Shaw as she came into the room, but she stood quietly in the background watching the 'doctor's' methods. He rubbed, kneaded, and pinched the patient while keeping up a monotonous incantation. Then he sucked the skin of the sick man. After a few moments he stopped, and took from his mouth small pieces of string and a piece of wire, which he claimed to have sucked out of the patient. 'That made 'im sick. He get better now', he said. The onlookers gazed open-eyed at the supposed results of the operation. When Mrs Shaw remonstrated with him, saying he had them in his mouth all the time, he became very angry and stalked out of the room. Some time afterwards Major's wife died, and he took to his bed and said he would not get better; that he was going to die. The doctor

'Old Major and Wife' at Coranderrk Aboriginal station. Source: Samuel Leuba collection in Clark 2016b; also published in Shaw 1949, p. 36.

could not find anything wrong with him. We tried to rally the old man, but it was of no avail; in a few days he was dead (Shaw 1949, p. 28).

William Thomas has also given us a detailed description of an Aboriginal doctor named Yammerboke, aka Malcolm, who belonged to the Mt Macedon clan:

The blacks have various kinds of doctors for eyes, bowels, head, &c., and, like white physicians, are noted in proportion to the remarkable cures said to have been wrought. But the highest pitch of the profession is flying. Among the tribes who have visited the settlement there has been but one, that has come to my knowledge, possessed of this power, whose name is Malcolm, of the Mount Macedon tribe. I have known this man to be sent for 100 miles. The blacks say that he has power to soar above the clouds, and to fly like an eagle; he also can, in some cases, recover the marmbula (kidney fat) when it has been stolen. I have a most singular account of one of his serial journeys, together with the solemnity of the encampment during his two hours' flight, but cannot trace it now. This Malcolm (aboriginal name Myngderrar) is said to have inherited this power from his father, who was famous before him (Thomas in Bride 1898, p. 92).

James Dawson has left a rare account of a female sorceress, known as the 'White Lady', who was the widow of the Mt Kolor clan-head (presumably Tuurap Warneen, discussed above):

The aborigines had among them sorcerers and doctors, whom they believed to possess supernatural powers. In the Kolor tribe there was a sorceress well known in the Western district under the name of White Lady, who was the widow of the chief, and whose supernatural influence was much dreaded by all. As an emblem of her power, she had a long staff resembling a vaulting pole, made of very heavy wood, and painted red. This pole, which she said was given to her by the spirits, was carried before her by a 'strong man' when she visited her friends or attended a meeting. On occasions of ceremony, it was dressed up with feathers of various colours, and surmounted by a bunch of the webs of the wing feathers of the white cockatoo. The pole-bearer, whose name was Weereen Kuuneetch, acted also as her servant. After ushering her to the meeting, he hid the pole at a short distance from the camp, while singing and amusements were going on, as it was too sacred to be exposed to common inspection. At bedtime he brought it into the circle by her direction, and held it upright before the fire, as a signal of retirement for the night. At her death the pole was carried off by the spirits, and no one has seen it since.

In order to support her pretensions to supernatural power, she would, on some moonlight night, leave the camp with an empty bag made of netted bark cord, and return with it full of snakes. These she said were spirits. No one, therefore, dare go near them or look at them. She described one as pure white, another black; the rest were young ones. She emptied the bag near the fire and made them crawl around it, by pointing with a long stick, and speaking to them.

On another occasion, having left the camp for a while on a moonlight night, she pretended, on her return, that she had been to the moon; and, in proof of her visit, produced a tail of a lunar kangaroo — an old fur boa which she had got from the whites. Besides this boa she had a number of charms round her neck, and, in her bag portions of the bones of animals, beads, pieces of crockery, bits of brass and iron, and strangely-shaped stones, each having its particular spell, and capable of producing good or evil, as suited her interests. This clever old witch was very much annoyed when any white person scrutinized and exposed the contents of her bag; but the natives, though the more sensible of them were not sorry to see her powers and mysterious charms ridiculed, were too much afraid of her to smile, or join in any mirth at her expense (Dawson 1881, pp. 55–6).

Practices of the medicine men and women

Assistant Protector William Thomas has recorded an example of a harming practice called Murrina Kooding or 'Strength lost', which involved stealing a lock of hair and/or head ornaments which had the immediate effect of draining the life away of those affected. He also described in detail the ceremony by which the doctors reversed the harmful effects and restored the strength of the victims. This example highlights the important role the doctors had in reversing harming practices caused by others.

In the encampment south of the Yarra, on the evening of [blank] were Goulburn, Mount Macedon, Barrabool, Yarra, and Western Port blacks. The Goulburn lubras, quite naked, stole upon seven young men. No sooner had the women their hands on the heads of the young men than the latter appeared helpless; they cut from each young man a lock of his hair. As soon as the hair was cut the young men fainted; the women took the ornaments from the men's heads and decamped. The young men's friends came about them to comfort them, but life apparently could scarcely be kept in them. Their friends sat with them the whole of the night.

On the following morning, the doctors assembled; a fire was made about a quarter or half a mile [400–800 m] from the encampment, and the seven young men were brought, each borne by two friends bearing pieces of lighted bark in their hands, to the spot; the young men were placed round the large fire at some distance, and before each was the bark brought by the friends. The doctors, mumbling and humming, with a piece of glass bottle commenced scraping off all the hair from the crown of the head to the feet, and then rubbed them from head to feet with werup (red ochre). The young men lay speechless during the whole of the time the ceremony was being performed, and every muscle of their faces seemed to be keenly noticed by the doctors. This ceremony lasted from sunrise to three hours afterwards. I understand that these young men would have died had not this ceremony been performed. Strength left them as the lock fell from their heads. (Is not this some semblance to Samson's case?) (Thomas in Bride 1898, pp. 92–3).

The Chief Protector George Robinson witnessed an Aboriginal doctor treating a sick man in Melbourne:

I witnessed with them a most ludicrous scene by a native doctor who was exorcising the sick man. He was squeezing the sickness from the patient into an old rag. The poor man submitted to the operation and looked wishfully on. A small fire was placed on one side of him &c. and to describe the whole would occupy too much time (Robinson 7 May 1839 in Clark 2014a, p. 31).

James Bonwick describes various methods employed by 'medicine men':

Medicine men. They have an odd fashion of relieving pain by sucking out from the part affected a piece of wood or bone, the cause of the suffering, and which is always exhibited to the patient; the charlatanism of some modern quacks is about as clumsy as this, but often equally efficacious upon believing minds. A description is given of the operation of the native physician upon an ague subject, which is interesting and suggestive. The woman was stripped of her blanket, and carried by two men to and fro over a fire. Then the doctor put a string round her waist, and holding the ends, gently pulled her toward him, repeating a charm. He then gazed earnestly in her face, catching her eye, still uttering a charm, to produce a magnetic influence. Going up to her, he began to employ friction on her body. Others came, and joined in the universal shampooing. Then catching her up, they rolled her in rugs, laid her before the fire, and left her to an undisturbed repose (Bonwick 1856, p. 83).

Haydon has recorded the circumstances of the death of a Woiwurrung warrior. Arriving at the Aboriginal camp, he learned that one of the Aboriginal men was close to death. Haydon went to the man:

> The sick man was stretched out on his back, and appeared gasping for breath the hand of death pressed heavily upon him. I was concerned to find in him an old friend who had often accompanied me on shooting excursions. He looked up at me, on one of his wives saying something to him, and held out his hand, I took it in my own, which he grasped with all that peculiar energy consequent on being in great pain he could scarcely articulate except in broken sentences. I asked him where was the seat of his sufferings? He told me in his loins, and said 'that some wild black fellow had put a hot clay brick inside him, which was burning him and melting his fat.' Had this not been a most solemn occasion, I could not have avoided laughing in his face, but I endeavoured to comfort him by telling him he might yet recover, although I had little hope, for from his description I imagined he was suffering from inflammation of the kidneys. It was a warm evening, and he was perfectly naked, presenting a most humiliating sight; every muscle appeared stretched to its greatest tension, his eyes presently became fixed, and after a severe struggle and a few deep sighs, the soul of one of the greatest warriors in the Woeworong tribe departed into the world of spirits (Haydon 1846, pp. 109–10).

Europeans and their treatment by Aboriginal doctors

Several accounts have survived of Europeans undergoing treatments by Aboriginal doctors. Sometimes the Europeans were feigning illness so that they could experience the healing procedures. Assistant Protector of Aborigines responsible for the Goulburn River district in Port Phillip, James Dredge, allowed an Aboriginal doctor named Moonim Moonim, to treat his tic douloureux (trigeminal neuralgia or inflammation of the fifth cranial nerve):

> I sat in a chair, outside the hut, and he commenced operations by feeling my face, where I described the seat of pain to be. His object appeared to me to be to ascertain if the parts were swollen. He then rubbed his right hand briskly under his left arm and immediately applied it to my face, muttering something at the same time. He next applied his mouth to the parts and continued sucking and blowing alternately for several seconds making a whizzing noise with his mouth and continuing the muttering still. When he withdrew his mouth, he spat on the ground as if to empty it of the baneful contents which he had extricated from the affected parts. This operation was performed twice, first on the upper part of my neck, and next on my temple, when the ceremony ceased

for this time, the operator affirming that 'moloko all gone very bad'. Wur-rum also expressed his sorrow at my indisposition, and said he would perform for me then 'all gone bad'. War-ra-wulk too expressed the same sentiments (Dredge 1845b, letter of 9 October 1839).

Christopher Hodgson recounted a time he shammed illness so that he could experience the treatment from an Aboriginal doctor:

> Out of curiosity, I one day pretended to be unwell, and shammed so exquisitely, that the venerable doctor thought I was ill. I waited patiently for the result: and as I imagined there was but one, universal cure, and that not an agreeable one I was all anxiety. The part infected was my ear, no pulse felt no tongue put out such forms were despised; but, one hand seizing my chin, the other was applied with great vigour to the part affected spittle was the ointment; hand was the means of applying it; and if ever a man wanted a stiff neck, I cannot recommend him a surer recipe; for I certainly thought that there was some truth in the homeopathic system, and that he would only end his operation by inflicting upon my poor ear the very disease I pretended to have; but fortunately had not. Though a physician for others, he was unable to cure himself; for, having one day left him my horse to take care of, while I was otherwise engaged in the camp, I found he had ventured on his back; and the animal not being accustomed to such a freight, had thrown him off and caused a compound fracture of his arm; he was useless, and the only answer I obtained from him was cries and groans. A few pieces of bark were hastily stript, and bandaged firmly round with my neck handkerchief; the bone replaced as well as my knowledge of the art of surgery allowed, and within three weeks he was able to use it as well as before. Many a time afterwards he wished to repeat his cure upon me, but it would not do. On being asked if my ear was well, I always answered in the affirmative; and he seemed mightily pleased at the thought of having alleviated the pain and sickness of a White Man (Hodgson 1846, p. 204).

Anonymous (1865b, p. 3), was another who feigned illness to test the skill of a native doctor:

> … some years ago [I] pretended to be ailing, on purpose to test the skill of a famous doctor on the Lower Murray, and two large pieces of stone were pretended to be sucked out of his chest. After which Doctor Peter said he would be sure to feel better, and have a good night's rest (Anonymous 1865b, p. 3).

Leckie recounted a story involving William Barak in which a white official challenged the belief that it was possible to harm a person using a lock of their hair:

> Before he reformed Barak, last chief of the Yarra Yarra tribe, was given to over indulgence whenever he could obtain intoxicating liquor. After one bout he became ill, and he declared the cause to be sorcery. He believed that the blacks

of Gippsland were using his spears to do him evil and he ordered Punty a Gippsland black, who was in the district to return to Gippsland. Punty refused. Barak sneaked behind Punty, cut off a lock of his hair, and threatened to burn it if he did not do as he was told. A fight ensued which necessitated official interference. A white official ordered the hair to be given to him, and, in an effort to enlighten the aborigines, he offered Barak a lock of his own hair. He defied Barak to do him harm. Barak would not accept the challenge on the ground that he had no power over white men but when shortly after, Punty died nothing would convince Barak that it was from natural causes (Leckie 1932, p. 7).

In 1863, the Presbyterian missionary Rev. John G Paton, undertook numerous field visits in Victoria hoping to meet with Aboriginal doctors or sacred men or priests. In February 1863 at Wonwondah Station west of Horsham he met two doctors. After some negotiation Paton was permitted to see the contents of a doctor's bag, especially the objects that they said made people sick and well. Paton negotiated the purchase of some of the contents:

> After much talk among themselves, he took the money; and in our presence Mr. Paton selected a stone idol, a piece of painted wood of conical shape, a piece of bone of human leg with seven rings carved round it, which they said had the power of restoring sick people to health, and another piece of painted wood which made people sick; but they made him solemnly promise that he would tell no other black fellows where he got them (Paton 1891, p. 269).

Several days later Paton was visiting Hexham Station, where he had the opportunity to examine another doctor's bags, and was permitted to purchase three items:

> … in the bag 'were kangaroo tusks or bears' tusks, pieces of human bone, stones, charred wood, etc, etc. She described the virtues attributed to the different articles. If any evil was wanted to befall one of another tribe, the 'doctor', after muttering, threw such a stone in the direction he was supposed to be, wishing he might fall sick, or might die, etc. The spirit from the idol entered his body, and he was sure to fall sick or die. Another piece of charred wood, that the 'doctor' rubbed on the diseased part of any sick person, made the pain come out to the spirit in the wood, and 'doctor' carried it away. All this time the other blacks were in evident dread of the things being seen and handled, repeating, 'No white man ever see these before'. Mr. Paton got three specimens from them, viz., an evil and a good spirit, and a piece of carved bone' (Paton 1891, pp. 271–2).

When Robert Hood, the local station holder, asked the Aboriginal people how he had never heard of or seen these things before, despite living among them for several decades, he was told: 'Long ago white men laughed at black fellows, praying to their idols. Black fellows said, white man never see them again! Suppose this white man not know all about them, he would not now see them. No white men live now have seen what you have seen' (Paton 1891, p. 272).

Causes of death and ill-health

In Aboriginal societies across Australia illness and death was often perceived to have a supernatural agency, and was blamed on sorcery, breaches of religious sanctions and social rules of behaviour, intrusions of spirits and disease-objects, or loss of soul. The writings of the Chief Protector, George Robinson, are replete with references to illness and death. For example, on 4 November 1839, Robinson was told that an old Aboriginal man named 'George', had died in Melbourne because four Barrabal blacks (Wathawurrung), had seized and bound him, cut a piece of flesh out of his buttock and eaten it, and then sewed up the skin which occasioned his death (Robinson 4 November 1839 in Clark 2014a, p. 66).

Robinson also learned of a disease called Korr or blood, which was different from the Mindi (smallpox) disease, which the Daungwurrung and Woiwurrung considered came from the Banebedora clan of the Bangarang people. The east Kulin feared that this would be released in response to the arrest of the large group of Aboriginal men as a result of the Lettsom raid in Melbourne in October 1840. In response to their arrest, Yammerboke aka Mr Malcolm, a native doctor (see above), spent much of his time engaged in performing incantations to avert the coming calamity (Robinson 20 October 1840 in Clark 2014a, pp. 222–3).

> Korr or blood: a disease that attacks the crown of the head, covered the head like [...], bleed at the nose, bleed at the posterior, the posterior opens like a pannican, the stomack is empty, blood at the penis. This disease comes from the Banebedorro country. 1. Worgunbul, 2. Ginegine, 3. Teerregunnuc: the names of the blacks that will bring the korr (Robinson 15 October 1840 in Clark 2014a, p. 220).

With the release of the prisoners, Robinson was advised that the contagion known as correr, bloody issue, was averted, and the Banebedora in Melbourne were preparing to return to their home country with haste to stay the malady (Robinson 17 November 1840 in Clark 2014a, p. 232).

In 1844 Robinson learned of the death of a Daungwurrung youth in retribution for the death of the aged Mr King.

> Chief Constable informed that Natives were fighting at Merri Creek. Visited the camp pm, about 500 Natives present, the fight had terminated. The cause of quarrel was the murder of the Goulburn youth killed at Manton's. The circumstances are that [blank] alias Mr. King (aged) belonging to the tribe of Aborigines Western Port died on the Long Beach east side Hobsons Bay in March of 1843, a natural death as was fully ascertained but as the Natives seldom believe death to ensure from natural causes they endeavoured to find out from among their enemies the supposed individual Natives or tribe which had been the cause of his death (they believe it to be done by incantation) - for some time Bilbillerere, chief of the Waverongs - lay under the bone and his life was threatened and he was kept in alarm (Robinson 8 February 1844 in Clark 2014a, p. 532).

At Omeo in 1844, Robinson learned that a magician named Cor ro mung, alias Slarney, 'professes to smell the approach of white or black persons coming to hut and other marvellous things are ascribed to him. The Omeo blacks are said to possess power of causing a hard substance to enter the throat and stomack of people and causing death' (Robinson 22 June 1844 in Clark 2014a, p. 579). At the Loddon protectorate station, Robinson learned that: 'Native magicians are able to extract and put in sinews into man and cure them from rheumatics, take out sinews from lower animals and put them into man' (Robinson 23 September 1847 in Clark 2014a, p. 764).

Ethnographer John Mathew gave a detailed account of sorcery as was practised in Gippsland. Here he discusses the sorcery rituals believed to cause sickness:

> In Gippsland the practice of sorcery was called Ngurra Ngurra, or Ngurrabin. It was practised in a multitude of ways. One method was as follows: - If the death of a person were aimed at, a lock of his hair was stolen and kept intact by being tied with a piece of cord. This, apparently, represented the person that it was taken from. Then it was warmed at the fire, and afterwards removed from the camp into the bush, where the operator and a number of his friends would sing to it. The person whose death was designed would sicken and die in about a fortnight.
>
> Singing was also resorted to for good ends. Thus Brauwin (whatever that might be) could be charmed away from sick people by singing, and they would recover. The use of singing in working magic reminds one of the incantations employed in European countries and elsewhere in former days.
>
> To carry out another method an instrument called murrawan was made of iron bark wood. At one end it had a hook like that on the wommera, the lever for spear-throwing, and at the other a bunch of eaglehawk feathers. The murrawan was placed surreptitiously between the proposed victim and his camp fire. The result was that he became ill and died.
>
> A still easier method was simply to put a bone upon a person's footprint, call out his name, and his feet and legs would swell. To counteract the effects of this magic, a specially gifted, aged woman, would procure some long grass, like that which they made baskets of, and twist it into a cord. Then she would sit on her knees before the fire, with the patient at her side, away from the fire. The old hag would draw the cord of grass from side to side across her own gums, and spit blood and saliva into a wooden bowl. The blood was supposed to be drawn from the swollen legs. This mode of phlebotomy was applied to cure other kinds of ailments (Mathew 1925, p. 66).

Leckie noted that cursing was as important to a native doctor as was curing:

> Cursing is no less a part of a Wer-raap's practice than curing. Death curses are cast in a variety of ways. Here is a popular prescription. Take a piece of green bark in the left hand and scoop up about a pound of hot ashes. Cast the ashes

in the direction of the enemy and singing the curse prescribed by the Wer-raap invoke the spirits and all the birds of the air to carry the ashes and hurl them on to the doomed man. Then, according to the Wer-raap he will wither up like the scorched bark and die. Pain is inflicted on a distant enemy by making a model of the part of his body to be afflicted and beating it slowly before a fire. A favourite method of inflicting death, or pain, is to obtain some of the enemy's property, preferably weapons or hair, and burn them slowly in a bag enumerating, meantime, the particular pains desired (Leckie 1932, p. 7).

Edward M. Curr (1883) noted that there were three forms of sorcery: *millin*, *ngathungi*, and *neilyeri*. These are Ngarrindjeri terms from the Lower Murray.

> *Millin.* A big-headed club called 'plongge' is used entirely for millin. A mere touch from it is the cause of disease and death. They sometimes knock down an enemy, then tap his chest with the plongge, hit him with it on his shoulders and knees, and pull his ears till they crack. He is then called plonyge wategeri. He is now, by this means, given to the power of a demon called Nalkaru, who will create disease in his chest, or cause him to be speared in battle or bitten by a snake. Very frequently the plongge is used upon a person sleeping.

> *Ngathungi.* This form of sorcery is practised with bones, or remains of animals which have been eaten. A bone of some bird or beast which an enemy has eaten is obtained. This is mixed with grease, red ochre and human hair. The mass is stuck in a round lump on the end of a skewer of kangaroo's bone, and is then called 'ngathungi.' When injury is intended against the man who ate the animal from which the bone came, the ball above described is put down before the fire, and as it melts disease is supposed to be engendered in the person so bewitched, and if it wholly melts he dies. Anyone who knows that another person has ngathungi capable of injuring him, buys it if he can and throws it in the river or lake. This breaks the charm.

> *Neilyeri.* This is practised by means of a pointed bone. It is scraped to a very fine point. Sometimes an iron point is used. This is poisoned by being stuck into a dead body. Any one wounded by it usually loses a limb or dies. Sometimes this wound is inflicted secretly, when the person is asleep. The bone is kept moist by being wrapped in human hair soaked in liquor from a dead body. The natives are so terribly afraid of neilyeri that they dread even to have the weapon pointed at them, looking upon it as having a deadly energy, even when thus used.

Hill and Hill (1875, p. 117) confirmed that the Narrinyeri 'think that no persons die naturally; but that illness and death are always the result of sorcery, in which they have a profound belief. Poisoning, by the insertion of putrid matter taken from a corpse, is a not uncommon mode of revenge among these tribes.'

Dawson noted that in response to the threat of sorcery, Aboriginal people had developed habits of sanitation to ensure that they were safe from harming practices:

> It is worthy of remark that nothing offensive is ever to be seen near the habitations of the aborigines, or in the neighbourhood of their camps; and although their sanitary laws are apparently attributable to superstition and prejudice, the principles of these laws must have been suggested by experience of the dangers attendant on uncleanness in a warm climate, and more deeply impressed on their minds by faith in supernatural action and sorcery. It is believed that if enemies get possession of anything that has belonged to a person, they can by its means make him ill; hence every uncleanness belonging to adults and half-grown children is buried at a distance from their dwellings. For this purpose they use the muurong pole (yam stick), about six or seven feet [180cm–2 m] long, with which every family is provided. With the sharpened end they remove a circular piece of turf, and dig a hole in the ground, which is immediately used and filled in with earth, and the sod so carefully replaced that no disturbance of the surface can be observed. Children under four or five years of age, not having strength to comply with this wholesome practice, are not required to do so; and their excreta are deposited in one spot, and covered with a sheet of bark, and when dry they are buried. It may be as well to say here, that, besides this sanitary use of the muurang pole, it is indispensable in excavating graves and in digging up roots, and is a powerful weapon of warfare in the hands of the women, who alone use it for fighting. The aborigines believe that if an enemy get possession of anything that has belonged to them — even such things as bones of animals which they have eaten, broken weapons, feathers, portions of dress, pieces of skin, or refuse of any kind — he can employ it as a charm to produce illness in the person to whom they belonged (Dawson 1881, p. 12).

> They are, therefore, very careful to burn up all rubbish or uncleanness before leaving a camping-place. Should anything belonging to an unfriendly tribe be found at any time, it is given to the chief, who preserves it as a means of injuring the enemy. This wuulon, as it is called, is lent to any one of the tribe who wishes to vent his spite against any one belonging to the unfriendly tribe. When used as a charm, the wuulon is rubbed over with emu fat mixed with red clay, and tied to the point of a spear-thrower, which is stuck upright in the ground before the camp fire. The company sit round watching it, but at such a distance that their shadows cannot fall on it. They keep chanting imprecations on the enemy till the spear-thrower, as they say, turns round and falls down in the direction of the tribe the wuulon belongs to. Hot ashes are then thrown in the same direction, with hissing and curses, and wishes that disease and misfortune may overtake their enemy (Dawson 1881, p. 54).

Phillip Davis Rose, a settler in the Grampians district of western Victoria, writing in 1853, recalled the use of spatulas (gunigalk or excrement sticks) to bury human excrement was commonplace when he first arrived in the district: 'A few old men still use a spatula to dig a small hole, and cover their evacuations like the Israelites of old. In 1843, at the Grampians, all did this' (Rose 1853 in Bride 1969, p. 324).

Joseph Parker, whose father was Assistant Protector responsible for the Loddon district of the Aboriginal Protectorate from 1839 until 1851, recalled that the Djadjawurrung believed in two forms of death: one natural and one superstitious:

> They did not believe in death from natural causes, except in the two extremes of life - old age and infancy; but assert that there were two forms of death - the Moo-char-moo-roop (literally, 'take the spirit') and the Boor-kur-moo-rar (or, 'break the kidney-fat'). The first of these terms, Moo-char-moo-roop, applied only to those who had been removed by a very sudden death without any apparently previous illness. The spirits of those who died in this way were said to have gone to the West (Whar-ree-whin-knam-mytch-oo), and would re-appear at some future time brighter and more perfect beings. Under this belief, mothers, whose infants had died in the way here described, have been known to carry the dead body on their backs for months after death, affirming that the longer they carried the corpse in this way the more future happiness would their child and themselves enjoy, and believed that its departed spirit had the power of granting to its parents the influence of 'witchcraftism.'
>
> The other form of death - Boor-kur-moo-rar - a horrid superstition, was the cause of perpetual murder and blood-shedding. If one of their number died from natural causes, or was killed, no matter how, the moment life was extinct the body was tied up and prepared for interment, and carried to a piece of clear soil; two of the companions of deceased would then dig a small trench round the body, generally of an elliptical form, which would be carefully swept and minutely examined in order to find a hole. Should they succeed in finding one, they would place a straw in it, and carefully mark the bearing it pointed to. They would then proceed in the direction indicated by the straw, and take a similar life to the one they had lost; that is to say, should a man die, the life of another man would be taken, and if that of a woman or child, a similar life would also be taken.
>
> In the year 1846 an incident occurred which painfully illustrated this superstition. The Melbourne or Coast natives lost a man of their tribe, generally supposed from natural causes. A number of the deceased's friends resorted to the usual mode of trench-digging, and, strictly in accordance with the strawpointing, proceeded to Knee-rarp -now known as Joyce's Creek......and there at mid-day attacked a party of the Ja-jow-er-ong natives, who were at the time hunting, and killed a fine young man who was at the time unable to defend himself. The friends of this young man, although eye-witnesses to his murder, and in the full knowledge of who the guilty parties were, proceeded in the usual way to tie up the body and dig the trench. The straw pointing in the

direction of the Goulburn, a strong party, consisting of eighteen men, were then equipped with spears, &c., and in about a week from the Knee-rarp tragedy a similar life was taken by this party in the locality named. Thus it was that this horrible superstition kept these people in constant fear of molestation, and caused them to be continually moving from place to place (Parker in Smyth 1878, vol. 2, pp. 155f).

Harming practices

Many Europeans were exposed to Aboriginal beliefs about harming, but often did not understand them. The following example from c.1855 demonstrates this. Charles D Ferguson, a gold miner in the Happy Valley district south-west of Ballarat, had, on one occasion, offered to assist a wounded Aboriginal man. He expressed surprise at the horror displayed by the Aboriginal man's friends and family when he, Ferguson, signalled his intention to cut the patient's hair so that he could treat the wound. Clearly, Ferguson was unaware of the Aboriginal concern that pieces of the human body, such as the hair, in the hands of traditional enemies could be used for harming practices. This explains why the Aboriginal women very carefully picked up every piece of hair so that it did not fall into enemy hands.

> The man had a terrible cut on the head, the gash being nearly three inches [7.5 cm] long and laying open the skull. I had often heard of the thickness of the skull of the blacks, but had never before seen one laid open, nor did I ever believe that it was half so thick as this man's. I had brought some court-plaster and some castile soap and a pair of scissors. It was necessary to cut away some locks of hair. The hair of these natives is as thick as a mat, is never combed, and is as coarse as a horse's tail, and as soon as I commenced to cut it the woman set up a louder and still more disagreeable howl. I stopped them, but found they did not want me to cut his hair. I explained the necessity thereof to save his life, and then they quieted down and appeared satisfied, but watched me and picked up every hair that was dropped. I plastered him up and left him, and came that night to see my patient and found he had become conscious, but did not believe he would recover (Ferguson 1888, pp. 308–9).

In 1891, Alexander Ingram, a surveyor in Hamilton, informed Rev. John Mathew that in the volcanic rock country, about a quarter of a mile [400 m] from Mt Eccles, near Macarthur, in western Victoria, there were two well-like holes of great, but unknown, depth, into which in former times the Aborigines were accustomed to drop a morsel of the excrement of an enemy who was supposed thereafter to pine and die (Mathew no date, *Papers* Ms. 950 (j)).

A further example of this belief is provided by the sequel to the death in 1841 of Eurodap, alias Tom Brown, a clan head of the Gurngulag, the Djargurdwurrung clan belonging to Lake Colongulac and Mt Myrtoon. On 24 April 1841 Eurodap had agreed to accompany the Chief Protector, George Augustus Robinson, during his 1841 tour of the Western District to Port Fairy, Portland, and the Grampians. On 25 June in the Wulluwurrung dialect region of the Dhauwurdwurrung tribe, Eurodap was killed by Pongnorer, alias Joe, a 'fighting man' or warrior of the Wandidj gundidj, the Dhauwurdwurrung clan belonging to the Tahara and Murndal Stations on the Wannon River. In September 1843 Reverend Francis Tuckfield, one of the missionaries at Buntingdale, informed Assistant Protector Thomas (in Smyth no date, *Papers* Ms. 8781) that the Gurngulag people eventually returned to the country of the Wandidj gundidj, where by unspecified means they obtained a lock of Pongnorer's hair. The hair was wound around three small sticks three inches [7.5 cm] in length, and was kept close to heat or fire at all times. By this means they believed Pongnorer would get 'a hot head' and become ill with fever. On their return to Lake Colongulac they sent the package to the Gulidjan at Buntingdale Mission, their eastern neighbours, with instructions to send it on to the Wathawurrung at the Barrabool Hills and thence to the Boonwurrung, keeping it hot at all times. The Boonwurrung were finally to bury it in a mudflat in their country; and when the hair on the amulet rotted, Pongnorer would perish.

In August 1841 Robinson was presented with a parepole, a charm or amulet, by Linebonearrermin, a member of the Ngutuwul balug (the Djabwurrung clan belonging to Mt Langi Ghiran). A parepole consisted of the flan or fat of a human subject from the kidney or near the heart tied up in a piece of skin or rug. The fat is procured from a victim belonging to a tribe with whom they were at enmity. The possessor of the parepole believed that in consequence of his retention of this fat, the enemy tribe would die or whatever he desired would happen to his enemies. Robinson believed this was why the fat was sometimes taken from white men when they were killed by Aboriginal people.

Traditional Aboriginal beliefs about harming practices continued into the early 20th century in south-east Australia as the following exchange between Anthony Anderson and Natalie Robarts, the last matron at Coranderrk Aboriginal station, over the disposal of the contents of a cuspidor or spittoon, demonstrates.

> He has a cuspodor [sic] which he uses he tells me but he threw away the cardboard filler as it was full, I told him he should [have] burnt it all, 'oh no no burning no good, I'd soon kick the bucket if I put it in the fire, I know from olden time, Chuck him out, that's all right but not burn him, I know, I know all about it!' (Robarts 8 August 1910 in Clark 2016b, p. 71).

At Lake Boga, local baker, Alexander Stone, who maintained an active interest in the lives of the Aboriginal people he employed in his bakery, recorded that: 'Most sicknesses were believed to be the result of evil machinations of enemies, and no person died of any sickness (old age excepted) that was not caused by enemies' (Stone 1911, p. 440).

Dangerous places and sources of dangerous powers

Biernoff (1978) in an analysis of traditional Aboriginal society in eastern Arnhem Land has highlighted that Aboriginal land is replete with meaning at several levels of reality. It is both the container and focus of ambivalent and dangerous powers. Knowledge of dangerous places has a 'marked effect in restricting and patterning movement, access to land, and land use' (Biernoff 1978, p. 92). The focus of danger may be spatially confined to such entities as springs, rocks and trees, or diffuse areas such as marshes, plains or creeks. 'Rarely localities exist which must be avoided by everyone, more commonly all but those protected by ritual knowledge and position must practice avoidance.' Biernoff identified four levels of danger in relation to the permissible duration of exposure to a dangerous place:

> One can walk over the land with safety but cannot halt; One can stop for a short time but must not sleep; One can camp and sleep but not for more than one night; and Locality specific restrictions may exist which affect the use of particular resources: water, plant foods, animal foods, use of fire wood, honey, mineral pigments – 'can't eat', 'can't make holes' etc.
>
> Trespass will result in death or serious illness and close friends and relatives are not immune from the danger. Accordingly, before travelling permission is asked and knowledge of safe routes and places is sought, and one does not travel in unknown regions without a guide who has knowledge of and responsibility for the area.

Albert Le Souëf noted that: 'The constant treachery practised prevents the different tribes from often leaving their own territory; if they do, they are never sure of their lives; for the same reason they do not like going about after dark, or leaving their camps, unless several are together; there may be a lurking foe behind every tree or bush' (Le Souëf in Smyth 1878, vol. 2, p. 223). This fear of travelling in the dark may also be a fear of harmful spirits.

Robinson observed a custom called pardewundingarroengar that his Aboriginal guides enacted whenever they were in travelling through someone else's country. They snapped their fingers on their pullerterlorn (woomeras) and then rubbed them against their noses and then shook it from them once. They told Robinson they did this 'to let the blacks of this country know they are there. They do this when they imagine the natives are talking about them' (Robinson 30 June 1841 in Clark 2014a, p. 383).

Biernoff classified dangerous places along a continuum from 'Dreamtime origin place' – places associated with important events in major origin myths; 'Dreamtime story place' – sites of incidents in less important local dreamtime myths; 'Secret ceremony place' – where secret ceremonies take place; 'Public ceremony place' – where major public ceremonies are regularly held; 'story places, long time ago' – where some catastrophe has produced a large number of deaths, through disease, massacre, magic, or the activity of supernatural forces; and 'Place of recent death or serious illness' – which must be avoided for a period of time.

Certain foods were also avoided – Robinson was told by a Daungwurrung youth that young men did not eat Emu meat because it made them fall from trees when they were climbing for possum and cause them to break their limbs or kill them (Robinson 1 June 1840 in Clark 2014a, p. 191).

In 1843 Robinson travelled with several members of the Native Police Corps who were members of the Woiwurrung from the Melbourne district. Near the Murray River, Robinson noted that while the Native Police were able to fish for yabbee they told him: 'they cannot fish for munye, Dick said he must not fish as country did not belong to him and if the natives saw him they would at some other time call him to account. Redman, the native police, said the natives of the country, he believed, would take his feet' (Robinson 31 March 1843 in Clark 2014a, p. 493). In July 1846, despite his hunger, Robinson's Aboriginal companion, Merringundidj, a Djadjawurrung, would not eat possum or wombat. Merringundidj explained to Robinson that 'if eat it, plenty very bad and die, belonged to another black country' (Robinson 17 July 1846 in Clark 2014a, p. 723).

Robinson noted that after a Daungwurrung man had died at a campsite in Melbourne, his people buried the body. They also buried with him all his clothes and implements. A fire was made on the actual spot where he died, and a spear belonging to the deceased was broken into two pieces and thrust into the fire. They also burnt his *willam* or *wuurn*. After the funeral the campsite dispersed and people removed to other sites (Robinson 30 December 1839 in Clark 2014a, p. 74).

The efforts of missionaries and protectorate officials to keep Aboriginal people in continuous residence on their stations were often hampered by Aboriginal people temporarily leaving the stations during periods of illness or when an Aboriginal death had occurred. For example, several deaths took place at an Aboriginal campsite at Arthurs Seat in late 1839, and Robinson noted that in consequence the Daungwurrung Aboriginal people told him they would leave in the morning. Robinson (30 December 1839 in Clark 2014a, p. 74) noted that they were 'very much alarmed and afraid they shall all die'. In November 1841, the Aboriginal residents at the Loddon Aboriginal station began to leave. They explained that they were leaving because 'too much sick at the station at Willam.be.parramul' (Robinson 20 November 1841 in Clark 2014a, p. 440).

Conclusion

This chapter has surveyed the south-east Australia literature concerning Aboriginal beliefs in the causes of illness and death. It has also examined the role of male and female doctors in using sorcery to induce death and illness and their ability to counteract sorcery-related illnesses. Treatments for ailments such as cuts, bruises, and stomach-ache, and the general Aboriginal pharmacopeia are the subject of the next chapter.

Chapter 12
Healing

Ian D. Clark

Plants feature prominently in Aboriginal remedies chiefly used to relieve symptoms such as fever, congestion, headache, skin sores, tired or swollen aching limbs and digestive problems. Treatment can involve drinks, washes, massages and aromatherapies. The drinks are made by heating water with plant additives, and in Aboriginal English are commonly referred to as 'tea'. Since European colonisation, washes are prepared by boiling plants, with the cooled liquid applied externally to the body. Some plants are heated, then rubbed or massaged into swollen parts of the patient's body. The aroma of plants is generally transferred to the patient through contact with steam and smoke (Clarke 2008b, pp. 12–13).

Introduction

The Aboriginal pharmacopeia is vast; far too large for a detailed description in this chapter (for an example of regional studies, see Clarke 1987). The diversity of herbal remedies served Aboriginal people well. As hunter-gatherers they had to move with the seasons through different habitat zones in the landscape, which meant that it was necessary for them to possess knowledge of a broad range of remedies. It was important for Aboriginal people to know the seasonality of each plant species as well, some of which may not be as effective or even available at certain times of the year (Clarke 2008b, p. 13).

Early commentators in south-eastern Australia were impressed with Aboriginal healing practices; James Bonwick, for example, noted: 'They have cures for many complaints which exhibit no small intelligence' (Bonwick 1856, p. 82). Rev. Daniel Mathews considered their use of plants was effectual:

> Their knowledge of surgery is very scant, but they have medicinal remedies in roots, plants, flowers, and shrubs, which, before the introduction of Epsom salts, pills, and drugs, were accredited as being effectual. Their belief in the healing properties of eucalyptus leaves is possibly as old as the race, and I have known of many remarkable recoveries in consequence of its use (Mathews 1899, p. 45).

Stanbridge also discussed some of the practices of Aboriginal doctors and the remedies they used to treat illness:

> The medical doctor occasionally administers a decoction of a fleshy-rooted geranium [probably Austral Cranesbill, the native geranium, *Geranium*

solanderi], the only root used medicinally, and he has been known to bleed in the arm with a sharp flint; but incantation is the panacea for all the ills to which flesh is heir, whilst it is also regarded as the cause. The patient is seated in front of the operator, who utters a monotonous chant, makes passes by drawing his hands downwards over the part affected, and at intervals rubbing and blowing upon it. At the conclusion, supposing the disorder to be rheumatism, hot ashes are applied, but as incantation loses its power by the presence of a third person, it is very seldom and only by accident that the ceremony is witnessed. There is but small field for the doctor's art, as the only contagious disorder appears to be a mild form of hooping-cough, but many persons in advanced life are met with who are pitted as by small-pox (Stanbridge 1861, p. 300).

An exhaustive description of medical procedures employed by Aboriginal doctors throughout south-east Australia and Aboriginal treatments for ailments is beyond the limited scope of this chapter. However, the following discussion examines some of the procedures known to have been practised in the study area, and some Aboriginal treatments for ailments such as snakebite and toothache.

Isolating the sick

One of the earliest accounts of Aboriginal healing treatments and remedies in south-east Australia is provided by Dr WH Baylie, who served as medical officer at the Goulburn River Aboriginal protectorate station in the early 1840s. As a medical practitioner his comments on Aboriginal medicine are insightful. For example, he noted that Aboriginal people were aware of the need to isolate people who were sick:

> They are sensible of the infectious character of disease, and they use the precaution of removing their sick from the encampment by day to some distance, where on finding a shady place they erect a few brambles and form a shelter to the great comfort of the invalid (Baylie 1843, p. 88).

General treatment of wounds

Bonwick explained that there were numerous treatments for wounds, depending on their nature and severity:

> Wounds are covered with a plaster of mud and ashes. Flesh wounds heal with rapidity under their simple treatment … Splints and bandages are employed in accidents … For certain disorders the excrements of animals are rubbed over the head (Bonwick 1856, p. 82).

Stone confirmed that: 'The general treatment for wounds consisted in the application of a plaster of wet red pipeclay, bound on with opossum fur rope, and this rough and rude treatment met with general success' (Stone 1911, p. 440).

William Thomas was impressed with the efficacy of Aboriginal treatments applied to wounds:

> Wounds of whatever kind which do not affect a vital part are more readily cured than white people. I have seen most desperate wounds inflicted by their weapons (that would have kept Europeans for months invalids) healed in an incredible short time, to the astonishment of medical men. Wounds, whether by accident or otherwise, are immediately attended to by their doctors; if in the fleshy part of the body, they suck the blood from the wound, and continue sucking until blood ceases to be extracted; if little blood comes from the wound they know all is not right, and will put the patient to pain by probing the wound with their lancet (a sharp bone), or place the body in that position so to compress the opposite part to force blood; they know well the consequence of stagnant blood or matter, especially in the upper part of the body; when the wound is thoroughly, clean, they leave the rest to nature, and place a lump of 'pridgerory' (a kind of wax oozing from trees) on the wound; should there follow a gathering, they open the wound afresh, and see all right, and again cover it over with pridgerory (Thomas in Victoria 1861; Anonymous 1861, p. 6).

In 1870, Dr Andrew Ross, of Molong, New South Wales, published a paper in the *New South Wales Medical Gazette*, in which he detailed the recovery of an Aboriginal man from a spear wound in the abdomen. Ross had treated the man in late 1864, though he resorted to traditional treatments which led to a complete recovery: 'bathing the wound and swollen abdomen with a few of the most tender undershoots and leaves of the red gum tree [*Eucalyptus camaldulensis*], wrapped together in the form of a wreath, and every now and again dipped into a billy (a tin vessel capable of holding from half a gallon to a gallon of hot water, which was constantly kept warm by the fire' (Ross 1871, p. 17). Ross's account is testimony of the antimicrobial properties of compounds found in Eucalyptus, and interest in eucalyptus oil as a possible antiseptic.

> In submitting so remarkable a case to the notice of the medical profession, I do so more to stimulate research in the matter, and to attract special attention to the effects of so novel and simple a remedy, than with the view of pluming myself with any degree of credit for the recovery of the patient — a circumstance more indebted, I fully believe, to the efficacy possessed by this species of the Eucalypti than any services which I may have rendered the unfortunate sufferer. Whether it may be found to act so obediently in similar cases (abdominal wounds) in the European subject, is a matter which must be left to be confirmed by future experience. Neither do I pretend to explain its *modus operandi*; but as a topical agent in cases of severe punctured wounds, it certainly appears to possess some useful and valuable qualities, which I think are worthy of being further investigated, and thus my sole desire to give it publicity through the medium of your columns. The matter, of course, as may be observed, was

accidentally, but impressively, brought under my notice while engaged attending to this important and serious case; and even the most sceptical cannot, I think, on reading the details, readily doubt the statement. In fact, the leaves of the various species of the Eucalypti, it is well known, possess a peculiar strong aroma, especially when bruised, somewhat resembling the odour of camphor. To those who may have chewed them, this fact is palpable, and scarcely needs corroboration; for they are not only observed to possess a bitter but rather pungent taste, followed by a kind of benumbing or cooling sensation in the mouth, which very soon disappears. It is, probably, due to these aromatic ingredients, combined with some kind of astringent mucilaginous extract or principle, that it was [sic.] such extraordinary therapeutic properties — properties, I believe, that are not only sedative, antiphlogistic, but even catalytic or antiseptic. As to whether it will act so internally or in other wounds is another question, for in this case it was only used externally (Ross 1871, p. 17).

Mathews (1899, p. 47) was another who noted the efficacy of the treatment for a spear wound: 'The young man made light of it, although the blood flowed freely. Taking a little moist clay from the bank, he pressed it on the wound, and in a few days he was quite restored'. Le Souëf noted that for cuts or wounds 'they apply bandages and often earth poultices, which, by-the-by, often have a marvellous effect' (Le Souëf 1878 in Clark and Cahir 2016, p. 84).

One newspaper correspondent in 1909 recalled the efficacy of an Aboriginal treatment for a flesh wound his mother had witnessed in the 1840s. In the absence of Western remedies he followed the Aboriginal treatment and noted that the treatment produced immediate benefits:

> Sir,- In the forties the blacks were very numerous on the Loddon Valley and the Murray. My mother (then in her teens and a resident at Thorpe's Station) was an eye witness to a pitched battle between these two tribes. Among other casualties, a lubra was seriously wounded on the thigh by a 'boomerang'. The aboriginal doctor took her in hand, and freely sprinkled cold wood ashes on the wound, immediately washing them off by using his mouth as a syringe. This he repeated for a long time. The lubra recovered. A few days ago a friend of mine was badly scalded on the wrist by steam from the spout of a boiling kettle. Not having the usual remedies at hand, I thought of the wood ashes, applied them and, to our surprise and pleasure, found that the ashes absorbed the moisture, excluded the air, and gave the patient immediate relief. Continuous treatment in the course of a few days produced a new skin, the patient in the meantime continuing his duties without any inconvenience. A brown mark the size of a walnut is now the only indication of what had been a very serious burn. Perhaps some of your readers will be able to throw more light on this very interesting subject. -Yours &c. T. MARTIN. Ballarat, Aug. 18. (Martin 1909, p. 8).

Blood-letting

Aboriginal people practiced blood-letting or phlebotomy. Stone explained that: 'Blood-letting was sometimes practised, and was carried out by sawing an opossum fur cord backwards and forwards over the spot from which blood was desired. Of course in most ailments the doctor was the chief person officiating, and managed to imbue the patient with his great ability' (Stone 1911, p. 440). Le Souëf confirmed they are 'skilled in the art of bleeding; they open a vein with a piece of sharp flint or shell; they often rub and knead with their knuckles the affected part (Le Souëf 1878 in Clark and Cahir 2016, p. 84).

John Mathew recalled, in a public lecture, his observation of a 12-year-old boy undergoing the operation of bleeding:

> A line made from the tail of an opossum was tightly tied round his waist. At a little distance the doctor sat with a portion of the line drawn through his mouth. Occasionally he spat blood from his cut lip into a billycan placed at a little distance further on. This blood was supposed to be drawn by the doctor from the body of the boy. A boy with a pain in his side consulted the medicine man, who was skilfully supposed to draw from the place a piece of newspaper, glass or some other substance that somebody who desired to do him an injury had aimed at him from somewhere (Mathew 1918, p. 2).

Steam baths and cold water bathing

When patients did not respond satisfactorily to simple treatments, a steam bath was sometimes resorted to. Stone discussed the steam bath, or 'Burree' as it was known in the Lake Boga district:

> [It] was prepared by making a large fire, and after it had burnt out the ashes were raked away, and a piece of bark was laid over the heated spot, a thick layer of mistletoe bushes and leaves were heaped upon it, and then the patient was rolled in an opossum rug and laid upon the bushes. A profuse perspiration was generally induced, which often had the effect of greatly improving the condition of the patient (Stone 1911, pp. 440–1).

As a general treatment for fever, Aboriginal people employed the naturopathic treatment of hydrotherapy or cold water bathing. Dr Baylie noted, 'I have known them to plunge into the water when high inflammatory action was going on, it has often been a means of checking the circulation of blood throughout the system, and thus has allayed the fever' (Baylie 1843, p. 88). Thomas discussed this treatment in more detail:

> The aboriginal doctor's treatment in fevers is strictly the cold water system, no matter what kind of fever it may be, accompanied with prohibition of animal food. The doctors have a quantity of water by them, fill their mouths full and spurt it over the whole of the patient's body, back and front, and for a considerable time on the navel, then with their hands throw it over face and breast, then lay

the patient on the back, breathe and blow on the navel, incantating continually while operating. If the patient be young, the doctor will carry him and plunge him into the river or creek; the adult patient will voluntarily plunge himself in three or four times a day. The blacks obstinately persist in this mode of treatment, although they find death generally the result. I was not a little surprised to find many years back that this was also the mode of treatment adopted by the natives of the South Sea Islands. I was called to witness their habits, when a party of them were enticed over by the late Mr Boyd. They were located at Mr Fennel's (Mr Boyd's agent) on the banks of the Yarra; as soon as fever attacked them they crept to the banks of the Yarra and plunged themselves in three or four times a day (Thomas in Victoria 1861; Anonymous 1861, p. 6).

Use of ligatures

Ligatures were used to strengthen muscles and to suppress hunger pangs and headaches, presumably to cause hypertrophy or a reduction in blood flow. Robinson noted that some were worn so tight that 'after worn a few days it cuts through the skin' (Robinson 6 June 1841 in Clark 2014a, p. 361). In 1843, Billy Billy aka Urquor, the head of Djabwurrung clan centred on Mount William, was arrested for his alleged involvement in sheep stealing on Ledcourt Station. In July 1843, William Thomas visited him in the Melbourne gaol where he was being held pending his trial. Thomas noted in his private journal, 'I gave him some bread & meat, he had his belly tied up, a general custom with all other Blks I have met with to keep off hunger' (Thomas 12 July 1843 in Stephens 2014, vol. 1, p. 524). Christopher Hodgson noted the use of ligatures to suppress hunger:

> I have known them fast for days when on some expedition, only appeasing their appetites by a tighter and tighter fastening of their opossum belts round their decreasing stomachs. This allays the pangs of hunger; but woe to the unfortunate kangaroo or animal that first falls to their chance! (Hodgson 1846, pp. 224f).

Friedrich Gerstaecker, was another writer who observed their use and enquired into their purpose:

> the Murrumbidgee and neighboring blacks ... Nearly all of them go perfectly naked, and many have a thin cord tied very tightly round their waist. I was told by some shepherds that they intended this for a kind of medicine, to stop the free circulation of the blood; and this seems in so far probable, that I have found several other blacks who wore strings, tied in a similar way, round an arm or a leg, also round their heads. The latter they told me, was to prevent headache (Gerstaecker 1853, p. 440).

Rheumatism

Aboriginal people treated rheumatism through a range of treatments including the use of ligatures, massage, hydrotherapy and taking a whale bath. Robinson noted that at the

Loddon Aboriginal station the native doctors cured rheumatism by implanting animal sinews into the patient: 'Native magicians are able to extract and put in sinews into man and cure them from rheumatics, take out sinews from lower animals and put them into man' (Robinson 23 September 1847 in Clark 2014a, p. 764).

In 1843, Dr Baylie noted that rheumatism was treated by using ligatures made from the intestines of kangaroo or Emu:

> Rheumatism is very general, as may be inferred from their being so much subjected to the variation of temperature, often assuming a most acute degree of inflammatory action. They employ as a means for their relief a number of tightened cords above or over the parts affected, thereby checking the circulation and allaying the symptoms or paroxysm of pain; these cords are made from the intestines of the kangaroo or emu, neatly twisted like our fine twine, and possess a great degree of strength and elasticity; in cholic or spasmodic action of the intestines, they apply this cord and check the violence of the pain by pressure (Baylie 1843, p. 87).

William Thomas noted that friction or massage was a general treatment for rheumatism. However, for rheumatism in the legs or thighs the patient was placed in hot ashes nearly up to their knees:

> Their general remedy [for rheumatism] is friction; if very severe about the thighs or legs, the doctor gets a good mound of hot ashes prepared, made solely from bark which is without grit; the patient is laid on his belly, and the doctor rubs most unmercifully the hot ashes on the part affected as a butcher would in salting meat; if in the thighs or legs, the patient is put into the mound of heated ashes nearly up to his knees, where he sits whilst the doctor is rubbing with hot ashes the parts affected. During this process the doctor is incantating, blowing occasionally a portion of the dust into the air with a hissing noise; when sufficiently operated upon, the invalid is wrapped up in his blanket (Thomas in Victoria 1861; Anonymous 1861, p. 6).

Dawson elaborated on the treatments available for rheumatism, including an infusion of bark from the Blackwood tree (*Acacia melanoxylon*):

> In cases of pain in one spot the skin is scarified, and the blood allowed to flow freely. When the pain is general, and arises from severe cold or rheumatism, a vapour bath is produced by kindling a fire in a hole in the ground, covering it with green leaves, and pouring water on them. The sick person is placed over this, and covered with an opossum rug, and steamed till profuse perspiration takes place. He is then rubbed dry with hot ashes, and ordered to keep warm. Another cure for rheumatism is an infusion of the bark of the blackwood tree, which is first roasted, and then infused while hot. The affected part is bathed with the hot infusion, and bandaged with a cord spun from the fur of the flying

squirrel, or ringtail opossum, with a piece of opossum rug as a covering (Dawson 1881, pp. 56–7).

Albert Le Souëf noted that for rheumatism: 'they plunge the patient into cold water' (Le Souëf 1878 in Clark and Cahir 2016, p. 84). Bonwick (1856, p. 82) confirmed that rheumatisms 'are relieved by shampooing, and inflammations by bleeding and by cold water applications; no female is bled'.

At Twofold Bay, in southern New South Wales, the Aboriginal people used a whale-cure for rheumatism where sufferers would sit inside the festering carcasses of whales in order to relieve their symptoms. If patients could endure the smell and sit in a hole made in the carcass for up to 30 h, they were promised 12 months of relief from the pain. The warmth and fumes generated by the rotting carcass were believed to have healing properties. RH Mathews (1904a, pp. 252–3), observed the Aboriginal people of Twofold Bay undertaking the whale cure for rheumatism. After the meat and blubber were removed, he wrote, the Aboriginal people would lower themselves into the carcass and cover themselves with whale fat to treat their pain. Hyde (2013, p. 60) believes that it 'was not long before knowledge of this treatment was transferred to their European neighbours, and Eden's local businesses were more than happy to accommodate those willing to pay for the experience'. Louis Becke has left a detailed account of the whale cure:

> The 'cure' is not fiction. It is a fact, so the whalemen assert, and there are many people at the township of Eden, Twofold Bay, New South Wales, who, it is vouched, can tell of several cases of chronic rheumatism that have been absolutely perfectly cured by the treatment herewith briefly described. How it came to be discovered I do not know, but it has been known to American whalemen for years.
>
> When a whale is killed and towed ashore (it does not matter whether it is a 'right,' humpback, finback, or sperm whale) and while the interior of the carcase still retains a little warmth, a hole is cut through one side of the body sufficiently large to admit the patient, the lower part of whose body from the feet to the waist should sink in the whale's intestines, leaving the head, of course, outside the aperture. The latter is closed up as closely as possible, otherwise the patient would not be able to breathe through the volume of ammoniacal gases which would escape from every opening left uncovered. It is these gases, which are of an overpowering and atrocious odour, that bring about the cure, so the whalemen say. Sometimes the patient cannot stand this horrible bath for more than an hour, and has to be lifted out in a fainting condition, to undergo a second, third, or perhaps fourth course on that or the following day. Twenty or thirty hours, it is said, will effect a radical cure in the most severe cases, provided there is no malformation or distortion of the joints, and even in such cases the treatment causes very great relief. One man who was put in up to his neck in the carcass of a small 'humpback' stood it for sixteen hours, being taken out at two-hour intervals. He went off declaring himself to be cured. A year later he had a return of the complaint and underwent the treatment a second time.

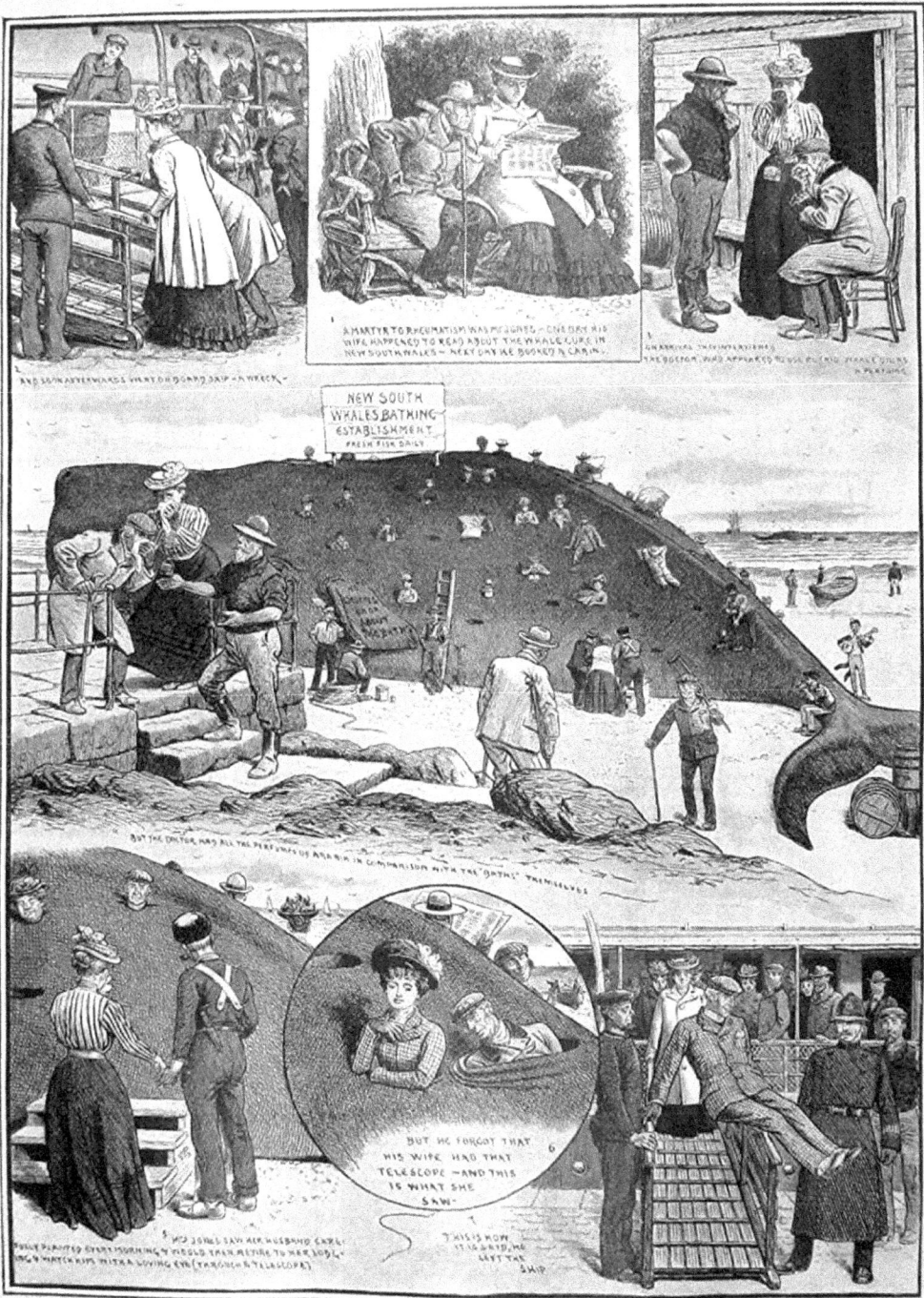

The Whale Cure for Rheumatism in Australia. Source: W Ralston, *The Graphic* 31 May 1902.

All the 'shore' whalemen whom the writer has met thoroughly believe in the efficacy of the remedy, and by way of practical proof assert that no man who works at cutting-in and trying out a whale ever suffers from rheumatism.

Furthermore, however, some of them maintain that the 'deader' the whale is, the better the remedy. 'More gas in him,' they say. And anyone who has been within a mile of a week-dead whale will believe *that* (Becke 1904, pp. 239–40).

Joint pain

Dawson noted that for pains in the joints, the fresh skins of eels were wrapped round the place, flesh side inwards. He commented that the same cure was very common in Scotland for a sprained wrist (Dawson 1881, p. 57).

Skin infections

Skin infections such as boils were treated with lotions made from wattle bark and Wild Marshmallow:

> The blacks treat boils and swellings thus: when hard, they lotion the part well with decoction of wattle bark; when obstinate, they boil wild marshmallow [*Althea officinalis*], and poultice - if it softens and does not break, they apply their sharp bone lancet (Thomas in Victoria 1861; Anonymous 1861, p. 6).

Thomas has written of a skin disease called by the eastern Kulin of central Victoria *bubberum*. Marie Fels (2011, p. 78) has undertaken a detailed study of Thomas's writings on the origins of Aboriginal disease and she believes that *bubberum* may have an association with dogs. The treatment included a poultice of red ochre and wattle bark.

> The aborigines are deeply afflicted with a disorder called by them bubberum, white men call it itch, but it is in no way like it; it appears as raised dark scabs, and spreads, joining each other, until they cover almost all the lower extremities. It seldom affects the head or upper parts, but I have known it almost cover the thighs and legs, so that the afflicted one could with difficulty move about. The native cure for this distemper is every night and morning to grease the parts affected with wheerup (a red ochre), mixed with decoction of wattle bark. I knew one instance of this disease becoming most distressing to a white man, in a respectable position, who was continually cohabiting with black lubras (Thomas in Victoria 1861; Anonymous 1861, p. 6).

Burns

Burns were treated by 'dabbing the parts over with melted fat, afterwards dash the parts affected with a puff made of opossum fur and the dust of wheerup' [red ochre] (Thomas in Victoria 1861; Anonymous 1861, p. 6).

Venereal disease

Across the study area at least two treatments were applied to venereal diseases. One method involved the application of heat and ashes; the other involved the application of a wattle bark lotion:

The natives when at times when any one of their number is afflicted with venereal bury them up to the shoulder. Mr Crook's stockkeeper saw a woman thus afflicted. The natives made a large fire, then scraped away all the fire and cleared away ashes then dug up the ground with stick, made a hole same as for damper and laid the woman in it and covered her all over with the ground except her head she screamed. It must be a severe operation from the heat (Robinson 26 June 1844 in Clark 2014a, p. 581).

Bonwick (1856, p. 82) confirmed that venereal ulcers were sprinkled with alkaline wood ash. William Thomas also detailed the use of a wattle bark lotion in this complaint:

Though this [venereal] disease in the first instance must have been contracted from the whites, the native doctors have proscribed a cure which though simple has proved efficacious; they boil the wattle bark till it becomes very strong, and use it as a lotion to the parts affected. I can state from my own personal knowledge of three Goulburn blacks, having this disease so deeply rooted in them, that the then colonial surgeon, Dr Cousen, on examining them said life could not be saved unless they entered the Hospital and an operation performed, which they would not consent to; after eighteen months these three blanks returned to Melbourne among the tribes (two were young and the other middle-aged) perfectly cured, and the blacks assured me that they had only used the wattle bark lotion. Dr Wilmot, our late coroner, also saw these three blacks whilst in this state, and after their soundness, and in his report upon the aborigines stated: 'However violent the disease may appear among aborigines that it could not enter into their system, as it did in European constitutions' (Thomas in Victoria 1861, Appendix 3; Anonymous 1861, p. 6).

Dr Baylie has described a skin infection that he called 'excrescence'. It may refer to a form of syphilis but requires more research to be certain.

I arrived at the station about the middle of November, 1841, and immediately inspected them; here, indeed, I must present a picture truly painful to any feeling mind. They were almost all afflicted with disease, particularly with a malady (sui generis) which I believe has existed for a considerable period amongst them, to which, from its pulpy appearance and form, I have given the name excresence [sic]. It has originated in their constitution and is very contagious, delibilitating the sufferer, and spreading with rapid violence through the system, often preventing him from searching after his daily food; starvation therefore ensues, and death is the painful consequence (Baylie 1843, p. 87).

Baylie noted that the treatment for excrescence involved the application of lotions made from astringent barks:

As to medicine they never knew that such a powerful means of relief existed until the arrival of the white population; true, they use some of the astringent barks macerated in water as a lotion, which they apply to the excresences [sic] before alluded to: beyond this, I have never known them to use anything pertaining in the least to drugs (Baylie 1843, p. 88).

Dysentery

Dysentery was treated by drinking a decoction made from the leaves of the native currant and other plants (Bonwick 1856, p. 82). Thomas noted:

> The aborigines of Australia are very subject to dysentery, but not to the fatal extent as Europeans their remedy for this disorder is drinking plentifully of the decoction of wattle bark and eating gum the day, and pills night and morning made by themselves of wattle bark and gum (Thomas in Victoria 1861; Anonymous 1861, p. 6).

Chest infections

Thomas noted that chest infections were monitored very carefully and treatments depended on the seriousness of the infections. Thomas considered the treatment meted out to those expectorating blood to be extreme:

> The blacks study much the color of the spittle in those affected in the lungs, and know well its stages. When the patient begins to expectorate blood, much attention is paid him; should this increase, which is generally the case, the doctors hold a consultation, and when once a consultation is held the doctors will not allow the patient to take any more medicine from the whites. The invalid is laid on his back and held firm by three or four black's, whilst the native doctor keeps continually pressing with his feet, and even jumping on his belly. I need scarcely state that this cruel practice brings on premature death (Thomas in Victoria 1861; Anonymous 1861, p. 6).

Insect repellents

Clarke (2015b, pp. 237–8) has suggested that in the south-east of South Australia there was a practice of throwing Coastal Ballart (*Exocarpos syrticola*) foliage onto camp fires in order to repel insects (Bonney 2004, p. 66). In the Coorong area, foliage from the Coastal Daisy-bush (*Olearia axillaris*) was burnt to keep insect vermin away from the camp (Clarke 2012, pp. 96–7), as well as being rubbed on the body as a mosquito repellent (Bonney 2004, p. 69; Clarke 2015b, p. 237). Stannard, who had experience with the Djabwurrung and Djadjawurrung peoples, noted that they greased their bodies with animal fat, 'to prevent the flies and mosquitoes from tormenting them', but she found the grease had a 'most offensive odour' (Stannard 1873, p. 97).

Headache, earache, toothache and stomach-ache

In terms of headache, we have three early accounts of the treatments employed by Aboriginal healers:

If [the pain is] of long standing the patient is compelled to lay on the back, the native doctor places his foot on the patient's ear and presses this organ until water literally gushes from the patient's eyes; however rough the treatment, I have known this operation to give relief, and the patient cured (Thomas in Victoria 1861; Anonymous 1861, p. 6).

In the Wiimbaio of north-west Victoria, 'A severe headache was sometimes treated by digging out a circular piece of sward, and the patient laying his head in the hole, the sod was replaced over his head, on which the Mekigar sat, or even stood for a time, to squeeze out the pain. Even under this practice the patient sometimes declared himself relieved' (Howitt 1904, p. 381).

Dawson explained that severe headaches of long continuance, required strong remedies, and were cured by burning off the hair and blistering the skin of the head (Dawson 1881, p. 57). In south-west Victoria, earaches were treated by pouring water on hot stones placed in a hole in the ground, and holding the ear over the steam (Dawson 1881, p. 57).

Dawson has provided a detailed description of three methods used to treat toothache – an application of gum from a white gum tree; the second by wearing a basket rush cape, and the third, involving the actions of an Aboriginal doctor or healer:

Sow thistles are eaten raw to soothe pain and induce sleep. The gum of the eucalyptus, or common white gum tree [presumably *Eucalyptus viminalis*], is a cure for toothache. It is stuffed into the hollow of the tooth. Teeth are never extracted unless they are loose enough to be removed by the finger and thumb (Dawson 1881, p. 57).

To cure toothache, a cape made of the basket rush is worn over the shoulders and round the neck, and is laid aside when the pain is gone — its name is weearmeetch. Another remedy is the application of a heated spear-thrower to the cheek. The spear-thrower is then cast away, and the toothache goes with it in the form of a black stone, about the size of a walnut, called karriitch. Stones of this kind are found in the old mounds on the banks of the Mount Emu Creek, near Darlington. The natives believe that when these stones are thrown into the stream at a distance from their residence, they will return to the place where they were found; and as they are considered an infallible remedy for toothache, they are carefully preserved. They are also employed to make an enemy ill, and are thrown in the direction of the offending tribe, with a request to punish it with toothache. If, next day, the stones are found where originally picked up, it is believed that they have fulfilled their mission. Not far from the spot where these stones are plentiful, there is a clump of trees called karriitch — meaning toothache — and the natives of the locality warn their friends never to go near it, for if they do they will be sure to get toothache. Stones of a similar description are found in the sand hills on the sea coast, and are put into a long bag made of rushes, which is fastened round the cheek. The doctor always carries these stones in his wallet, and lends them to sick people without fee or reward (Dawson 1881, pp. 59–60).

Leckie has described how Aboriginal healers treated stomach-ache:

> Another form of 'cure' was sometimes undertaken by the Wer-raaps. The patient, having complained of a pain in the stomach, excited the sympathy of the tribe by writhing on the ground. The Wer-raap calling on Len-ba-moor for assistance clutched the patient's flesh with his hands and released it with expressions of disgust. At last he was successful. He produced a handful or stones and stick which he declared the wild spirit had placed in the patient to make him sick. The patient, fearing to contradict the Wer-raap pretended to be cured and, if the illness were merely indigestion he recovered naturally in a few days. Should the illness prove to be more serious it was the evil power of the wild spirit which was blamed, and not the ignorance of the Wer-raap (Leckie 1932, p. 7).

Delayed lactation in new mothers

James Dawson noted that delayed lactation was remedied by bathing in lime-water made from the ash of burnt fresh-water mussel shell:

> Women unable to nourish their newly-born infants have their breasts bathed with lime-water, which is made by burning the shells of fresh-water mussels and dissolving them in water. Every married woman carries several shells in her basket, which are commonly used as spoons (Dawson 1881, p. 57).

Treating colic

Dr Baylie noted that when children had colic, their mothers placed their children upon their stomachs during the colicky episode, and kept trampling upon them until they benumbed the pains (Baylie 1843, p. 87).

Snakebite

Snakebite was treated by applying ligatures, incising the wound and sucking the poison from the bite (Haydon 1846, p. 73; Stone 1911, p. 44). In December 1870 correspondents to the *Hamilton Spectator* discussed the merits of various methods of treating snakebite that were employed by Aboriginal peoples in the district:

> I once saw quite a different operation performed by the blacks belonging to the Wannon tribe, and which proved remarkably successful. In the case alluded to, a lubra was bitten on the ankle, a ligature was immediately tied tightly round the leg above the wound, and the part incised. A circular piece of leather, with a hole in the centre, was placed over the wound, the leather being kept constantly wet with warm water, when one of the coolies applied his mouth, and kept up the sucking process for more than an hour, during which time he extracted the poison along with a large quantity of blood (Hardwick 1870, p. 3).

W. Carr-Boyd, of Lilydale, detailed a variation of this treatment that involved the use of ligatures and blood-letting but did not involve sucking venom from the wound:

> Perhaps the public would like to know how the blacks in more than one part of Australia that I have existed in cure the most venomous snake-bites, provided that the bite is on a portion of the body (for example, the leg or arm) that a string can be tightly secured a few inches between the bite and the trunk of the body within a few minutes after the snake has bitten. I have seen three blacks at different times bitten by the most venomous snakes - two men and one woman. The men were bitten on the leg below the knee, and both recovered, but the woman was bitten on the breast, and died in less than half an hour. The way the blacks were cured was as follows: They tied a piece of human hair string (which they made up as strong and fine as the best whipcord) about three or four inches above the bite, as tight as they could make it: then with one of their sharp stone knives cut a circle round the bite, about the eighth of an inch deep, and about a quarter of an inch from the two fang punctures; then, just below the bite; they slit the largest vein I could see, which, in a very few minutes, seemed to let every drop of blood out of the leg below the string; then they got one of my water-bags, and kept a stream of water running on the leg, just above the bite, the while they kept rubbing the leg down between their hands as hard as they possibly could. This they kept up for about 20 minutes, when every drop of blood seemed to be got out of the wounded part of the leg. They then got a piece of sharp thin wood and twitched up the slit veins, dabbed some dirt on the wounded parts, undid the string, and half an hour after [they] … were bitten they were as well as ever I saw them (Carr-Boyd 1892, p. 9).

Some colonists such as Samuel Clutterbuck (c.1855) noted that whether or not an Aboriginal person was willing to suck the venom from the snake bite victim, depended upon how much they liked the victim: 'if black snakes bite you you're sure to die, unless you cut the wound out with a sharp knife, and then get a native to suck the venom out, which they will sometimes do if they are particularly friendly with you. You can't persuade them to do it for money unless they like you very much, it makes their lips swell to an enormous size.'

Ophthalmia

Gordon Leckie recounted a story about Simon Wonga (1824–1874), a leading Aboriginal elder in the Melbourne district, who suffering from ophthalmia was operated on in the Melbourne hospital. However, when the operation failed to restore his sight, immediately, he turned to the services of an Aboriginal doctor, and the following day, his sight returned. According to Leckie, Wonga thereafter credited the Aboriginal doctor with his recovery.

> An example of the blacks' faith in their doctors was provided by Wonga a member of the Yarra Yarra tribe. Wonga attended the Melbourne Hospital where he was operated upon for ophthalmia. The operation was successful but

he was discharged still blind for as is usual with this operation it takes several days for the sight to return. Wonga did not understand this and he consulted Tall Boy a noted Wer-raap who pretended to extract three straws from the back of Wonga's head. On the next day Wonga could see and on the day after his sight was almost normal. To his death he treasured the three straws and the belief that Tall Boy, and not the [hospital] surgeon, had regained his sight for him (Leckie 1932, p. 7).

Refreshing drinks and tonics

Aboriginal people consumed tonics to maintain their general health and body function. To invigorate themselves, Aboriginal people in southern South Australia described taking 'blood medicine' that could be made from the stems of Sow-thistles (such as *Sonchus hydrophilus*) or Pale Flax-lily (*Dianella longifolia*) roots. Tonics are taken to maintain good health rather than as a remedy for an existing ailment, so they are strictly speaking not medicines (Clarke 2008b, pp. 7–8).

Robinson (27 May 1840 in Clark 2014a, p. 186) noted that Aboriginal people made a sweet beverage by tying small boughs (teerang) of Ironbark (yeerip) (*Eucalyptus sideroxylon*) together (which was called *time*), soaking them in water and then squeezing out the mellifluous portion.

Dawson noted that a tonic was given for indigestion. It was obtained in this manner: 'the small roots of the narrow-leafed gum tree, or the bark of the acacia, are infused in hot water, and the liquor drunk as a tonic' (Dawson 1881, p. 57). When a child had overeaten, Dawson wrote, 'its mother gathers yellow leeches from underneath dry logs, and bruises them up along with the roasted liver of kangaroo, and sow thistles, and compels it to eat the mess, which is called kallup kallup. It acts as a strong emetic. Adults, when ill from overfeeding, are sometimes induced to take this dose, in ignorance of its composition; and it affects them strongly, but beneficially' (Dawson 1881, p. 57).

European uses of bush medicine

On the frontier, explorers and settlers gained knowledge of the bush through observing Aboriginal hunter-gatherers. Europeans incorporated into their own 'bush medicine' a few remedies derived from an extensive Aboriginal pharmacopeia. Differences between European and Aboriginal notions of health, as well as colonial perceptions of 'primitive' Aboriginal culture, prevented a larger-scale transfer of Indigenous healing knowledge to the settlers. Since British settlement there has been a blending of Indigenous and Western European health traditions within the Aboriginal community (Clarke 2008b, p. 3).

The first British colonists came to Australia from an industrialised nation, bringing with them knowledge of newly developed Western medicines and their own folk remedies (Hagger 1979; Pearn and O'Carrigan 1983). Hard-pressed settlers in remote regions were forced to rely upon the local bush for many essential things, such as 'bush medicines', as supplies from Europe were scant and infrequent. Western European understandings of the causes of poor health shaped the settlers' immediate response to the Australian environment. The British persisted in using flannel underwear in spite of hot summers, because it was believed to prevent colds and rheumatism (Clarke 2008b, p. 14).

British colonists of the late 18th and early 19th centuries did not consider that hunter-gatherer societies possessed highly developed systems for managing their health and wellbeing. While some Europeans were in awe of the capacity of Aboriginal people to naturally recover from serious physical injuries, particularly spear wounds, most relegated the practices of recognised Aboriginal healers to the arena of trickery, magic and sorcery (Clarke 2008b, p. 14).

Writer, artist and architect, George Henry Haydon is an example of an early colonist with an interest in Aboriginal medicine:

> The black wattle, (*acacia affinis*) yields a gum which would perhaps be found to answer for many purposes as well as gum arabic; it is of an amber colour, and when fresh from the tree has a pleasant taste. It is used as a medicine by bushmen when attacked by dysentery, and is seldom found to fail in producing salutary effects. This gum forms an article of great consumption with the natives, but in consequence of the reckless manner in which the trees have been stripped of their bark for the purposes of tanning and for exportation, many parts of the country are now quite destitute of this, to them useful and important article of diet, forming as it does, or did, their principal vegetable food (Haydon 1846, pp. 82–3).

James Dannock (c.1855) wrote of entrusting himself to Aboriginal medicine: 'I took bad with the dysentery and the black lubras got me wattle gum and when I did not get better they said 2 days that fellow go bung [dead] so I thought I had better clear out and got the blacks to put me over the river in a canoe'. George Rowe, a miner at Bendigo and Castlemaine, was another who was conversant with Aboriginal [Djadjawurrung] medicines for dysentery, noting that the 'wattle gums give off a great deal of gum which is very similar to gum Arabic and good as medicine for the dysentery so is the decoction from the bark' (Rowe 1854).

On mining fields and towns, mishaps or accidents had fatal consequences for miners or bush dwellers because medical attention was unavailable or difficult to obtain. Subsequently Aboriginal advice about the healing properties of herbs and plants and some of their methods were adopted. In the Dimboola district when Horatio Ellerman accidentally wounded his companion, the Wergaia stemmed the flow of blood by packing the wound with a poultice composed of the fresh contents of a sheep's stomach, probably an adaptation of a traditional medical procedure. Ellerman rode 150 miles [241 km] to fetch the nearest doctor at Carngham near Ballarat, returning within three days to find the victim still alive (Longmire 1985, pp. 28–9).

Colonist and author Dame Mary Gilmore stated in her reminiscences of growing up in rural New South Wales that:

> ... the white forgets the uncounted ways in which he [was] ... unintelligent (and still would be unintelligent) but for what the blacks taught. As parallels to

the treatment of snake-bite by sucking, take the use of eucalyptus, the application of weak wattle tan-water for burns and blisters, of clean mud as poultices, of native gums in dysentery, the eucalyptus beds and steam pits for colds and rheumatism, and ask was it a black or a white intelligence that was first to find and apply these (Gilmore 1934, p. 232).

Gilmore stated that a whole industry in making medicines owed its existence to Aboriginal practices:

> It is true there is a eucalyptus extract industry now; but the knowledge that led to that was originally derived from the natives, who used eucalyptus leaves in steaming, and for wounds. For rheumatism steam pits were made, heated by fires, raked out, lined with leaves and then possum-rugs laid over the top. Another use of the leaves was as a strapping for wounds that needed closing in order to heal. These uses came to the pioneers from the blacks (Gilmore 1934, p. 226).

The historical records are incomplete on the origin of most colonial healing practices. Few settlers would acknowledge that Aboriginal knowledge was the source of their remedies. Although Europeans came to use some of the same healing remedies as Indigenous people, it is not known how many of them came about through the direct acquisition of Aboriginal knowledge. Colonists experimented with plants that appeared similar to European species they were familiar with, although the resemblance was often superficial (Campbell 1932, pp. 77–80; Clarke 2008a, chapter 2; Cribb and Cribb 1982, chapter 3).

The native lilac or False Sarsaparilla (*Hardenbergia violacea*) was used as a tea-based medicine due to the similarity of its leaves to true sarsaparilla, although its effectiveness is doubtful. Botanist Frederick M. Bailey remarked that the 'roots of this beautiful purple flowered twiner are used by 'bushmen' as a substitute for the true sarsaparilla, which is obtained from a widely different plant. I cannot vouch for any medicinal properties' (Bailey 1880, p. 8).

Europeans targeted wild aromatic plants to make herbal teas. Fragrant oils and drinks were made out of the Native Pennyroyal (*Mentha satureioides*) and other Australian mint species (Bailey 1880, p. 19; Cribb and Cribb 1982, pp. 78–79; Lassak and McCarthy 1983, pp. 15, 19, 77, 88, 175; Low 1989, p. 177).

The discovery of new sources of medicine potentially had economic benefits for the fledgling colony. Surgeon Dennis Considen in New South Wales wrote to English botanist Joseph Banks in 1788 and stated that:

> … this country produces five or six species of wild myrtle [species of *Melaleuca*, *Kunzea* and *Leptospermum*], some of which I have sent you dried. An infusion of the leaves of one sort is a mild and safe astringent for the dysentery (D. Considen cited Campbell 1932, p. 80).

A bitter taste was a strong indication for a plant having a potential use as a tonic or for the treatment of indigestion. In early New South Wales, a type of 'acid berry', identified as the Sour Currant-bush (*Leptomeria acida*), was used to treat sick convicts who had arrived from England suffering scurvy (Cobcroft 1983, pp. 18, 27, 29–30, 32; Campbell 1932, p. 83; Powell 1990, p. 94).

The influence of Indigenous plant use practices is apparent in the origin of a few colonial remedies. In the early 19th century, Aboriginal women living with European sealers in the southern region from Kangaroo Island to Bass Strait used their plant-based remedies and medical charms to treat their families (Clarke 1996, p. 62; Leigh 1839, pp. 160–1). In these communities, 'teas' made from 'bush ti-tree' (*Leptospermum* and *Melaleuca* species) were used to medicinally 'purify' the blood (Anonymous 1844a, p. 6). In south-eastern Australia the settlers used the Aboriginal remedy of applying a poultice made from the creeper known as Old Man's Beard (*Clematis microphylla*) onto aching joints of the legs and arms (Clarke 1987, p. 6; Lassak and McCarthy 1983, p. 56).

Conclusion

In Aboriginal Australia the traditions for maintaining health were based upon cultural beliefs of the causes of sickness and the powers of their remedies. The causes of serious illness were generally attributed to supernatural sources. While Western European scholars see the charms, medicines, tonics, narcotics and stimulants Aboriginal people use as falling into distinct categories of use, for the users the distinctions are either blurred or non-existent. Many of their everyday remedies were derived from plants. The primary role of Indigenous healers is to mediate religious forces in order to restore health in seriously ill patients. As a 'doctor', their techniques did not always utilise substances that Europeans considered to be medicines. Any discussion of Aboriginal healing needs to distinguish between healing lore, which as Indigenous Biocultural Knowledge was known to most people in Aboriginal social groups, and healers' lore, which concerns specialist Aboriginal healers combating sorcery and religious taboo breaking, and the like.

On the frontier of British colonisation the level of hardship that Europeans experienced would have determined when they utilised bush medicines, some of them based upon Aboriginal remedies. Colonists used bush medicines that were easily collected and required little processing. They were more likely to use a wild plant as a medicine if it had some physical, albeit superficial, resemblance to European species. Perceived similarities were not restricted to physical appearance alone, but also involved characteristics like taste and smell. Aromatic wild plants were particularly attractive as medicines to settlers, who were already primed by their Northern Hemisphere experiences.

Aboriginal healing practices were placed under severe pressure through European colonisation. While the Aboriginal pharmacopeia coped with the treatment of general ailments experienced by a dispersed population of hunter-gatherers, it was less successful when dealing with the introduction of European exotic diseases. The availability of new technologies since European colonisation has brought about a significant modification of Aboriginal healing practices. A level of coexistence has developed between Indigenous and Western health systems in remote Aboriginal communities. Here, the success in maintaining health will continue to be influenced by Indigenous views on the causes of ill health and the ways to cure it. Since the late 19th century, Aboriginal medicines have been used as a guide to finding plants useful for pharmacologists (Clarke 2008b, p. 20).

Chapter 13
Trade

Ian D. Clark

Introduction

Aboriginal trade involved physical objects such as stone and ochre and cultural objects such as song, dance, stories and ceremonies (Clarke 2003a). Pioneering research into stone axe production and exchange in south-eastern Australia (McBryde 1984a, 1984b) and the investigation of intercultural trade for possum skin cloaks (Cahir 2005), provide a platform for this chapter's exploration of intra-cultural and intercultural trade and exchange. Critical analysis of primary 19th century ethnographic works (such as Dawson 1881, Smyth 1878, Mathews 1894 and Howitt 1904) and primary source documents, such as the journals and papers of George Augustus Robinson (Clark 1990b, 2001a, 2001b, 2014a, 2014b) and William Thomas (Stephens 2014), that concern south-eastern Australia, establish clearly the existence of trade and exchange between individuals and groups across the study area as well as trade with European colonisers.

Smyth, for example, when commenting on Aboriginal trade, noted the absence of a medium of exchange such as tokens and money, and commented on Aboriginal sagacity when transacting with Europeans:

> They traffic only by exchanging one article for another. They barter with their neighbours; and it would seem that as regards the articles in which they deal, barter is as satisfactory to them as sale would be. They are astute in dealing with the whites, and it may be supposed they exercise reasonable forethought and care when bargaining with their neighbours. The natives of some parts, however appear to be reckless traders (Smyth 1878, vol. 1, p. 180).

Smyth's comment suggests that the relationships established through trade are more than simply an exchange of goods. The goods 'traded' were often distinct to particular 'countries', that is, they were related to resources or economic needs, and were inter-tribal in character. The exchange often took place at large gatherings and inter-group meetings, especially at initiation ceremonies (McBryde 1984a, p. 132). Classic descriptions include Howitt's (1904, pp. 718–19) accounts of the Yuin barter in south-eastern New South Wales, and of the Wotjobaluk meetings in north-western Victoria, and Dawson's (1881, p. 78) account of meetings at Mt Noorat in the Western District of Victoria. In the Murray Basin, trade linked Aboriginal groups who were dispersed along the River, including others living far beyond the boundaries of the region. The Darling and Adelaide regions were also linked to the Lower Murray through trade. The social contact necessary to enable bartering helped promote good relations between neighbours (Clarke 2009c, p. 156).

There are detailed accounts of the items of exchange:

> The Yuin ceremonies of initiation were attended by people from a district included by Shoalhaven River, Braidwood, the southern part of Manero, and Twofold Bay. At the termination of these ceremonies, when the novices had gone away into the bush for their time of probation, and when the people were about to separate, there was held a kind of market, at which those articles which they had brought with them for exchange were bartered. It was held at some clear place near the camp, and a man would say, 'I have brought such and such things,' and some other man would bargain for them. A complete set of articles is one Ngulia or belt of opossum-fur string, four Burrain or men's kilts, one Gumbrum or bone nose-peg, and a complete set of corrobboree ornaments. It was the rule that a complete set went together. Weapons might be given in exchange, and a complete set of these is 'two hands,' that is ten, fighting boomerangs (Warangun) being the straight-going ones; the same number of grass-tree spears (Jjumma); one of each kind of shield, namely the Bejnata, used for stopping spears, and the Millidu, used for club fighting; one club (Gujerung or Bundi), and one spear-thrower (Meara). The women also engaged in this trade, exchanging opossum rugs, baskets, bags, digging-sticks (Tualt), etc. Not only were these things bartered, but presents were made to friends and to the Headmen by the other men. The women also gave things to the wives of the Headmen. A Headman who was held in great esteem might have as many things given to him as he could well carry away (Howitt 1904, pp. 718–19).

Trade was a means by which resources could be redistributed, driven by the recognition that some raw materials were superior to others.

> Bartering was also practised by the Wiimbaio, with the blacks from higher up the Darling River, who occasionally brought down wood of the mulga tree for spear points, slabs of stone, and hard and heavy pestles of granite for pounding and grinding seeds and tough tubers. These they exchanged for nets, twine, or fish-hooks (Howitt 1904, p. 718).

Dawson's account of trade in south-western Victoria identified the source locations of valued resources such as ochre, spears, gum and kangaroo skins:

> At the periodical great meetings trading is carried on by the exchange of articles peculiar to distant parts of the country. A favourite place of meeting for the purpose of barter is a hill called Noorat, near Terang. In that locality the forest kangaroos are plentiful, and the skins of the young ones found there are considered superior to all others for making rugs. The aborigines from the

Geelong district bring the best stones for making axes, and a kind of wattle gum celebrated for its adhesiveness. This Geelong gum is so useful in fixing the handles of stone axes and the splinters of flint in spears, and for cementing the joints of bark buckets, that it is carried in large lumps all over the Western District. Greenstone for axes is obtained also from a quarry on Spring Creek, near Goodwood; and sandstone for grinding them is got from the salt creek near Lake Boloke. Obsidian or volcanic glass, for scraping and polishing weapons, is found near Dunkeld. The Wimmera country supplies the maleen saplings, found in the mallee scrub, for making spears. The Cape Otway forest supplies the wood for the bundit spears, and the grass-tree stalk for forming the butt piece of the light spear, and for producing fire: also a red clay, found on the sea coast, which is used as a paint, being first burned and then mixed with water, and laid on with a brush formed of the cone of the banksia while in flower by cutting off its long stamens and pistils. Marine shells from the mouth of the Hopkins River, and freshwater mussel shells, are also articles of exchange (Dawson 1881, p. 78).

Beveridge, when discussing the Watiwati (Watty Watty) people of the Swan Hill district and their neighbours, referred to their exchange as a 'primitive kind of commerce':

> The articles of commerce which the aborigines exchange with each other consist of reeds for spears, red ochre and chalk for painting purposes, stone for tomahawks, fibre for nets and cord, opossum cloaks, wood for weapons etc. Some of these articles are peddled backwards and forwards, even as far as the Tropic of Capricorn, each tribe gladly exchanging its local productions — of which it has abundance — for such commodities as are the produce of other tribal territories, and in which their own locality is altogether lacking. At first, this doubtless seems a very primitive kind of commerce, but really, it was ample for all the simple requirements of these savage tribes, the advent of the civilised race gave to them tastes and wants which, until then, were altogether foreign to their nature (Beveridge 1883, p. 20).

Isaac Batey, was told in 1862 by an Aboriginal stockman from the Lachlan that stone for hatchet heads came 'from a hill down in the Melbourne country', which is presumably a reference to the Mt William quarry. Batey learnt 'our aboriginals went inland carrying stone implements with other things. What they brought was exchanged with remote tribes for what they produced, hence it appears that … [they] … have what I shall call the commercial instinct' (Batey no date, pp. 129–30).

McBryde (1984a, p. 133) has noted that the 'ecology and resources of different regions are seen as major factors determining the nature of exchange and the goods exchanged'. For McBryde, 'Exchange itself encompasses a diversity of activities and processes', and is used in an entirely open sense as 'reciprocal traffic, exchange, movement of materials or goods through peaceful human agency' (Renfrew 1979, p. 24 in McBryde 1984a, p. 134). 'Goods' are taken to include both tangible material items and the intangible items such as services, knowledge and rituals. In some cases, it also included fire. It was noted that in winter if the Narangga people of the Yorke Peninsula lost their fire, and their wood was too wet to ignite the Grass Tree fire-sticks, they sent someone to the Murray River people to ask for gifts of fire (Clarke 2009c). The Kaurna of the Adelaide Plains traded skin cloaks, quartz flakes, and red ochre for fire from the Murray River people. Red ochre is a mineral resource that had significant cultural value – it was used as a pigment for artistic and ceremonial purposes, and was used in rock painting, to decorate artefacts such as shields, boomerangs and clubs, and also for medicinal purposes (Jones 2007). Certain ochre deposits were more highly valued than others and were important in Aboriginal mythology (Peterson and Lampert 1985). In South Australia, for example, red ochre from a mine in the Flinders Ranges was believed to be the blood of a sacred emu, and was highly sought after in Central Australia (Jones 2007).

It was noted that at times the exchange involved items of great value:

> Not only were articles which the people made themselves bartered, but also things which had some special value, and had perhaps been brought from some distant place. Such an instance I heard of at one of these meetings many years ago. An ancient shield had been brought originally from the upper waters of the Murrumbidgee River, and was greatly valued because, as my informant said, it had 'won many fights.' Yet it was exchanged, and carried away on its farther travels (Howitt 1904, pp. 719–20).

Some trade practices were associated with elaborate rituals, and the social contact necessary to enable bartering probably helped engender positive relations between neighbours. One example was the *ngia-ngiampe* custom of the Murray Basin that involved the exchange of ornaments made from human umbilical cord strings (Berndt *et al*. 1993; Clarke 2009c). Aboriginal people used the existence of the *ngia-ngiampe* trading relationship to seek temporary harbour after a dispute.

Items for exchange

The ethnographic sources stress that material goods were essential items of inter-group exchange, especially those products that were distinct to particular group territories. These goods included raw materials such as ochre and finished products, such as stone, stone hatchets, cord or net bags, bundles of reeds or spears (Curr 1883, p. 273; Krefft 1862–65, pp. 366–7; Mathews 1894, p. 303; 1897, pp. 150–1; Stanbridge 1861, p. 296; Taplin in Woods 1879, p. 40). McBryde (1984a, p. 134) observes that exchange acts as an adaptive mechanism, compensating for ecological inequalities between Aboriginal 'countries'. The items exchanged included consumables (food), tangible goods and intangible items such as songs, dances, names or services, as well as the living (women in marriage exchange). Marriage is regarded as an exchange; its arrangement also involves forms of transfer of goods and gift-giving between the families concerned. William Thomas specifically mentions the 'purchase of a wife' and lists the bride price (Bride 1898, p. 67).

European items such as clothing, handkerchiefs, tomahawks, twine, sugar, flour and rice join Aboriginal items such as hatchet heads, spears, opossum rugs, nets and mats, and feature in the records of exchange. As an example, Katherine Kirkland, a squatter at 'Trawalla' Station in western Victoria in the early 1840s, discussed inter-cultural trade: 'We sometimes got some skins of the opossum and flying-squirrel, or tuan, from the natives. It was a good excuse for them to come to the station. I paid them with a piece of dress, and they were very fond of getting a red pocket-handkerchief to tie round their necks' (Kirkland 1845, p. 20).

McBryde (1984a) notes that comments by Howitt (1904) and Stanbridge (1861, p. 295), as well as Krefft's (1862–65, p. 386) descriptions of the feverish craft activity preceding a meeting at Yelta, on the Murray River, in July 1857, suggest that groups ensured they had a sufficient stock of items for trade. However, she qualifies that this drive need not be directed to economic gain from the transaction. Similarly records of the 'ownership' of certain resources and the strict control of their use need not imply protection of their commercial potential as such. One of the best examples of resource control and access restriction is provided by the conventions pertaining to the Mt William stone quarry (see Brumm 2010). Resources important for subsistence were also often regarded as 'owned' by specific local groups, e.g. fishing rights along Salt Creek at Lake Bolac in the Western District of Victoria. Members of other groups could only catch fish or eels with the permission of the land-owning group.

Brumm (2010) has examined 'The falling sky' belief held by the Kulin people of central Victoria that the sky was a dome propped up by wooden poles in the mountains in north-east Victoria (Howitt 1884, 1904). A man or supernatural being maintained the props, and if they should fail, the Kulin people believed the sky would collapse and the clouds burst and the Earth would be destroyed. These props were maintained by stone tomahawks, and the ethnographic record confirms that some time before 1835 the Wathawurrung learned of their imminent collapse and they frantically gathered a supply of stone axes and sent them as 'payment' to the supernatural being in the north-east (Langhorne 1837; Morgan 1852, pp. 57–8). Brumm's (2010, p. 189) analysis is that the axes were likely to have been sourced from the Mt William quarry.

The context of exchange

Goods were acquired by exchange, by gift, or as the result of special expeditions to the source, where negotiation with its owners may be involved. Large inter-group gatherings were a prominent feature of the life of societies in south-east Australia, and meetings were carefully organised, and scheduled to coincide with times of abundance in local food resources; participants were invited by messengers who travelled from group to group. These messengers often became men of importance.

> At certain seasons of the year, usually in the spring or summer when food is most abundant, several tribes meet together in each other's territory for the purpose of festivity or war, or to barter and exchange such food, clothing, implements, weapons, or other commodities as they respectively possess; or to assist in the initiatory ceremonies (Eyre 1845, vol. 2, pp. 218–19).

Eyre witnessed such meetings between the Moorunde groups and people from the Lake Bonney district. Seasonal abundance of eels at Lake Bolac and seasonal 'treats', such as the exudate from eucalypts in the Wimmera/Mallee, called La'ap, and the Bogong Moth (*Agrotis infusa*) in the southern uplands, also provided foci for such meetings (Clark 1990b, 2001b, 2014b). Robinson commented on the social and political importance of these gatherings.

> These masses are a collection of representative tribes and the eeling and whaling seasons are wisely taken advantage of by them for holding their great social and political meetings … But, of all the places I saw, Lake Boloke [Lake Bolac] was the most interesting. This spot, celebrated for its eels and its central situation, appears to have been fixed upon by general consent for the great annual meeting of the tribes of the interior, and it is for the same reason the sections in and near the coast assemble at Tare-er [south of Tower Hill] during the whaling season (Robinson April 1841 in Clark 2001b, p. 18).

Robinson discussed the Festival of the La'ap in an 1845 report:

> The Mallee (*Eucalyptus dumosa*) a species of gum scrub in the vicinity of Lake Hindmarsh extends longitudinally with the Murray over a large tract of country; the wood is famed among the Aborigines for making spears, being flexible tenacious and hard. The Mallee is celebrated also for the sacharine [sic] quality of its leaves, of which the natives during the season of February and March make a luscious drink called la'ap; the Festival of the La'ap is an occasion of great interest to the natives, when they assemble in large numbers to settle their disputes and adjust other matters connected with their tribes. The laap is found in a crystalline or candied form on the underside of the leaf from which it exudes, and is a white colour at least what I saw had this appearance (Robinson 1845 report in Clark 2014b, p. 484).

McBryde's (1984a) analysis is that meetings held primarily for exchange were rare because exchange was ordinarily subsidiary to social and ceremonial activities. Those that did take place commonly fell into one of two kinds: (1) those concerned with fulfilling marriage arrangements, the occasions on which girls betrothed in infancy joined their husbands. Curr (1883, pp. 128–34) described such a meeting held near Tongala in the 1840s. (2) Those for exchange of food, such as the meetings Buckley attended (Morgan 1852[1980] pp. 56–9).

> Buckley mentions that a messenger (bihar) came from the Wudthaurung to propose that the latter should exchange eels for roots. The place of meeting was about fourteen days' distance to travel. The exchange was made by two men of each party delivering the eels and roots on long sheets of bark, carrying them on their heads from one party to the other until the bargain was concluded. When the tribes separated an agreement was made to meet again for barter (Howitt 1904, p. 718).

Meetings were often concluded with an exchange of goods as a means of confirming and cementing 'friendships'. Howitt gave the following account of the conclusion of the *Bunan* initiation ceremony held by coastal groups in southern New South Wales:

At the end of the Bunan, when the boys have gone off by themselves and before the different lots of peoples return to their own localities, a kind of market was held, in some clear space near the camp where the people laid out the things they had brought with them. A man would say 'I have such and such things', another would bargain for them. A complete set of one ngulia (belt of possum fur cord), four Burrian (kilt), one gumbun and one complete set of Kul butgun, that is corroboree things. It was the rule that complete sets of things went together, weapons might be given in exchange. A complete set of weapons was ten fighting boomerangs (straight going ones) warangun, ten grass tree and jagged wood pointed spears (gumma), one of each shield — for stopping spears (bembaia) — for club fighting (Millidu), one Knib dub (gug-ju-rung) or (Bundi) and one spear thrower (wommera). The women also engaged in this trade exchanging possum rugs, bags — yamsticks (tuali) (Howitt no date a, *Papers*, Box 8, Folder 5, Paper 12).

Similar exchanges marked the end of meetings of the Wotjobaluk, the Jupagalk and members of the Kulin group of tribes. On such occasions, as well as exchanging goods of various kinds, participants made gifts to visitors.

When the people who attended the great tribal meetings of the Wotjobaluk were about to depart to their homes there was an assembly at the Jun, or men's council-place, where they exchanged the articles which they had brought for the purpose. These articles were such as the following: sets of spears, respectively called Guiyum-ba-jarram, or jag spear, and reed spears; opossum skin rugs, called Jirak-willi (opossum skin); men's kilts, called Burring-jun, made of the skin of the kangaroo-rat (Goiyi), or padi-melon (Jalla-gur); armlets worn round the upper arm, called Murrumdat-yuk; wooden bowls called Mitchigan; in fact, all the implements, utensils, arms, and ornaments used by these people. It was to such a meeting that the Jajaurung man, Tenamet-javolich [Sergeant Major], before mentioned, carried stone from the quarry at Charlotte Plains to be made into axe-heads. The same was the case with the meetings of the Jupagalk tribe (Howitt 1904, pp. 717–18).

According to William Barak, the Woiwurrung *ngurungaeta*, this exchange of gifts was designed 'to make friends' (Howitt 1904, pp. 717–18).

The same practice of barter occurred when there were great tribal meetings in the Kulin nation. Such a meeting was held about the year 1840 at the Merri Creek near Melbourne; at which people came from the Lower Goulburn River, from its upper waters, and even from as far as the Buffalo River. Not only was barter carried on, but, as Berak said, people made presents to others from distant parts 'to make friends' (Howitt 1904, p. 718).

Presumably 'making friends' explains many of Robinson's experiences in the Western District in 1841. On leaving a camp shared with local Aborigines for some time the two groups exchanged gifts. 'This is their custom', commented Robinson, noting that: 'much feeling was exhibited on this occasion' (Robinson 12 May 1841 in Clark 2014a, p. 331). Curr (1883, p. 125) noted that among the Bangerang, 'Neither kissing, shaking hands, nor any other salutation of the kind was in use amongst the tribe, though frequently men of different tribes made exchanges of arms or articles of dress in token of good-will; and women, in like manner exchanged ornaments.'

Mathews attended initiation ceremonies in various parts of New South Wales in the 1890s, and in his writing makes general reference to this context for exchange (Mathews 1894, p. 303). According to McBryde (1984a, p. 140), one aspect he recorded which eluded other observers was the formal exchange of gifts between the initiates' families and sponsors. He describes these exchanges for both Wiradjuri and South Australian groups, for whom tooth evulsion was an important part of initiation rituals:

> After a time, which may be only a few months duration, or it may be a much longer period, the headman who took the tooth away sends messengers to the tribe to which the owner of the tooth belongs, stating that it will be brought back at such a time. On receipt of this message, preparations are made to meet the strange people at the time appointed. On these occasions it is the custom for each tribe to make presents to each other, which takes the form of exchange or barter. Supposing, for example, that there is plenty of suitable stone for making hatchets and whetstones in the country belonging to one tribe, they will exchange these commodities with the men of another tribe, in whose country there may be suitable wood for making spears and other weapons. People who have coloured clays will exchange them for skins of animals not plentiful in their own country. Others will have string made of the bark of certain trees, richly coloured feathers of rare birds, reeds for making light spears, and so on, which they exchange for other articles. It may be that some of the men and women exchange exactly similar articles with the people of another tribe merely as mementos of their meeting. At these gatherings, the hosts arrange themselves in a line, with their presents and other commodities lying on the ground near them. The visitors advance and form into a row opposite the hosts, and display their presents in a similar manner (Mathews 1897, pp. 150–1).

McBryde (1984a, p. 141) has noted that one of the earliest records of an inter-group meeting at which goods were exchanged comes from the journal of William Bradley, who settled near Tinpot Creek in the Costerfield district, near Heathcote, Victoria, and established good relations with local Aboriginal groups. On 12 November 1838 he recorded in his diary:

> Today two groups of blacks met at the encampment by the deep hole in the creek, they at first appeared to act as strangers but it soon became apparent that there were indeed individuals in either group who were familiar with each other. Both groups it would appear were happy to see each other. The stranger

group as I will call them had travelled from the south and they had carried with them a number [of] … stone hatchets … Some of these hatchets were polished while others were still quite rough and I imagine still require further work. The group of blacks who are camped on the creek were eager to obtain these hatchets and in return for one polished axe they gave two of their opossum skin covers. For a hatchet still in a roughened state they gave in return a number of their light bamboo spears. This bartering as I shall call it went on for some time, but only amongst the menfolk, the women and children stayed behind the men with only some of the older women having anything to say (Bradley in McBryde 1984a, p. 141).

Bradley later learned that the hatchets came from 'near Lancefield' (i.e. the Mt William quarry). The 'Bamboo spears' are presumably reed spears from the wetlands of the Murray and its tributaries.

Role of messengers in exchange
Messengers were often the medium through which groups learned of the desire of other groups for items for exchange. Peter Beveridge discussed the role and function of messengers or postmen (*ngallow wattow*) among the Murray River tribes near Swan Hill, who were able to travel from tribe to tribe with impunity, relaying news between groups:

These men also negotiate all barter and trading required by their respective tribes. At the first blush it would almost seem that the aboriginies [sic] could not have very much in the shape of goods to dispose of, but that would be an erroneous conclusion to arrive at; as the districts inhabited by the different tribes produce each their own particular class of commodities, and those alone; therefore, as the aboriginal requirements over the whole of the colony are very similar, the only manner in which many of their wants can be supplied is by means of barter. For example, the tribes inhabiting the mountainous regions have an abundance of stone suitable for making axes of; and the tribes which roam over the vast depression forming the Murray river valley have many miles of reed beds, from whence reeds are procured for making spears, and not having any stone for axes on their own territory, they procure it by exchanging reeds with the inhabitants of the stone country, where reeds stout enough for spears do not grow. The same mercantile relations obtain in the matter of gums, resins, ochres, etc. Thus a single glance will suffice to show that the *Ngallow Wattows* have always abundance of work cut out for them, going from tribe to tribe on trading expeditions (Beveridge 1889, pp. 165–6).

One of Howitt's Djadjawurrung sources was Sergeant Major, a Djadjawurrung messenger.

My Jajaurung informant, whose father married a woman of the Jupagalk tribe, and whose maternal grandmother was of the Leitchi-leitchi tribe, was one of those men who were sent on important messages. He was free of three tribes, first on account of his father, who lived with the Jupagalk, also on account of his mother and grandmother, as well as of his own tribe the Jajaurung. Thus he became a messenger and intermediary between these tribes. His mother's sister was married to a Jajaurung man who lived at Charlotte Plains [now known as Moolort Plains, near Maryborough], and her son took care of a stone quarry at that place, from which the tribes to the north-west were supplied with axe-heads. My informant on one occasion brought down word that the Wotjobaluk, at Lake Hindmarsh, were in want of stone for axes, and this material was obtained from the quarry, and carried up to the next great meeting of the tribes on the Wimmera for barter by my informant's father (Howitt 1904, p. 689).

Gifts

Goods also changed hands as direct gifts. In the literature of the late 19th century prestation, or the offering of gifts, receives scant attention compared with that given the direct and immediate reciprocity of barter, though recorded observations of the 1840s give it prominence (McBryde 1984a). Such gifts could represent delayed reciprocity or exchange of goods for services, such as healing, and knowledge. The European recorder may not have observed, understood or perceived the transaction in its entirety. Gifts often involved prestige or valued goods such as possum rugs, hatchets, hatchet stone, or the rare 'Bundit' spears from the Otway forests. This type of spear was considered: 'so valuable that it is never used in fighting or hunting, but only as an ornament. It is given as a present in token of friendship or exchanged for fancy maleen spears from the interior' (Dawson 1881, p. 88).

Gift-giving formalised particular occasions such as the arrival or departure of a visiting group. Robinson's journals and reports provides numerous examples of this form of gift giving:

> I explained the object of my visit and deputed them to deliver my message to their tribe and my wish to meet the Boolucburrers [people of Lake Buloke] at Kilambete. I gave them some trifles I had in my valise and eels and native weapons were given to me. The strangers and my natives exchanged raiment, a custom prevalent among the Aboriginal tribes (Robinson 1841 report in Clark 2001b, p. 15).

> I desired … alias Tom Brown to give the big man his blanket and in return the one gave me a large jagged spear. The woman gave us three eels and Brown exchanged his shirt for a kangaroo skin mantle. We then parted with the best possible feeling and with an understanding that we were to meet the following day (Robinson 2 April 1841 in Clark 2014a, p. 285).

> Preparing to depart. The shirts I had given to my natives they gave to those strangers and received spears or other trifles in return, as I remember. This is their custom and it matters little to me as all the natives were under my protection and guardianship. Some gave their blankets and a few gave away trousers and other articles they had obtained. Much feeling was exhibited on this occasion (Robinson 12 May 1841 in Clark 2014a, p. 331).

Robinson understood that the gifting of spears and other weapons at meetings was a token of friendship (27 May 1841 in Clark 2014a, p. 350). On 23 June 1841 Robinson presented a group of Aboriginal people he met with 'a new rug and a knife and also medal to each and five pannicans, beads and rug, and the Governor's letter, flour, tea and meat. They were astonished at my manifoldness ... After the formality of our meeting was over Mingbim sent one of his wives, he had three, for some native weapons and gave two each to my natives, a leangle shield and [blank] or wommera. This is a sign of good understanding' (Clark 2014a, pp. 374–5).

The practice was not confined to the groups Robinson met in the Western District. Batman had distributed presents before the formal meeting he held in June 1835 with leaders of the Kulin clans living near the site of modern Melbourne to discuss purchasing a tract of their country. On the day following the formal discussions he recorded in his journal:

> Just before leaving, the two principal chiefs came and brought their two cloaks, or royal mantles, and laid them at my feet, wishing me to accept the same. On my consenting to take them, they placed them round my neck and over my shoulders and seemed quite pleased to see me walk about with them on (Bonwick 1973, p. 99; Billot 1979, p. 118).

Thomas described 'a ceremony of mutual friendship' between groups from the Devil's River (Daungwurrung) and from the Yarra (Woiwurrung) at his camp just outside Melbourne in 1847:

> ... some Devil's River blacks arrive ... Sunday March 21, 1847. Early proceed to the encampt [encampment] which I find much excited. One Bunhile (Davy's son) had lost his lubra who had eloped. Am much affected by a ceremony of mutual friendship, the Devil's River gave the Yarras 1 koogra [possum skin cloak] and a number of spears. The Yarra blacks give 5 splendid koogra and bundles of pocket handkerchiefs, sashes, &c &c European articles all new given by Old Borenuptune, brother to the late chief [Billibellary] (Thomas 20–21 March 1847 in Stephens 2014, vol. 2, p. 234).

Gifts were also given to 'create friendship' between groups and individuals (Curr 1883, p. 268). Those who acted to settle disputes were presented with gifts, though they may also have been payment for services rendered. Stahle, a Moravian missionary with extensive experience in Victoria, said of a senior man in south-western Victoria: 'The men of the tribe were under an obligation to provide him with food, and to make all kinds of presents to him, such as kangaroo and opossum rugs, stone tomahawks, spears, flint knives etc' (Fison and Howitt 1880, p. 277).

Similarly those who exercised power through magic and ritual received gifts from 'those who benefitted and those who feared' them (Howitt 1887, p. 47). Dawson (1881, p. 75) refers to a leader of the Spring Creek clan, who arbitrated in disputes: 'In reward for his services he returned home laden with presents of opossum rugs, weapons and ornaments.' Gifts may repay obligations, as well as promote social or political influence. Buckley perceived this aspect of Aboriginal social relations among the Wathawurrung. On finding a cask washed up on shore he removed the iron hoops which he 'knew were valuable to the natives': 'Having broken up the iron hoops into pieces, I some days after divided them amongst those who were most kind to me, and by these presents added greatly to the influence I had already acquired over them' (Morgan 1852[1980] pp. 113–14).

Gifts were usually made to those who sponsored young men in initiation. Howitt (no date b, p. 8) records that Barak presented Billibellary with an opossum rug following his (Barak's) investiture 'with the insignia of manhood'. Mathews noted exchanges between sponsors and the families of initiates after the Wiradjuri *Burbung* and after the *Kannetch* ceremony in South Australia (Mathews 1897, pp. 150–1).

Marriage arrangements also involved the giving of gifts between the families and individuals involved (Morgan 1852[1980], pp. 95, 165–71). The arrangement itself was a form of exchange, whose agreement was binding. There is a record by Thomas of a 'bride price': two large possum rugs, two or three dozen possums, and other trifles (Thomas in Bride 1898, p. 67). In most instances, however, the reciprocity was a continuing obligation (Thomas in Bride 1898, pp. 67–8; Beveridge 1883, pp. 23–5; Smith 1880, p. 3; Smyth 1878, vol. 1, pp. 76–7, 79; Curr 1883, p. 248). Sometimes the return was in kind, such as another marriage arrangement. The importance of gifts and exchanges of this kind seems universal throughout the south-east, with the possible exception of the Kurnai (McBryde 1984a).

Robinson cautioned the authorities in Port Phillip that: Aboriginal people wearing the clothing of Europeans who had been killed by Aboriginal people was not necessarily evidence that the wearers had committed the crime, as items of European clothing were regularly traded. In the case of the murder of a shepherd near Mt Napier in 1847, and the subsequent killing of two Aboriginal men, Robinson noted: 'It was said as a matter of concern that they were bad characters and had the murdered man's clothes on at the time, now that is expected and admitting they had on things is no proof they were the murderers as the natives exchange articles with each other or send them to their friends of other tribes' (Robinson 2 September 1847 in Clark 2014a, p. 762).

Exchange and gift-giving involved obligation, and the non-fulfilment of obligations often bore serious consequences, ranging from loss of status and humiliation to death. One of the incidents which triggered the Rufus Creek massacre in south-western New South Wales seems to have been the failure of European shepherds to keep a promise to supply certain goods in return for the services of Aboriginal women. Clearly, they had been expected to conform to the rules governing such exchanges (Massola 1969). Robinson suspected that failure to honour sexual exchange obligations was the cause of the deaths of some Europeans on the frontier. The following example from Robinson's journal highlights the consequence of the failure of a hutkeeper in western Victoria to fulfil his obligations:

> Bayley had a man waddied by natives. He directed an enquiry to be made by natives into the cause … the particulars were that the hut keeper agreed with the man to leave his gin three days at the end of which time he would give the gin a blanket and the man a blanket. He kept her two days, beat her and turned her away. The native sought his revenge and watched at the water hole in the

bushes. The man went down three days, each time with his musket for he expected the black would be revenged. The man knew he should be shot if he attempted anything. The fourth day the man went for water without his gun and the native waddied him severely. This was justice (Robinson 26 June 1841 in Clark 2014a, p. 380).

Special expeditions

Acquisition of goods was usually arranged at inter-tribal or inter-group meetings through exchange and prestation. Goods were also obtained by making special expeditions to their source, often in alien, if allied, territories (McBryde 1984a). These were usually to acquire items important for subsistence or technology, especially raw materials not available locally such as fine-grained hard greenstone for hatchet heads like that quarried from the outcrops on Mt William, or reed stems for spear shafts (Bride 1898, p. 109; Morrison c.1971, p. 22; Smyth 1878, vol. 2, p. 154: Krefft 1862–65, p. 366). In many instances these expeditions involved exchange, for goods must be bartered to acquire the commodity from its owners. For example, at the Mt William stone quarry strangers were not permitted to work the outcrops, but must negotiate a price for their needs with those who had the right to do so. Reed spears from the Goulburn reed beds and opossum rugs were recorded exchanges for Mt William stone (Howitt 1904, pp. 311–12, 340–1, 689–90, 718; Smyth 1878, vol. 1, p. 161; vol. 2, pp. 298–9). The exchange rate was one possum rug for three pieces of Mt William stone.

Intercultural trade with Europeans

From the commencement of British colonisation in the Port Phillip region in 1835 there were attempts to open up formal trading networks with the Indigenous people (Cahir 2005). Squatters from Van Diemen's Land (Tasmania), who had occupied land around Indented Head (on the Bellarine Peninsula), sought to employ local people in making baskets (Billot 1979, p. 62). Members of the Port Phillip Association hoped that if a significant bilateral business relationship could be established, then inter-racial relations would be more conciliatory at Port Phillip than they had been in Van Diemen's Land.

Exchange between Europeans and Aboriginal people also involved foodstuffs. Charles Hall, for example, when at the Grampians in the early 1840s, noted that the kidney fat of a sheep would purchase a dozen brace of quail (presumably Stubble Quail *Coturnix pectoralis*). 'One old villainous - looking black of my acquaintance used to catch large bundles of quail, which he would barter freely for suet. The kidney fat of a sheep would purchase a dozen brace' (Hall 1853 in Bride 1969, pp. 270ff). Yabbies (*Cherax destructor*) were bartered for tea and sugar and other supplies. Hall noted:

> The lubras fished up crawfish from the shallow muddy water-holes with their toes and yam-sticks, and exchanged them for the dainties of civilized life. A large tin-dishful might be obtained in barter by a small expenditure of tea and sugar, and when treated with a certain degree of gastronomic science formed a not unwelcome change of diet from mutton chops or salt beef, which in those

days was the almost unvaried food of the 'cormorant squattocracy' (Hall in Bride 1969, pp. 270ff).

Robinson noted that the Burung baluk (Djadjawurrung) people were at Charles Ebden's Carlsruhe Station exchanging tuan [Brush-tailed phascogale] skins for flour (Robinson 21 February 1840 in Clark 2014a, p. 111), and there was a regular trade for tuan skins at Joseph Docker's Bontherambo Station, north of present-day Wangaratta (see Robinson 11 February 1841 in Clark 2014a, p. 258).

The demand by pastoralists and their servants for possum skins and especially possum skin rugs was widespread. The rugs were so popular that a small number of white entrepreneurs established a lucrative trade in the skins in the 1840s. Robinson observed how several pastoralists and merchants in Melbourne and the 'settled districts' had become wealthy from the considerable intercultural trade in artefacts, possum skins and lyrebird tails.

> The natives state that white men in the country and residents in Melbourne supply them principally for the purpose of shooting bullen-bullen ie native pheasants and squirrels, the skin of the latter and lyre tails of the former being given as an equivalent for the use of guns and ammunition. These skins and tails are I understand of valuable consideration and have by some been turned to very profitable account (Cannon 1983, p. 726).

Possibly the first record of the sale of possums in Victoria comes from the Melbourne area. John Pascoe Fawkner, the first European to occupy land in the vicinity, recorded in February 1836:

> Mr Henry Batman sent blacks out to get parrots, got [William] Buckley to abuse William Watkins for buying squirrel skins for me and I find him forbidding the natives to sell us any skins or birds. He wants them all himself (Billot 1979, p. 41).

George Langhorne, an early missionary in Port Phillip, also noted that a substantial monetary trade was well established by 1838 for lyrebird feathers:

> A considerable number of the blacks obtain food and clothing for themselves by shooting the Menura pheasant or Bullun-Bullun for the sake of the tails, which they sell to the whites (Cannon 1982, p. 229).

By 1844, the trade in possum skins was so lucrative that large volumes of skins were being offered for sale to white settlers (Robinson 4 November 1844 in Clark 2014a, p. 638). Assistant Protector William Thomas (1844) reported: 'Loddon blacks arrive, bringing in some thousands of skins for sale'. Similarly, Dr James Horsburgh (1849), the Medical Officer In Charge of the Goulburn Protectorate Station (1846–53) noted that the 'natives also obtain both money and food for opossum skins, ~2 [sterling] pounds of the former article being laid out in my presence to hawkers'.

George Gilbert, a bullock driver at the Goulburn River Protectorate station, witnessed 'large quantities of skins' being procured from the Aboriginal people in exchange for flour

(Gilbert 1842). EB Addis, Commissioner of Crown Lands, considered that the 'Barrabool tribe' [Wathawurrung] was attracted to the Geelong township chiefly because of the ease with which they were able to trade possum skins and lyrebird tails for European foodstuffs:

> … the town of Geelong attracts them greatly, partly from curiosity and otherwise by the facility they procure offal meat from the sheep and cattle killed at the butcheries, and rice, flour or sugar, in exchange for birds and skins … (Addis 1842).

Thomas recorded in May 1840 that the people in his Western Port District were eager to work on his district's station and to exchange 'Aboriginal manufactures' for food rations. Thomas (1840) recorded that the Boonwurrung people traded 17 possum and kangaroo skins and seven baskets for flour and other unspecified goods.

At times, Aboriginal people sought to acquire European weapons through exchange. The following example documents an unsuccessful attempt at a station near Geelong:

> [A.M.] some natives called on their way to Geelong they did not stop long as Mr. R gave them nothing fearing he should be favoured with more … They wished to exchange some spears for an old musket but were told they wanted to kill white men. To do this they replied 'no no borak white men but kangaroos. Borak kill white man plenty bang black fellow' (Adeney 1842–43, 18 February 1843, p. 369).

James Nealer, a shepherd at Buninyong reported that a group of Wathawurrung offered to exchange some possum skins for a sheep: 'On the 25th of July [1838] four natives came to [me and] my flock of sheep and wanted one, offering some squirrel skins' (Cannon 1983, p. 307). GF Read, a pastoralist also at Buninyong, was the subject of a visit in April 1838: 'A great many natives came here today and exchanged skins for flour' (Griffiths 1988, p. 4). The rate of exchange for items varied, but Charles Griffith, a pastoralist near Bacchus Marsh, reported on one occasion that two unidentified Wathawurrung men received flour and sugar for a kangaroo tail and skin, and on another occasion noted that he was busy tanning 'several oppposum skins and touan skins, the latter is the flying squirrel … which we have got from the natives in exchange for flour' (Griffith c.1845–c.1859, pp. 287–8). In March 1845, John Cotton, a squatter on the Goulburn River, west of present-day Yea, had adopted a similar trade and exchange rate:

> … they know very well that we never give anything unless we receive something in return, so they generally come provided with opossum skins, for which we give them rice, sugar, bread or anything of the sort that we can spare; they generally prefer rice and tobacco (cited in Clark 2013, p. 95).

The trade of Aboriginal manufactures such as baskets, kangaroo and possum skins, and buckets was commonplace in the 1840s. Officials such as Chief Protector Robinson frequently obtained such items for his own collection or sold them on to George Lilley, a produce merchant in Melbourne who also had a stall at the Melbourne market (see Clark and Heydon 2004, p. 65).

Travellers throughout south-eastern Australia also traded with Aboriginal people, as the following examples illustrate. Friedrich Gerstaecker paddled down the Murray River from Albury in March 1851, in a self-made canoe, until it broke, after which he walked *c.* 1100 km to Adelaide. On one occasion during his rambles he came upon an Aboriginal camp and he recorded the following transaction of some tobacco for a boomerang.

> Coming one day to a camp of blacks, who wanted smoke, I refused at first; one of them threw, while I was with them, his boomerang at a little walloby that jumped up close to us, while we were opposite each other. The weapon missed the animal but coming back to us with arrow speed, I had just time to dodge away from under it. Even then it grazed my arm, leaving, as I afterward found, a deep blue spot for a remembrance. While the blacks laughed and jumped at the fun of the thing, I turned to the little black fellow as ugly a black as I had ever seen and offered him tobacco for his boomerang. We soon made a trade, and I followed my road unmolested (Gerstaecker 1853, p. 454).

At Aldinga, in South Australia, sisters Florence Davenport Hill and Rosamond Davenport Hill, who travelled to Australia in 1872, attempted to purchase a basket from an Aboriginal girl, but refused to pay the asking price:

> Here we stopped to negotiate the purchase of a basket from a singularly handsome half-caste girl combining the lustrous eyes and dazzling teeth of her aboriginal descent, with the straight nose and thinner lips of her white parent who was walking along the road with an old native woman, equally remarkable for her ugliness. We objected to the price asked as too high, to which the elder woman replied, by saying 'tucker was very dear,' tucker being food, in Australian slang (Hill and Hill 1875, p. 126).

Upon reaching Goolwa, the sisters attempted to purchase 'native mats and baskets' but were told that they would have to leave their order with an agent as the local Aboriginal people were busy picking native currants:

> At Goolwa we inquired again for native mats and baskets, but found we could only leave our order for them with an obliging agent, who told us that all the aborigines of the neighbourhood were busy picking native currants (Hill and Hill 1875, p. 134).

During the gold rushes, Aboriginal people were quick to grasp the economic opportunities presented to them by the miners flooding to the gold diggings. This was especially the case with miners seeking to trade with Aboriginal people to obtain manufactures such as possum skin rugs (see Castlemaine Association of Pioneers and Old Residents 1972, p. 220; Pepper and De Araugo cited in Clark and Cahir 2003, p. 133). Miners and others writing in this period have left glowing reports about the benefits of obtaining possum skin rugs from Aboriginal people. George Henry Wathen, an English geologist visiting the Victorian goldfields, extolled the virtues of possessing a possum rug and acknowledged that the settlers considered them to be the most highly valued intercultural trade item in Victoria (Wathen 1855, p. 131). With thousands of miners congregating in towns across Victoria, the volume of trade in possum skins increased exponentially (see Cahir 2005).

Reports from several Aboriginal Station managers across Victoria in the latter part of the 19th century described the lucrative trade being conducted by Aboriginal station residents. In December 1870 the manager of the Condah Mission in Western Victoria wrote: 'Some of them earn a little money by making and selling baskets and mats, and occasionally an opossum rug' (Shaw in Clark 1990a, p. 49). According to John Green, the manager of Coranderrk, the Aboriginal station in Healesville, the high quality of the rugs, and the speed with which the Aboriginal people could manufacture them, combined with their ready sale, enabled some Indigenous Victorians to achieve a degree of economic independence (Wiencke 1984, p. 52). Andrew Porteous, a local guardian in the Carngham district (1860–77), reported that the demand by Europeans for Indigenous manufactured goods continued to be economically sustainable. In 1866, he reported that a good living could be made from selling possum rugs, and that the Aboriginal women enjoyed making baskets and nets which they sold to Europeans (see Clark 2008).

Newspaper reports both at home and abroad also reveal a strong interest in Indigenous manufactured goods, particularly in possum skin rugs. An 1865 report in the *London Times* noted a request by a Welshman for a possum rug to be made by Aboriginal people of the Ballarat district so he could show his country people what 'the pioneers of the goldfields frequently used to sleep in' (Anonymous 1865c, p. 5). A Wathawurrung couple obliged and were paid 30 shillings. In 1861, the *Ballarat Star* carried a satirical article supposedly attributed to 'A Blackfellow' which beseeched the Colonial Government to provide market protection for the Indigenous trade in possum skin rugs (Curoc 1861, pp. 7–8).

Conclusion

Trade as practised by Aboriginal peoples involved physical objects such as stone and ochre and cultural objects such as song, dance, stories and ceremonies. Trade was a means by which resources that were widely dispersed could be redistributed, driven by the recognition that some raw materials were superior to others. Aboriginal Indigenous Biocultural Knowledge was used when taking advantage of a seasonably abundant source of material, and turning it into an advantage through the formal establishment of reciprocal relationships between Aboriginal groups. For this reason, trade practices were often associated with elaborate rituals. The social contact necessary to enable bartering would have helped engender positive relations between neighbours.

Chapter 14
Space

Philip A. Clarke

In Aboriginal Australia, early European observers recorded that the inhabitants held cosmological beliefs dominated by concepts of the heavens as a Skyworld. This was considered to be the upper part of a total landscape that possessed a topography that included Earth and the Underworld. Historical accounts from Aboriginal people described the Skyworld as being inhabited by human and spirit ancestors, who lived in a land similar to Earth, and it was believed that these entities still had influence over all living things.

Introduction

In the Australian ethnographic literature, 'the Skyworld' refers to a concept of the heavens as a country with distinct zones and regions, within which human soul spirits and Creation ancestors (see Chapter 1) existed alongside all the plants and animals that also lived on Earth. Overviews of astronomical traditions describe the main elements of the Skyworld; its physical structure, the influence of its occupants over earthly events, and the existence of genealogical relationships between the celestial bodies (Haynes 1992, 2009; Clarke 1997, 2008b, 2009a, 2014b; Johnson 1998, 2005, Morieson 2003; Norris 2007, 2016; Norris and Hamacher 2009, 2014; Hamacher 2011; Norris 2016). While on the level of mythological detail there is considerable variation across Aboriginal Australia, the Skyworld was nonetheless experienced through broadly shared systems of cosmology and aesthetics.

For academic cultural geographers, the cultural landscape is a concept that encompasses both the physical and cultural aspects of the human construction and perception of space (Sauer 1925 [1963]; Clarke 1994; Baker 1999). Along with what modern Europeans see as the terrestrial landscape are parts of a larger country that Aboriginal people saw and experienced as the Skyworld (Heavens) and Underworld (Land to the West). Indigenous interpretations of the sky and the weather it generates must be understood in terms of the cosmological traditions of powerful spiritual Ancestors that explain how the world was originally put together.

As discussed in the ethnographic examples below, Aboriginal people believed that they lived in the centre of a finite world, which comprised a curved but relatively level Earth, with a Skyworld above and an Underworld below (Clarke 1997, 2003a, 2009a, 2014b, 2014c, 2015a). The sky was often perceived as a solid vault that sat on top of what was described as 'a flat limited surface' (Howitt 1904, p. 433). In the Wimmera district of western Victoria, anthropologist Alfred Howitt recorded that:

A Wotjobaluk legend runs that at first the sky rested on the earth and prevented the sun from moving, until the magpie [Australian Magpie, *Cracticus tibicen*] (*goruk*) propped it up with a long stick, so that the sun could move, and since then 'she' moves round the earth (Howitt 1904, p. 427).

The Magpie's use of a stick is significant, as it appears to be based upon the observed bird behaviour that ornithologists have noted with the Australian Magpie – it is one of the relatively few avian species that use tools (Lefebvre *et al.* 1998; Iwaniuk *et al.* 2009). The theme of the sky being held up by wooden poles or living trees is present in other south-eastern Australian accounts (Clarke 2003a; Massola 1968; Morgan 1852 [1980]). Aboriginal logic held that the sky region began at the height of a tall tree or at most a hill (Clarke 1997, 2009a, 2014c, 2015a).

Within Aboriginal tradition, the Skyworld and the Underworld were intimately connected. It was recorded that: 'In the south-eastern portion of Australia, the old men used to say that the forms or spirits of the dead went to the westward towards the setting sun' (Smyth 1878, vol. 2, p. 224). In the Gippsland area of eastern Victoria, however, it was recorded that Aboriginal people believed that the soul spirit travelled directly to the east. One account noted:

> It is very singular, says Mr. Bulmer [Lake Tyers missionary], that the natives, who have no form of religion, should have a distinct idea of a spiritual existence. They think that the soul, as soon as it leaves the body, goes off to the east, where there is a land abounding in sow-thistles (*Thallah* [*Sonchus oleraceus*]), [upon] which the departed eat and live. The spirits are sometimes prevented from reaching the happy land by the moon, which devours them if they encounter it, and indeed feeds on stray mortals and spirits of departed men and women. When the moon is red, they see proof that it has eaten plentifully of its favorite food (Smyth 1878, vol. 1, p. 274).

The belief that soul spirits of the recently deceased followed the path of the Sun may help explain why Aboriginal people in south-eastern Australia would bury the bodies of their dead with the head facing towards the rising or setting Sun (Smyth 1878, vol. 2). At Lake Boga in central northern Victoria, the interred body generally had the head facing the direction of where the Sun sets on the horizon (Stone 1911). A variation among the Dhurga (Thoorga) people of south-east New South Wales was that the head of the deceased was faced towards the rising Sun in order to be warmed (Mathews 1904a).

The total landscape was viewed as containing places where dangerous beings dwelt. For instance, at Lake Tyrrell in north-western Victoria it was recorded that Aboriginal people believed that the Nurrumbunguttias were a class of ancestral beings who existed in the Creation and 'still possess spiritual influences upon the earth; whether of darkness, of the storm, or of craters, all the evil spirits are of them' (Stanbridge 1857, p. 138). Similarly, in the area west of Ballarat in Victoria colonist William Stanbridge recorded that:

> The tribes in the neighbourhood of Fiery Creek have two other malignant spirits
> - Neulam-kurrk, the evil spirit which inhabits craters and caves, and in the form

of an old woman steals children and eats them; and Colbumatuan-kurrk, the spirit of storms, which kills or injures people by throwing limbs of trees upon them or in their way for them to fall against at night. They have also a good spirit named Barn-bungil, who is the reliever of pain (Stanbridge 1861, p. 299).

The dangerous spirit entities were believed to be normally restricted to particular parts of the landscape (see Chapters 2 and 3), so it was considered a serious matter if for any reason the separateness of the Earth, Skyworld and Underworld regions was compromised. Aboriginal people looked for omens in the heavens (Clarke 1997, 2009a; Johnson 1998; Hamacher and Norris 2010a, 2010b, 2011). For instance, in northern Victoria, the Aurora Australis was seen as a sign that the blood from a massacre was rising into the Skyworld (Howitt 1904). At the time of British colonisation, it was widely believed in south-eastern Australia that the eastern prop that held up the vault of the Skyworld had decayed and started to collapse, as shown by British expansion in the region (Howitt 1904; Massola 1968; Clarke 2009a; Brumm 2010). They reasoned that unless formal gifts of possum skins and stone hatchet heads were sent to the old man who ritually looked after it, then people on the terrestrial plain on Earth would be crushed by the falling vault of the heavens. It was also believed that spirits of the deceased, once set loose from the Skyworld, would have caused much havoc for the living.

The Skyworld

The records that are relevant to the study of Aboriginal astronomy are imperfect, particularly for south-eastern Australia where British colonisation began early and was most intense. For reconstructing the Indigenous star lore here, we rely heavily upon the anecdotal accounts from settlers and colonial officials. The names of the individuals who were the Indigenous sources of the ethnographic data were generally unrecorded, although it is known that in the case of a colonist in south-western Victoria much of his information came from 'Weerat Kuyuut, the sagacious old chief of the Moporr tribe [Moperer gundidj clan of the Gundidjmara], and from his very intelligent daughter Yarrum Parpur Tarneen, and her husband, Wombeet Tuulawarn' (Dawson 1881, p. 99).

Translation difficulties for scholars were significant, not just in terms of recording vocabularies from many diverse languages, but with poor understandings of the ways in which space was perceived in Aboriginal Australia, which was fundamentally different from that of modern Western Europeans (Morphy 1995, 1999; Strehlow 1970; Sutton 1988, 1998b). While the early record is imperfect, it is clear that Indigenous people once possessed a large body of astronomical knowledge that concerned the connectedness of everything in their universe. This information was highly valued, as it was recorded that much of the published Aboriginal star lore from western Victoria had come from 'the Booroung [Borung] Tribe, who claim and inhabit the Mallee country in the neighbourhood of Lake Tyrell, who pride themselves upon knowing more of Astronomy than any other tribe' (Stanbridge 1857, p. 137).

According to Aboriginal tradition, the Skyworld was similarly organised to that of Earth, to the extent that the celestial bodies as ancestors were subject to the same laws as people and animals. Landscape objects and topographic features on Earth were recognised as continuing or being replicated in the Skyworld above. In the Aboriginal worldview, they were reflections of each other, with points of connection on the horizon (Clarke 2014c, 2015a). Terrestrial and celestial spaces were not just seen as having similar geographies, as in some cases particular topographic features were seen as being continuous from one to another. The prominence of rivers as topographical features in south-eastern Australia is reflected in them being identified as major features in the Skyworld. For instance, the Ngaiawang people living along the upper reaches of the Murray River in South Australia recognised the Milky Way as being part of the same river, with the stars on either side seen as men hunting in the Mallee (Tindale cited Clarke 1997). Similarly, in south-western Victoria the Milky Way was known as *barnk*, meaning 'big river' (Dawson 1881, p. 99).

For Aboriginal groups in some parts of south-eastern Australia, the forests dominated the Skyworld. For instance, it was recorded that the Wurunjerri (Woiworung) people on the northern side of Melbourne in Victoria believed that they 'had a sky country, which they called *Tharangalk-bek*, the gum-tree country' (Howitt 1904, p. 433). Here, the land was named after *tharangalk* trees, which are Manna Gums (*Eucalyptus viminalis*) that were the source of the highly prized edible sugar lerp, which is a creamy white flaky extract on the leaves that is produced by Psyllid leaf insects (Tindale 1966; Clarke 2007a). According to Howitt, other groups were said to have related beliefs concerning the Skyworld, and in eastern Victoria:

> By the Kurnai [Ganai] this place is called *Blinte-da-nurk*, or (freely translated) 'bright sky of the cloud,' also *Bring a-nurt*, or 'bone of the cloud.' The Ngarigo called the sky *Kulumbi*, and said that on the other side of it there is another country with trees and rivers (Howitt 1904, p. 433).

The traditional Aboriginal belief was that people were able to cross the gulf between terrestrial space and the sky more easily during the Creation period (Clarke 2014b, 2015a).

Skyworld as a home of spirits

The cultural construction of space in classical Aboriginal Australia was influenced by beliefs in the existence of spirits, who were those of Creation ancestors and deceased humans. Aboriginal people believed that the human soul fragments after death, leaving a ghost component that may linger on Earth, while the ancestral soul spirit merges with that of the Creation ancestor, often via a sacred place, and then ascends to an existence in the Skyworld (Clarke 1999b, 2007b). Much of what can be seen in the heavens was perceived to be either spirits or what the ancestors had produced. For instance, in north-western Victoria it was said that Warring the Milky Way galaxy was 'The smoke of the fires of the Nurrumbunguttias ['old spirits']' (Stanbridge 1861, p. 302).

The connections between human spirits and the Skyworld were broad, and involved both the beginning and end of an individual's life. It was widely believed that the Creation ancestors produced the spirit children (Tonkinson 1978). The Skyworld was also a destination for the soul spirits of deceased people, and it was an Aboriginal belief that they followed the same or similar routes that their Creation ancestors had once taken through the Aboriginal landscape. In south-western Victoria, colonist James Dawson recorded that:

On the sea coast, opposite Deen Maar [Julia Percy Island] … there is a haunted cave called Tarn wirring, 'road of the spirits,' which, the natives say, forms a passage between the mainland and the island. When anyone dies in the neighbourhood, the body is wrapped in grass and buried; and if, afterwards, grass is found at the mouth of the cave, it is proof that a good spirit, called Puit puit chepetch, has removed the body and everything belonging to it through the cave to the island, and has conveyed its spirit to the clouds; and if a meteor is seen about the same time, it is believed to be fire taken up with it. Should fresh grass be found near the cave, when no recent burial has taken place, it indicates that some one has been murdered, and no person will venture near it till the grass decays or is removed (Dawson 1881, pp. 51–2).

In another account from anthropologist Robert Mathews, it was explained that here:

On the shore of the mainland there are some rocks like stepping-stones stretching out into the sea towards the island. These rocks, some old black fellows told me, were purposely placed in that position, in the far away past, by the spirits of clever men, to enable their fellow-countrymen to reach their *maiogo* [spirit home]. The spirit of a deceased person walks on the land to the shore and springs out to a rock, then to another and another rock, and lastly makes a final bound over the intervening sea and alights on Lady Julia Percy Island, the native name of which is Denmar (Mathews 1904b, p. 70).

Many of the male ancestors who left Earth for the Skyworld at the conclusion of the Creation continued to hunt. In south-western Victoria, Dawson noted:

By some tribes the coal sack is supposed to be a waterhole; and celestial aborigines [ancestors and deceased spirits], represented by the large stars around it, are said to have come from the south end of the milky way, and to have chased the smaller stars into it, where they are now engaged in spearing them (Dawson 1881, p. 99).

In western Victoria there was a tradition that the Pointers were hunters that killed the Emu, seen as the Coalsack, with their spears stuck in a tree, represented by the Southern Cross (Stanbridge 1857; MacPherson 1881; Massola 1968). The Skyworld country was humanised in manner similar to that on Earth, as the ancestors had partitioned it into distinct areas by assigning celestial 'countries' to specific cultural groups (Maegraith 1932).

The identities of celestial bodies varied across Aboriginal Australia, although the corpus of cosmological traditions was united by some common elements (Clarke 2009a). Mathews remarked that:

Legends are more numerous concerning stars situated in the neighbourhood of the moon's path through the heavens, and in this way a zodiac may be said to exist. The stars near the ecliptic and the zenith change their positions in the sky more rapidly than those toward the poles, and therefore more readily arrest attention. Besides constellations at these high altitudes can be seen easily when the people are camped in thickly wooded country, whereas stars near the horizon would not be visible then (Mathews 1904a, p. 278).

Aboriginal people classified the cosmic bodies using the same system that they used for themselves. It was reported that:

> Conspicuous stars and star clusters all the way along the zodiacal belt, have well-known names and traditions. Moreover, each star figuring in the myths belongs to a phratry [moiety], section [marriage class], clan [descent group] or other subdivision, precisely the same as the people of the tribe among whom the tale is current. The names of the subdivisions, as well as the names of the stars, change amongst the people inhabiting different parts of the country. Sometimes the legends and nomenclature of the stars will be substantially the same among several adjoining tribes over an extensive region. In other instances, not only are the names of the stars different, but the traditions and the stars connected with them are altogether divergent (Mathews 1904a, p. 279).

The identity of cosmic bodies recognised in the Skyworld was not random. They generally saw a male ancestor and his wives, along with their family, weapons and dogs, as being in the same part of the night sky (Mathews 1904a). Ancestors as constellations that were seen to be chasing other ancestors across the night sky were understood to be repeating their previous actions on Earth, as Aboriginal traditions were cyclical.

Rock engraving, possibly of a sun ancestor, in a limestone shelter on the banks of the Murray River. PA Clarke, Devon Downs, South Australia, 1994.

In general, the Sun and Moon ancestors were seen to be of primary importance, due to their dominating influence over the night and day skies respectively (Clarke 2009a). In Victoria, the supreme male ancestor, Baiame, was credited with having created a fire that was seen as the Sun (Beveridge 1883; Haynes 1992). Here, the warmth of the day linked to the strength of his fire and how much fuel was left to burn. The Sun was seen as having a strong influence over people and the plants and animals of their country. Often the Moon was male and subordinate to the Sun, which was female (Griffin 1923; Clarke 1997). In western Victoria, for example, Nyaui the Sun clan had both the Moon and the planet Venus among its set of subordinate totems, which were mainly plants and animals (Mathews 1904a). The gender of the Moon is apparent in the Gippsland, where:

> In olden days the blacks used to sing to the full moon. They saw it in the figure of a man gathering and carrying a thistle-like plant called durluk. If the children were crying, the parents used to tell them about this man in the moon to pacify them (Mathew 1925, p. 66).

The identity of *durluk* is possibly the Sow-thistle (*Sonchus oleraceus*), as this species is an edible green (Gott and Conran 1991; Clarke 2015c).

Connections between Earth and the Skyworld

In Aboriginal Australia, it was considered that travel between all the regions of the Earth, Skyworld and Underworld was possible during the Creation period. Then, as the ancestors reached the Skyworld via the tops of large trees or the summits of high hills, something happened to leave them there (Clarke 2014c, 2015a). The means for ascent to the heavens was variously described as by climbing, using strings or by being taken up by whirlwinds, as shown with some of the examples discussed below. The stated reasons for the ascension by the ancestors after their creative feats on Earth were generally either to escape attack or calamity, or because they were tricked into it and became stranded. After the Creation period, access to the Skyworld by people was restricted to people with special powers, such as 'doctors' (Clarke 2009a, 2014b, 2015a; Elkin 1977; Smith 1880).

From the Wimmera district of central western Victoria there is a recorded myth that described the connection between terrestrial space and the sky (Bulmer 1855–1908 [1999]). In this myth, people living on Earth during the Creation period were easily able to gain access to the Skyworld in order to collect the abundant lerp manna from the trees. Other ethnographers from western Victoria recorded related accounts of a tree connection between Earth and the Skyworld. In the Kara-Kara district in western central Victoria it was an Aboriginal tradition there was once a 'regular highway between the earth and the upper regions' that was formed by a large pine tree (Southern Cypress Pine, *Callitris preissii*) growing on Earth and its branches crossed into the sky like a ladder (Mathews 1904a, pp. 281–2). Similarly, Howitt stated that in the Wimmera district of the same region:

> The Wotjobaluk had a legend of a pine-tree, which extended up through the sky (*Wurra-wurra*) to the place beyond which is the abode of Mamen-gorak

['father', 'ours']. The people of that time ascended by this tree to gather manna, which implies that trees grew there like the Eucalypt [Manna Gum, *Eucalyptus viminalis*], which in the Wotjobaluk country shed the so-called manna (Howitt 1904, p. 433).

In the Gippsland area there is a record of how the Muk-Kurnai, who were the 'ancestors of the Gippsland blacks', entered the Skyworld towards the end of the Creation period. It was said that:

> The tribe being engaged in fishing, Bulun, Baukan, and their son Buluntut, coming to the camp, took away all the fire, and began to ascend to the sky by way of Wilson's Promontory. Reaching the summit, Buluntut threw up a string, made of kangaroo [Eastern Grey Kangaroo, *Macropus giganteus*] sinew, which stuck fast to the sky. He then tested its strength by pulling on it, when it broke. He then tried a cord of the sinews of the Black Wallaby [Swamp Wallaby, *Wallabia bicolor*], which likewise broke. Finally, he threw up a cord of the sinews of the Red Wallaby [Red-necked Wallaby, *Macropus rufogriseus*], which held fast. Then saying to Bulun [and Baukan] …, 'Hold on round my neck,' he began to ascend the cord, Baukan carrying the fire (Howitt 1889, p. 54).

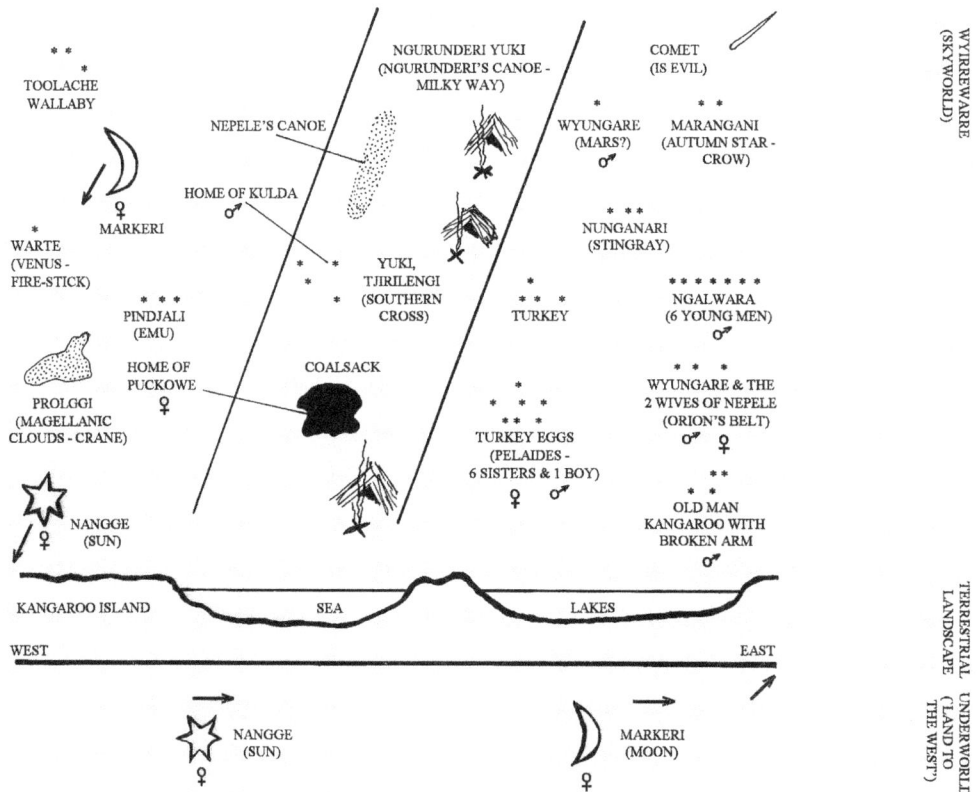

Skyworld, Earth and the Underworld, from the perspective of Lower Murray people in South Australia, reconstructed from historical records (PA Clarke 1994, Fig. 3.5).

The attempt of stealing fire was noticed by Wagulan the Crow (Australian Raven, *Corvus coronoides*]), who notified Borwon the Brown Hawk (probably Brown Falcon, *Falco berigora*) who swooped on Baukan and caused him to drop it. Bembrin the Robin (probably Flame Robin, *Petroica phoenicea*) found the fire on the ground and blew it into a flame and then smeared some of it onto his chest. In this manner, the Ganai (Kurnai) people on Earth kept the ability to make fire and robins gained their red colouring.

Underworld

Across Australia, it was a tradition that after the Sun ancestral being had travelled through the Skyworld to the western horizon it then returned by distant routes to the east (Clarke 2009a). Some Aboriginal peoples considered the path back as being just below the southern or northern edges of their country, while others thought it was directly underneath via an underground passage through the middle of the Underworld (Clarke 2003a, 2009a). The Wotjobaluk people of the Wimmera in western Victoria believed:

> … that the sun was a woman who, when she went to dig for yams, left her little
> son in the west. Wandering round the edge of the earth, she came back over the
> other side (Howitt 1904, p. 428).

The neighbouring Kulin people, however, believed that when the Sun disappeared in the west, she entered a cave at a place called Ngamat, which had the appearance of a hole in the ground like that left behind by a large tree burned in a bush fire (Howitt 1904).

The existence of the Underworld was also important in traditions concerning the Moon ancestor, as this is the area he travels through when not seen by people on Earth (Clarke 2003a, 2009a). The Moon was believed to have once lived in caves, as became apparent when late 19th century scholars were investigating the origin of Australian cave paintings that local Aboriginal people had claimed that they had not made (Smyth 1878, vol. 2). At a large cavern on the face of a granite cliff overlooking the Avon River in Gippsland, east of Melbourne, an Aboriginal man was asked about the red ochred hand and arm stencils and a large circular figure that had been drawn well out of reach on the roof, and 'He stated that his people believed that the Moon once dwelt in that cave, but becoming tired of the confinement, he ran up the roof of the cave, leaving his imprint at the top as he jumped up into the sky, where he has been wandering about ever since' (Smyth 1878, vol. 2, p. 222).

Caves feature in many Aboriginal traditions that concern the Sun and other ancestral beings (Clark 2007a; Howitt 1904). With an Aboriginal model of 'curved' space, it is possible that all of the lower geographic spaces outside the periphery of Earth were treated as the same. The Underworld was sometimes recorded as the 'Land to the West', where the spirits of the deceased went in order to eventually emerge in the east from where they could access the Skyworld by following the path of the Sun (Clarke 1991, 1997). For Aboriginal people the Underworld was a place of dangerous beings. It was recorded from a mission near Healesville in Victoria that:

> The Coranderrk blacks say that there is one man (*Kooleen*) under the ground
> (*Beek*) who has a long tail. He has a great many wives and many children. He is

a very bad man, and always laughs at the blacks because they have no tails (Smyth 1878, vol. 1, p. 429).

Orientation and appearance of the Skyworld

Aboriginal orientation in terrestrial space was based upon the observed movements of celestial bodies, particularly the Sun and the Moon, and the prevailing directions of the seasonal weather (Clarke 2009a; see Chapter 15). It is common in Aboriginal languages for the terms for west to refer to the 'direction to which the Sun travels', while the east is often associated with dawn or the moon, and in some instances the word for south is linked to cold (Nash 1992; Tindale 1974). In the south-east of South Australia the Bunganditj people referred to the groups living along the Coorong as the Wattatonga (Smith 1880, p. 62), which meant 'men of the evening' in reference to their country being to the west (Tindale 1974, p. 218). In the night sky celestial bodies were used for orientation. It is recorded that in south-western Victoria the constellation of the 'great hunter' Barrukill, seen as Hydra, was 'of great service to the aborigines in their night journeys, enabling them to judge the time of the night and the course to be taken in travelling' (Dawson 1881, p. 101).

Along the Murray River in South Australia the Meru people referred to another group who lived north of the North West Bend as 'Moon men' and their territory was said to be the 'country of the Moon' (Tindale 1953 [cited Clarke 2009c, p. 147]). The 'Moon-men' wore moon-shaped shell plates, which probably came to them along the documented ceremonial trade routes commencing at the northern Australian coast (Hercus 1980; McBryde 1984a). This association of a northern country with the Moon may also have been based on the belief that the Moon had a path that took it close to Earth in that region. It was a tradition among some people in south-eastern Australia that the Moon did on occasion reach the high places on Earth because it is recorded that the place Langi yan, known today as Mount Misery, west of Melbourne, was the 'resting place of the moon; as from a favourite camping-ground of the natives the moon appeared to rise over that hill' (Surveyor General of Victoria [cited Smyth 1878, vol. 2, p. 192]).

Aboriginal material culture has been shaped by a commonly held sense of aesthetics (Sutton 1988; Morphy 1998), which can also be seen to influence what was perceived in the sky. The orientation of stars was more relevant than the brightness of the individual elements, and colour was of fundamental importance when determining the significance of specific celestial bodies (Clarke 2009a; Haynes 1992; MacPherson 1881). In an Aboriginal perspective, celestial bodies that are not bright red or shiny white objects in the night sky are more likely to be seen as part of the background than as elements with individual identities (Clarke 2009a). Aboriginal people in south-western Victoria perceived the 'smaller stars, "narweech maering" as "star earth"' (Dawson 1881, p. 99).

In general, Aboriginal observers did not recognise shapes formed by connecting individual stars, although strings of stars representing clusters of ancestors were perceived as making tracks across the night sky (Clarke 2009a, 2014b). For instance, a 19th century newspaper correspondent outlining Aboriginal celestial lore remarked that: 'Stars that are in a line either horizontally or perpendicularly are generally honoured with a legend, in some cases extending over two or three of our constellations ... For the most part groups forming curves or angles are apparently ignored' (E.K.V. 1884). In summary, Maegraith claimed that: 'In general, the figured constellation is rare in aboriginal astronomy, a single star representing a whole animal or its tracks' (Maegraith 1932, p. 26).

Gender

In south-eastern Australia there are many structural similarities in Aboriginal beliefs concerning the Pleiades, Orion, Coalsack, Magellanic Clouds and the Milky Way, but there is considerable variation in accounts of the Southern Cross (Clarke 1997, 2014b; Stanbridge 1857, 1861). For their identity, particular constellations generally attracted a certain gender. For example, in most accounts of the Pleiades, or an asterism of it, there is a common theme of several female ancestors who are being chased by either a single male ancestor or by a group of them (Griffin 1923; Johnson 1998, 2011; Clarke 2009a). In south-western Victoria the Pleiades were called *kuurokeheear*, meaning a 'flock of cockatoos' in the dialect of the Kuurn kopan noot people (Dawson 1881, p. 100). In this area it was tradition of the Pirt kopan noot people that the Queen of the Pleiades was out with her six 'attendants' when she was taken away by Waa the Crow Ancestor [Australian Raven], who had disguised himself as one of the large edible white grubs that women were gathering using a small wooden hook (Dawson 1881). By this account Waa eventually became the bright star Canopus.

In many parts of Australia the Pleiades, as Sisters, were being pursued across the night sky by Orion, as young men (Clarke 1997; Johnson 1998, 2011). In north-western Victoria, Howitt recorded that:

> The Pleiades are, according to the Wotjobaluk, some women named Murkunyan-gurk, and the following account explains to some extent who they are. When they were on the earth, Boamberik was always running after without overtaking them. Now he is up in the sky, still chasing them, and still behind. I have not been able to identify the star Boamberik, but as it must be one not far from the Pleiades, it seems not unlikely that it may be Aldebaran (Howitt 1904, pp. 429–30).

At Lake Tyrrell in north-western Victoria it was recorded that 'Larnankurrk (Pleiades)' was 'a group of young women playing to Kulkunbulla', and 'Kulkunbulla (the Stars in the Belt and Scabbard of Orion)' was 'Several young men dancing. (A coroborree)' (Stanbridge 1857, p. 139). Howitt recorded from the area surrounding Melbourne that:

> According to the Wurunjerri, the Pleiades are a group of young women, the Karat-goruk, about whom there is a legend which recounts that they were digging up ants' eggs with their yam-sticks, at the ends of which they had coals of fire, which Waang, the crow, stole from them by a stratagem. They were ultimately swept up into the sky, when Bellin-Bellin, the musk-crow [Australian raven], let the whirlwind out of his bag, at the command of Bunjil, and remained there as the Pleiades, still carrying fire on the ends of their yamsticks (Howitt 1904, p. 430).

Many of the records for the Pleiades in south-eastern Australia are fragmentary, although there appears to be a significant overlap between them as they are generally assigned a feminine identity. At Lake Boga in central northern Victoria, the Pleiades (Sisters) were called *gorraitch gorraitch gourrk* and were seen as 'several young women' (Stone 1911, p. 451). On the border of north-east Victoria and south-east New South Wales, the Birdhawal people called the Pleiades *mamangalang*, from *mamang* meaning 'Elder sister' (Mathews 1907a, pp. 354–5). This constellation is often associated with cold weather (Mathews 1904a; Clarke 2009b). Aboriginal people cited cosmological events in order to account for the environmental rhythms of their country. Today the Pleiades in Australia are popularly referred to as the Seven Sisters (Clarke 2009a), although the stating of an exact number is probably a modern element because early Aboriginal counting systems would not have emphasised this larger figure (see Stanbridge 1861; Smyth 1878, vol. 2).

Skyworld influences over Earth

European recorders have acknowledged the role that heavenly bodies in classical Aboriginal tradition have had in the making of Earth as the terrestrial landscape. The Skyworld was seen as the source of the property of fire. In the Lake Condah area of south-western Victoria it was recorded that: 'A blackfellow threw a spear towards the clouds; to the spear a string was attached. The man climbed up with the aid of the string and brought fire to the earth from the sun' (Anonymous 1888c, p. 2). At Lake Tyrrell in north-western Victoria, it was reported that: 'War (Male Crow) (Canopus), the brother of Warepil [Sirius, Male Eagle]' was 'the first to bring down fire from (tyrille) space, and give it to the aborigines, before which they were without fire' (Stanbridge 1857, p. 140). It was recorded that 'Tyrell Lake' and 'Heaven' were both called *Derrell*, from which the European placename was derived (Stone 1911, p. 436).

Certain attributes of the plants and animals found on Earth were believed to have been derived from the Skyworld. For instance, the Glow-worm (*Arachnocampa* species) was said to have taken its light from the male celestial body Antares, which was called *Butt kuee tuukuung*, meaning 'big stomach' (Dawson 1881, p. 99). Similarly, a newspaper writer claimed that an Aboriginal myth, probably from the Australian east coast, 'states that the gum in the hearts of wattle trees [*Acacia* species] is made by shooting stars lodging there and breaking into bits' (Anonymous 1904a).

Through an understanding of the Skyworld and the role of the ancestors who dwelt there, Indigenous people believed that they had the ability to slow down the passage of time. For instance, a European named Locke wrote to a Melbourne-based newspaper to record his reminiscences of being a settler in the Kotupna district in north central Victoria and said that:

> I once went to the Moira accompanied by a blackfellow, and on our return I expressed to him my opinion that it would be dark before we reached home, whereupon he alighted from his horse, and, without saying a word, proceeded to cut a small sod of grass, which he placed in a fork of a tree, exactly facing the setting sun, remarking, 'Plenty quambee (stop), sun now, no pull away.' As it happened, we got home before it was dark, when Sambo exultingly exclaimed, 'No gammon [false] ground' (meaning the sod of earth) (Locke 1866, p. 3).

Aboriginal people considered that they had the ability, through spirit mediums, to manipulate the time and space relationship. For instance, one early Murray River resident in South Australia recalled seeing small flattened out piles of stone, resembling road metal, at intervals along routes used by Aboriginal people (Mallyon 1910). He was told by an Aboriginal person, 'Well, you know, fellow sit down under those stones who has control of the earth, and when blackfellow very tired him give presents to that fellow.' The first offering made would be a bough removed from a nearby tree. If the traveller was a man, this would be cast on, the thrower saying: 'That is your spear.' A second bough would follow, 'That is your waddy, make ground come shorter.' Similarly, women would offer the underground spirits two boughs, the first representing a possum rug, the second a net. In this way, Aboriginal people along the river made the ground shrink, and the length of their trip shorten. It is apparent from the above accounts that Indigenous people perceived time and space differently from Europeans.

Directions

Certain celestial bodies were named according to their observed proximity to the Sun. For instance, in south-western Victoria it was recorded that Jupiter had a feminine identity and was called *Birtit tuung tirng*, 'strike the sun' – as it is often seen near it at midday', and Venus was known as *Paapee neowee*, 'mother of the sun' (Dawson 1881, p. 99), probably as it is sometimes seen during the day. Similarly, at Lake Tyrrell in north-western Victoria, *Chargee Gnowee* (Venus), was seen as the sister of the Sun and wife of *Ginabongbearp* (Foot of Day), Jupiter (Stanbridge 1857, p. 138). At Lake Boga in central northern Victoria the 'morning star' was called *generpkoonberp*, meaning 'pulling up daylight' (Stone 1911, p. 451), and this may be another reference to Jupiter. While this planet is considered by many contemporary observers to be hard to see during the day, Aboriginal people were renowned for their keen eyesight.

The Skyworld was the destination of the soul spirits of deceased people, with specific areas of the night sky existing as places to live after their death on Earth (Mathews 1904a, 1904b). In western Victoria, Mathews recorded that:

> … every clan has its own spirit-land, called *mi-yur*, a native word signifying 'home' or final resting place, to which the shades of all of its members depart after death. The names of the miyurs are in all cases identical with the names of the clans. These miyurs are located in certain fixed directions from the territory of the tribe, some being situated toward one point of the compass and some another. Every man knows the direction of all the miyurs of his tribe in addition to his own (Mathews 1904a, p. 287).

In this area, the *miyur* compass directions for clans from the Gurogity moiety were given as follows:

> Dyalup, the miyur of which bears W. 5° S. Burt murnya, which has the same amplitude as Dyalup. Nyaui, E. 10° N. Kuttyaga, the same direction as Nyaui. Burt Wirrimul has a miyur bearing W. 25° S. Wartwurt bears N. 25° W. Wurt-pattyangal, E. 15° S (Mathews 1904a, p. 287).

Each clan was associated with its own set of totems, mostly animals such as birds and other game species, which were identified with the ancestral spirits who resided in the Skyworld.

For people in western Victoria, the *miyur* in the Skyworld was the source of unborn spirits. It was recorded that: 'The spirits of the dead congregate in the miyurs of their respective clans during their disembodied state, and from there they emerge and are born again in human shape when a favourable opportunity presents itself' (Mathews 1904a, p. 294). When game animals, such as kangaroos, snakes, possums and emus, were being cooked it was tradition that their heads were pointed towards the *miyur* of the clan with which they were totemically associated. Similarly, Mathews claimed that in south-western Victoria each clan possessed:

> … a spirit-home, which some of the tribes call *maioga* and others *mungo*. Every individual, when buried, is placed with the head pointing in that direction, and all the spirits go away to the same resting-place (Mathews 1904b, p. 70).

It is likely that the performances during the ceremonies that celebrated the clan ancestors were conducted in reference to the location of each *miyur*, which was probably located at the end of the celestial track of its ancestor.

Analogous beliefs to those just described appear to be held by other people in south-eastern Australia. Among the Kulin of the Gippsland, it was a belief that the *murup*, or soul, of the deceased went up into the Skyworld via the Underworld (Howitt 1904). It was said that: 'The *Murup* ascended to the sky by the *Karalk*, that is, the bright rays of the setting sun, which is the path to the *Tharangalk-bek* [Skyworld], and the *Karalk* was said by some to be made by the *Murup*s when in *Ngamat* [Underworld]' (Howitt 1904, p. 438). This account demonstrates the importance of the Sun ancestor's observed path across the sky for the Aboriginal worldview.

Wurdi Youang stone arrangement

The possibility that Aboriginal people used stone arrangements to record their knowledge of the Sun's tracks across the Skyworld is indicated by the scholarly analysis of the Wurdi Youang stone arrangement (Mount Rothwell archaeological site), which is located near Little River in western Victoria (Norris *et al.* 2012). Historical sources record the meaning of the Wurdi Youang as 'Station Peak, the big hill', from *wurdi* as 'big' and *youang* being 'hill' (Smyth 1878, vol. 2, p. 199), and the name for this area is now associated with the stone arrangement. Located here is an egg-shaped ring built of large stones that appear to have had some ceremonial function, although there is no early ethnographic evidence to suggest its significance in the local Indigenous cosmologies. It has been suggested that the overall shape of the stone arrangement is similar to an abalone shell and that this is suggestive of the place being used for increase rituals, held to increase the fertility of plants and animals (Morieson 1994 [cited Norris *et al.* 2012]).

In recent times scholars, who have been investigating the site with local Aboriginal people, have used mathematical tools to analyse the stone alignments (Norris *et al.* 2012).

Outlier stones at the Wurdi Youang site, which indicate the position of the setting Sun on the equinoxes and solstices. PA Clarke, You Yangs, Victoria, 2012.

They have confirmed that they do match the rising and setting of the Sun, and that this is something that is extremely unlikely to have occurred by chance. Given that the ethnographic evidence already presented above concerning high features like trees and hills being seen as points of connection with the Skyworld, the location of this particular stone arrangement and its alignments would together appear to support the notion that its makers perceived a possible intended connection between the Wurdi Youang site and the heavens, via the Sun.

The Wurdi Youang stone arrangement is a complex site and its structure is demonstrably not a naturally occurring feature. It does appear to have been made in ancient times. Ethno-astronomer Ray Norris described the site as follows:

> It measures about 50 metres along its major axis, which is oriented almost exactly East-West, and its constituent basalt stones range from small rocks about 0.2 m in diameter to standing stones up to 0.8 m high, some of which appear to be supported with trigger stones. At its highest point, the western apex, is a prominent group of three stones about 0.8 m high, with smaller stones, referred to as outliers, nearby (Norris 2016, p. 23).

Researchers have noted that the sight lines from a gap between the two largest boulders over the outliers marks the position on the horizon where the Sun sets on the solstices and equinoxes, while the straight sides of the stone arrangement leading from the point of the 'egg' indicate the directions to the solstices (Morieson 2003; Norris *et al*. 2012). Curiously, the three prominent stones at the western point of the arrangement, as viewed from the eastern point, delineate the spot where the Sun sets at equinox. As stated above, mathematical analysis of the configuration has dismissed any chance that it is a random alignment. The fact that the historic land tenure of the surrounding land has largely been held by one settler family, who have not developed the area, does not support a European origin for the stone arrangement. Unless the site was subject to an early hoax, which seems unlikely, we must accept the finding that the Indigenous builders of this stone arrangement had purposefully aligned the stones according to the positions of the setting Sun at the solstices and equinoxes (Norris *et al*. 2012).

Wurdi Youang is the only Aboriginal site in Australia that is known to have such a solar alignment, although analysis of the bora ceremonial ground alignments of central and eastern New South Wales suggest that these earthworks are orientated to the position of the Milky Way in the August night sky, when the galaxy's plane from Crux to Sagitarius is approximately vertical to the south-south-west (Fuller *et al*. 2013). It has been suggested that in Aboriginal Australia there are two main types of stone arrangements, which are those that Aboriginal people claimed were made by their ancestors, and those that were recognised as the creation of living people (McCarthy 1940). An archaeologist has suggested that the construction and location of many stone arrangements can be seen as strong evidence for a considerable investment of time and energy in activities that are not directly involved with subsistence, although having a ritual connection (Stead 1987). The ancestors' stone arrangements tended to be sacred sites and as such were totem centres and perhaps associated with religious power, while those said to have been constructed by people were done through the ongoing ritualised commemorations of ancestral feats during the Creation. Examples of stone arrangements seen as made by ancestors include those located in the Flinders Ranges of South Australia, and these were said to be used for rain-making ceremonies and were ritually sprinkled with animal blood when being used (Mountford 1927). The large earth rings, such as those found at Sunbury to the north of Melbourne, are thought to have been associated with initiations, and as such would have been recognised as the productions of people (Frankel 1982).

One of the few well-documented stone arrangements in Australia is at Partjar in the Clutterbuck Hills of the Western Desert in Western Australia (Gould 1969). The site is connected to Partjata the Western Quoll (*Dasyurus geoffroii*) ancestor, and while much of the ethnography gathered for the site relates directly to this ancestor it was recorded that here: 'The rocky slopes on the east side are men of the 'sun' side; the rocks on the west are men of the 'shade' side' (Gould 1969, p. 143). The 'sides' mentioned here are actually successive generation classes. The Sun also appears to divide the community in the Encounter Bay area of the Lower Murray, where Heinrich Meyer recorded from Ramindjeri people that:

> The Sun they consider to be female, who, when she sets, passes the dwelling-places of the dead. As she approaches, the men assemble, and divide into two bodies, leaving a road for her to pass between them; they invite her to stay with them, which she can only do for a short time, as she must be ready for journey the next day. For favours granted to some one among them she receives a present

of a red kangaroo skin; and, therefore, in the morning, when she rises, appears in a red dress (Meyer 1846 [1879]).

It is possible that the placement of the stones within the Wurdi Youang arrangement also related to the local Aboriginal social structure, in this case moieties. In the western Victorian/ south-east of South Australia region, the Sun and the Moon ancestors were members of opposed moieties (Fison and Howitt 1880; Mathews 1898).

The Wurdi Youang stone arrangement may be an example of a place where Aboriginal people acknowledged the estates of their ancestors, possibly when using a ceremony as a means of returning the soul spirit or 'shade' of a deceased person back to their totemic centre in the Skyworld. The importance of the Sun's position in relation to the horizon might be explained by early ethnographic records for south-eastern Australia, as given above, that indicate the widespread belief in souls going to the Skyworld along similar paths taken daily by the Sun ancestor. Aboriginal people would therefore have been keenly aware that different solar paths were taken through the course of the year, with the more distant paths when the Sun is low in the winter sky providing less daytime heat. As stated above, the spirit souls of the deceased were seen to enter the Skyworld via either western or eastern portals. The Wurdi Youang site may, therefore, have had less to do with increase rituals or foraging than it did with the celestial homes of the spirits. Future analysis of other stone arrangements may suggest that the imaging of the Skyworld on the ground was important during ceremonies that were held in honour of the Creation ancestors.

Discussion

In Aboriginal cosmology, the perceived Skyworld was a reflection of the terrestrial landscape. It was believed that human visitors to the Skyworld since the close of the Creation period were able to bring back to Earth 'new' rituals and songs they had learnt above. Across Aboriginal Australia, there were topographic features that were treated as portals linking the Earth with the Skyworld and Underworld. From Earth, the entry into the Skyworld during the Creation period was often perceived as being via the eastern horizon where the Sun rises, although the ancestors were thought generally to first travel to the western horizon where the Sun sets and then through the Underworld. In some cases, Creation ancestors were said to have made their ascent by climbing tall trees that connected Earth with the Skyworld. When the existing landscape was finally set at the close of the Creation period, certain tall trees remained as 'ladders' that allowed a variety of spirits and specially trained humans to travel both ways between these sections of the landscape.

The Aboriginal Biocultural Knowledge possessed by the Indigenous peoples of south-eastern Australia was framed by worldviews that organised space in vastly different ways from that of the Western Europeans who arrived as colonists in 1788. The actions of the ancestors during the Creation period established the order to the world as it had become. The characteristics of many species of plant and animal are accounted for in mythological narratives. With the imperfect recordings of the mythologies from south-eastern Australia, much of this detail has been lost, with the few detailed accounts available showing how Indigenous people would have passed this knowledge on within their community.

Chapter 15
Time

Philip A. Clarke

Indigenous concepts of time in Australia were cyclic, in contrast to modern Western European notions of it being linear. Aboriginal Australians believed much of the weather on Earth was generated by forces in the Skyworld, with the movements of celestial bodies being linked to seasonal changes. The ancestors were credited with having established the climate during the Creation and afterwards to have maintained some control of the weather from their abode above.

Introduction

The manner in which all people conceive and experience physical space and time is culturally determined. It was remarked that: 'While modern societies tend to depict these categories as types of independent entities, real things, or universal and objective categories, for most premodern and non-Western societies time and space remained embedded in their activities and events' (Iwaniszewski 2014, pp. 3–4). Early Aboriginal concepts of time differed from modern Western European perceptions as Aboriginal people did not see time as being linear or chronological (past–present–future) and generally structured their memory of events into a circular pattern of time, within which each individual saw themselves as being in the centre of a time-circle (TenHouten 1999; Janca and Bullen 2003). Events were placed in time according to their relative importance for the individual and his or her respective community, with the more important events being perceived as relatively close in time.

Indigenous Australians experienced time as the passage of cycles, including night–day, lunar months, annual seasons and travelling ceremonies that were held every few years (Morphy 1995, 1999; Silverman 1997). As hunter-gatherers, they were keen observers of any subtle changes within their environment, and these included the onset of seasons which were signalled by such things as the movement of celestial bodies, weather shifts, altered animal behaviour and the flowering of calendar plants (Dawson 1881; Clarke 2009b; Davis 1989, 1997; Morieson 2003). Such a system of determining the season was well suited to environmentally variable country, where the start and strength of a season is heavily subject to the El Niño climate cycle (see Green *et al.* 2010b; Meehl and Arblaster 1998; Tudhope *et al.* 2001).

For Aboriginal observers, much of the weather was generated in the Skyworld, with rain, hail, wind, thunder and lightning coming down from above and rainbows being a phenomenon that appears to connect the Earth with the sky. In the Lower Murray of South

Australia, thunder was regarded as the voice of male Supreme Ancestor Ngurunderi, while rainbows showed him urinating (Taplin 1874 [1879]). Along the Murray River in South Australia the Kombo (Rainbow) ceremony, which was chiefly concerned with arranging marriages, travelled annually downstream between Aboriginal groups (Tindale 1930–52, 1953 [cited Clarke 2009c]). Here, symbolically the outer arch of the rainbow was the male principle, while the inner arch was the female, and the whole rainbow represented the urine stream of a male supreme being. The Kombo was held in later winter, when the widespread presence of water made travelling easy.

The close connections between the Skyworld and Earth required that certain prohibitions in relation to the weather were practised. In north-western Victoria, it was reported that:

> According to the Wotjobaluk, the rainbow causes a person's fingers to become crooked or contracted if he points to it with a straight finger. This would prevent him from using his hand for making the markings with which the possum rugs are ornamented. Therefore, when pointing to a rainbow, the fingers must be turned over each other, the second over the first, the third over the second, and the little finger over the third, by which the evil is avoided (Howitt 1904, p. 431).

On Earth certain large trees, as portals to the Skyworld, remained as places to avoid due to their association with spirits and extreme weather. At Brown Lake near Mount Gambier in the south-east of South Australia there was once a Katal [spirit] tree, which was 'one imbued with spirits; its branches chafe together and it is said to 'talk'; some katal catch fire during storms, additional evidence to the natives that spirits inhabit them' (Tindale c.1924–c.1991 [cited Clarke 2015d, p. 276]). These large trees would have made unusual sounds during strong winds and possibly on occasion attracted lightning strikes.

Forecasting weather

European Australians have long recognised that Indigenous people had considerable experience and knowledge of local climates that could be used to predict future weather events (Clarke 2009b). For instance, a newspaper columnist remarked in an article concerning weather forecasting that:

> It is astonishing, however, how much weather wisdom has been developed in the world merely as the result of long-continued observations by unscientific people. The man whose life has been passed in certain localities has by reason of long intimate personal communion with nature become endowed with a 'gift' that is not to be despised. There are some who would even prefer to trust the instinct of the brute creation or the intuitive perception of aboriginals, whose traditions of the sky are not the least remarkable features of their native knowledge of the ways of nature. All this is only another way of emphasising the value of observation and deduction as the stepping-stones to knowledge (Anonymous 1904b, p. 4).

In 1876, Sydney-based meteorologist Henry Chamberlain Russell gave a detailed scholarly explanation of the wet and dry cycles within the Australian climate, and he utilised evidence that included the Aboriginal knowledge of past climate events, such as previous droughts (Russell 1876). Such use of Indigenous understandings of Australian weather, along with their oral history of the past, fits well with the statements of modern researchers who have described the value of incorporating Aboriginal Biocultural Knowledge into contemporary land management practices (Ens *et al*. 2012).

It is recorded that when predicting a change of weather in south-western Victoria the Aboriginal people took note of such things as animal behaviour, wind direction, the appearance of the sun and moon, colour of the horizon at sunset and sunrise and the time of day when rainbows occurred (Dawson 1881; Clarke 2009b). Many of the observations made concerning the weather were localised, and as such only relevant to a relatively small area. This is apparent in an account of Indigenous weather forecasting that appeared in 1876. An article from a newspaper correspondent that concerned sheep mustering in the Mallee country straddling the South Australian and Victorian border read as follows:

> 'Rain come bailie [good] this time!' said the blackfellow, 'because pelican [*Pelecanus conspicillatus*] fly away from billabong, and wallows all quambie [stop] along o' ground. Big one rain come bailie, me know, because curlew [Bush Stone-curlew, *Burhinus grallarius*] yabber [crying out] along o' day time.' The sky at that time was like a great canopy of lead, and the heat was thick, intense, and oppressive; and there was considerable agitation amongst the birds, as if they expected a catastrophe (Grassie 1876, p. 4).

In this region pelican observations were also made during dry times, as it was reported in a regional newspaper in 1885 that: 'There is a solitary pelican wandering about Hotspur River north of Heywood at present, and the aborigines at Condah [Lake Condah Mission] say that it is a sure sign of unusual drought in the Mallee' (Anonymous 1885a, p. 4).

Indigenous experiences of weather within a particular area would have produced sets of observations from which conclusions were drawn. In terms of animal behaviour, it was reported in a South Australian newspaper that for forecasting weather Aboriginal people possessed a 'set of weather indications read from the actions of animals, as, for example, their reputed forecast of heavy floods coming this season from an observation that the ants have removed their broods to the trees' (Elfin 1886, p. 3). In south-western Victoria, colonist James Dawson remarked that local Aboriginal people knew that:

> … when mosquitoes and gnats are very troublesome, rain is expected; when the cicada sings at night, there will be a hot wind next day. The arrival of the swift, which is a migratory bird, indicates bad weather. The whistle of the black jay [Grey Currawong, *Strepera versicolor*], the chirp of the little green frog [possibly Common Eastern Froglet, *Crinia signifera* or Growling Grass Frog, *Litoria raniformis*], the creak of the cricket, and the cry of the magpie lark [Mudlark, Murray Magpie, *Grallina cyanoleuca*] indicate bad weather; wet weather is more

likely to come after full moon. It is a sign of heat and fine weather when the eagle [probably Wedge-tailed Eagle, *Aquila audax*] amuses itself by towering to an immense height, turning its head suddenly down, and descending vertically, with great force and with closed wings, till near the earth, then opening them and sweeping upwards with half-closed wings to the same height. This movement it repeats again and again, for a long time, without exertion and with apparent pleasure. The aborigines call this movement 'warroweean,' and always expect warm weather to follow it (Dawson 1881, p. 98).

This colonist was greatly impressed with the knowledge and intellect of his informants and he remarked that:

> ... it is very questionable if even those [Europeans] who belong to what is called the middle class, notwithstanding their advantages of education, know as much of their own laws of natural history, and of the nomenclature of the heavenly bodies, as the aborigines do of their laws and of natural objects (Dawson 1881, p. iv).

In some parts of south-eastern Australia, Aboriginal people gave directions in reference to the sources of the weather and according to the paths of the Sun ancestor across the day sky. For example, at Lake Boga in central Victoria it was recorded that in the local language the term for north was *barrewill*, meaning 'where the hot winds come from', south was *boiecalling darn*, 'where the frost winds come from', east was *worwalling gnowie*, 'where sun rises', and west was *purticalling gnowie*, 'where sun sets' (Stone 1911, p. 451). The location of the country associated with other ancestors may have been referenced in other terms used for winds. For instance, in the Yaraldi language of the Lower Murray in South Australia, Berndt noted that the south-east winds were called *prolgimaii* (probably *prolggi mayi*), meaning 'native companion [Brolga, *Grus rubicunda*] winds' (Berndt *et al.* 1993, p. 239). This was possibly a reference to the Prolggi (Brolga) country, along with its myth track and associated sites, which were located to the south-east of the region in the area round Kingston (see Clarke 2016a). The ancestral paths of the Brolga may have continued to be used by the species after the Creation, with Brolgas known to migrate towards larger wetlands during drier times of the year (Walkinshaw 1973).

Seasonal calendar

Culture shapes the ways people divide the year and how they relate to seasonal changes within the landscape (Clarke 2009b; Ruggles 2014). In Aboriginal Australia, the understandings of annual cycles for different parts of the continent produced calendars that were recorded with seasons numbering from between four and nine (Reid 1995; Davis 1997; Clarke 2003a; Ryan 2012). Observations of such things as celestial movements, changes in animal behaviours, alterations in weather and the appearance of certain plants would together be taken to signal the onset of each season. A change in season would have had implications for where people placed themselves within their territory and for the specific foraging strategies that they employed there.

Aboriginal mobility was driven by both economic and religious considerations. It was remarked that while Aboriginal religion is primarily structured around 'places where ancestral events occurred, and their relative locations … time in the sense of seasons of the year or phases of the day and night carries much symbolic power in Aboriginal classical thought' (Sutton 1998b, p. 371). With reference to the seasonal calendar, Aboriginal people aimed to hold the ceremonies honouring their ancestors at certain mythological places over many days during a season when there was an abundance of food for participants, and to have it end with a full moon (Morphy 1999). The full moon would have provided extra illumination during the night for performers, as it was observed in Victoria that dances were 'performed nearly always at night, and not seldom when the light of the moon is sufficient to enable a European to read a book' (Smyth 1878, vol. 1, p. 175). Moonlight was also regarded as a deterrent for those dangerous spirits who were active during the night (Wyatt 1837 [cited Clarke 1997]).

The lunar cycle was also used as a measure of time, as shown by an account of initiations from south-eastern Australia that reported: 'For the space of one moon the youths are prohibited from seeing any one except the coradjes [senior men]' (Smyth 1878, vol. 1, p. 62). Similarly, in the case of burials, from south-western Victoria there is an account of relatives putting a man's body in a tree for the period of one moon before cremating it (Dawson 1881). In the 'dialects of the Yarra Yarra and Coast tribes' of Victoria, it was recorded that *carnboo meniyan* meant 'Lunation, the revolution of the moon, month' (Bunce [cited Smyth 1878, vol. 2, p. 137]). This derivation was based on *carnboo* as 'A, one, once, unit' and *meniyan* being 'Moon, lunar, relating to the moon' (Bunce [cited Smyth 1878, vol. 2, pp. 137, 141]). In the south-east of South Australia, the Bunganditj (Boandik) term for the Moon was Toongoom, which was also used to indicate the period of a month (Smith 1880). In the Encounter Bay area of the Lower Murray in South Australia the Ramindjeri people had a name for a half Moon, *marger-ald-narte*, which literally meant 'Moon of piece', while a full moon was *marger-ald rakkuni* and translated as 'Moon of round' (Meyer 1843, p. 78).

In south-eastern Australia, there are several recorded Moon myths that feature the lunar phases. For instance, in the Gippsland anthropologist Alfred Howitt recorded:

> The Kulin account is that the moon was once a man who lived on the earth. He wished to give the old Kulin a drink of water, so that, when they died, they could after a time return to life again; but the Bronze-wing pigeon would not agree to this, which made the moon very angry (Howitt 1904, p. 428).

Similarly, in north-western Victoria, he noted that:

> In one of the Wotjo legends it is said that at the time when all animals were men and women, some died, and the moon used to say, 'You up-again,' and they came to life again. There was at that time an old man who said, 'Let them remain dead.' Then none ever came to life again, except the moon, which still continued to do so (Howitt 1904, pp. 428–9).

In a record from the Encounter Bay area of the Lower Murray the Moon had a feminine identity, which was different from other accounts in surrounding areas. Missionary Heinrich Meyer said that here, not only was the Sun female, but that:

> The moon is also a woman, and not particularly chaste. She stays a long time
> with the men, and from the effects of her intercourse with them, she becomes
> very thin, and wastes away to a mere skeleton. When in this state, Nurrunduri
> [Ngurunderi] orders her to be driven away. She flies, and is secreted for some
> time, but is employed all the time in seeking roots which are so nourishing that
> in a short time she appears again, and fills out and becomes fat rapidly (Meyer
> 1846 [1879, pp. 200–1]).

In the south-east of South Australia, the Potaruwutj people considered the Moon, which
they called Mitjan the Eastern Quoll (*Dasyurus viverrinus*), was male and was wandering
alone, sometimes well fed and other times starving, because he had been driven away for
attempting to steal someone's wife (Tindale [cited Clarke 1997]). This myth records the
behaviour of individual male quolls defending large territories (Glen and Dickman 2006).
These myth narratives provide explanations of why the lunar cycles are different from those
of the Sun. In the case of the Encounter Bay account, there appears to be correlation between
the female menstruation cycle and the phases of the Moon, which accords with modern
medical understandings of human responses to the lunar cycle (Law 1986).

For the Aboriginal languages of south-western Victoria, the ethnographers recorded
many Indigenous terms for the weather, with storms classified according to the environmental
damage they would cause. For instance, in the Djabwurrung (Chaap wuurong, 'broad lip')
dialect a storm 'which destroys blossoms' was called *borran borran kulan chimmuk*, a storm
'which blows young magpies out of their nests' was known as *kang'aelap kang'aelap kaeaerae*,
while a 'Stormy day' was described as *muun muurt peetch* and a 'Storm, hurricane' as *puundaa
yirneen* (Dawson 1881, pp. xxxix–xl). In the same dialect, summer was listed as *kartii*, winter
as *moat moatt* ('cold') and spring as *bukkar yak eelang nor* ('summer coming') (Dawson 1881,
pp. xxxviii, xl, xlvi). There were related terms recorded from neighbouring dialects. In the
Kuurn kopan noot ('small lip') dialect summer was listed as *kaluun* or *peep kaluun* ('father of
heat') (Dawson 1881, p. xl). An early colonist stated that there were no terms recorded for
autumn in any of the south-western Victorian languages (Dawson 1881), although this
observation may just be a product of the poor fit between modern Western European and
local Indigenous calendars.

In temperate Australia, the majority of early European ethnographers collected terms
that could only be used in translations to the four seasons of north-western Europe, albeit six
months out of phase. In western Victoria, the first season of the year was *weeit* (autumn),
followed by *myer* (winter), *gnallew* (spring) and *cotchi* (summer) (Stanbridge 1857). The
colonist William Stanbridge observed that here:

> They have no name for numerals above two, but by repetition they count to
> five; they also record the days of the moon by means of the fingers, the bones
> and joints of the arms and the head. Their year commences at the end of March
> or the beginning of April, and is divided into four seasons (Stanbridge 1861,
> p. 304).

Given that most Indigenous Australian calendars have more complex divisions into seasons (Clarke 2003a, 2009b), Stanbridge's account was probably truncated. In south-western Victoria, the recorded Aboriginal languages were rich with different terms for clouds and winds (Dawson 1881; Stone 1911), and here in at least one language there were different terms for the lunar phases (Blake 2003).

For much of south-eastern Australia it appears that the different forms of weather and the individual seasons were perceived as totemic entities and therefore classified within the same worldview as people and their ancestors. Here, some Aboriginal groups had a social structure whereby, for the purposes of assigning marriage partners and for participating in ceremony, each individual belonged to one of two moieties (phratries), with membership determined through the mother's line (Fison and Howitt 1880; Mathews 1904b). This matrilineal system, with moieties variously described as Gamaty (Kumite) and Gurogity (Karaal) or equivalents of them, covered much of western side of south-eastern Australia. Within each moiety were at least several major totemic subdivisions or super families, each of which included several other totemic clans. For instance, in the south-east of South Australia the Wa ('Crow [Australian raven]') subdivision within the Kumite moiety contained the totemic clans of 'Rain, thunder, lightning, winter, hail, clouds, &c.', while the Karaal ('Black crestless cockatoo' [sic. Little Corella, *Cacatua sanguinea*]) subdivision of the Kroki moiety included 'Kangaroo, sheoak trees, summer, sun, autumn (fem.), wind (fem.)' totemic clans (DS Stewart [cited Fison and Howitt 1880, p. 168]).

Over much of eastern Victoria, a patrilineal system operated (Fison and Howitt 1880; Mathews 1904b), as it did in the Lower Murray region of South Australia (Howitt 1889; Mathews 1904b; Brown 1918). The patrilineal clans also had season and weather-related totems. In the Lower Murray, the Yaraldi-speaking Limpinderar clan had *waldi*, 'hot weather', as their *ngaitji* or totem (Brown 1918, p. 229), while another Yaraldi clan, Reinindjera, had *bani*, 'rain', as one of theirs (Berndt *et al*. 1993, p. 308). This inclusion of weather and the seasons within Aboriginal social structure would have reflected the Indigenous belief that their ancestors, many of them seen as stars, had at least some power over the Earth's climate.

Observable changes in the landscape, including the Skyworld, indicated the change of season. The links between the movements of heavenly bodies and the onset of seasons is widespread across the Australian continent (Stanbridge 1857). Colonist and anthropologist Robert Mathews noted in Aboriginal Australia it was generally recognised that:

> … the stars which occupy the northern sky in the cold winter evenings travel on, and are succeeded by others in the following season; and that these are again displaced by different constellations during the warm evenings of summer (Mathews 1904a, p. 279).

The actions of ancestors during the Creation, as visually documented in the heavens, were used to explain the seasonal cycle on Earth. At Lake Tyrrell in north-western Victoria the presence was recorded of:

Yurree (Castor), Wanjel (Pollux), two young men that pursue Purra [Capella, Kangaroo] and kill him at the commencement of the great heat, and Coonartoorung (Mirage) is the smoke of the fire by which they roast him. When their smoke is gone Weeit (Autumn) begins (Stanbridge 1857, p. 140).

Warmer seasons were associated with the availability of particular food sources. For example, again at Lake Tyrrell, it was recorded that:

Neilloan (Lyra), a Loan flying (*Leipoa ocellata* [Malleefowl]). The mother of Totyarguil, and discoverer of the loan eggs, which knowledge she imparted to the aborigines. When the loan eggs are coming into season on earth [in September], they are going out of season with her. When she sets with the sun the loan eggs are in season (Stanbridge 1861, p. 302).

In the Lower Murray region of South Australia, Ngarrindjeri people believed that the arrival of Waiyungari the Red Man, as a red star in the night sky, heralded the beginning of a breeding season for animals and a growth period for plants (Clarke 1999a). Based on detailed mythological descriptions, an astronomer has recently identified Waiyungari as the star Antares (D. Hamacher, pers. comm. 2017). For Aboriginal foragers, spring in south-eastern Australia was 'a time of rejoicing' and a season for 'fishing and hunting when the moon was shining' (Smyth 1878, vol. 1, p. 140). At this time possums, in particular, were hunted by moonlight (Smyth 1878, vol. 1).

The gathering of invertebrate foods was also dictated by the calendar. For instance, in the Mallee region of north-western Victoria the arrival of Marpeankurrk (Arcturus) in the northern night sky signalled the beginning of the Gnallew ('spring') season for gathering the larva of *bittur* the 'wood ant' (termite) (Stanbridge 1857; Johnson 1998). The arrival of other constellations signalled a change in water sources. In north-western Victoria, it was recorded that the constellation of 'Tourt-chinboiong-gherra (Coma Berenices, Berenice's Hair)' was 'A flock of small birds drinking rain-water, which has lodged in a fork of a tree' (Stanbridge 1861, p. 302). A qualification of this stated that Coma Berenices represented a tree with three main branches, and at the junction with the trunk there was a hollow where birds were drinking (MacPherson 1881; Massola 1968; Johnson 1998). The appearance of this constellation was symbolic of the dry summer weather, when such sources of drinking water would be crucial for human survival. For foraging bands, the calendar was a determining factor in their choice of subsistence strategies and their movement patterns.

In Aboriginal Australia, the regular patterns of celestial movement were linked to the known behaviours of ancestors, while less predictable events in the night sky were seen as malevolent omens and the product of sorcery (Clarke 1997, 2009a, 2014b). For instance, in south-western Victoria a meteor was called *Gnummae waar*, meaning 'deformity' and a comet *Puurt Kuurnuuk* was believed to be a great spirit' (Dawson 1881, p. 101). Similarly, in north-western Victoria there is a record of 'Porkelongtoute (Shooting Star); which portends evil to those that have lost a front tooth [through initiation], to avert which they stir the fire and cast about firebrands' (Stanbridge 1857, p. 140). Again in north-western Victoria, Howitt reported that:

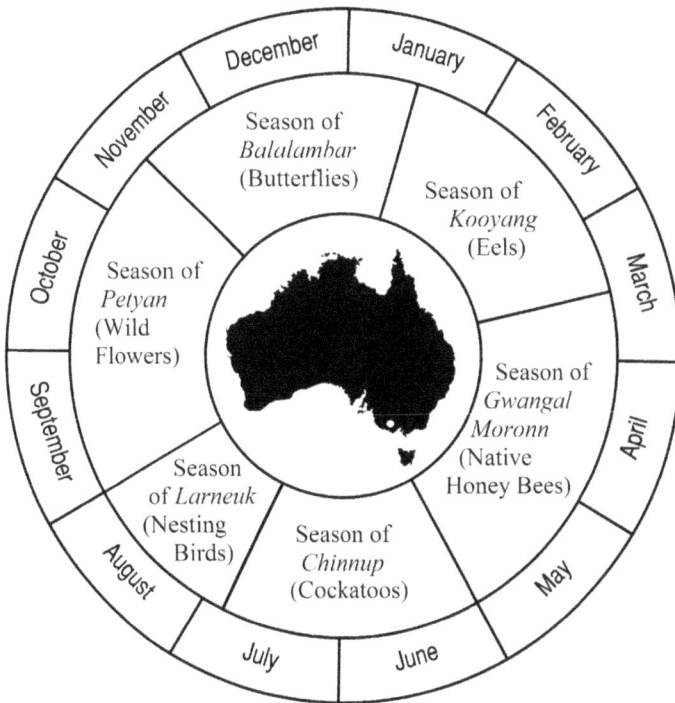

Seasonal calendar developed for the Gunditjmara people of south-western Victoria. PA Clarke 2017 (after Jones 2010; Jones *et al.* 2013). Note that the actual timing of the seasons would vary from year to year.

> The Aurora signified with the Wotjobaluk that, at some great distance, a number of blacks were being slaughtered, and that the Aurora colour is the blood rising up to the sky. When the Aurora was seen by the Kurnai [Ganai], all in the camp swung the Brett (dead hand [a magic object]) towards the alarming portent, shouting such words as 'Send it away; do not let it burn us up.' The Aurora is, according to one of their legends, Mungan's fire (Howitt 1904, p. 430).

Aboriginal people are not unique in their fear of irregular celestial phenomena. A late 19th century newspaper correspondent noted that in English folklore: 'any remarkable display of meteors, or the presence of a comet in the heavens, has always been regarded as a forecast of disaster' (Elfin 1886, p. 3).

For most of south-eastern Australia, the recording of Indigenous calendars is poor, to the extent that there are none that are complete in terms of relating to the time when Aboriginal foragers still foraged as they did before British colonisation (Clarke 2009b). In some areas, contemporary Indigenous communities, with help from academics, have used their knowledge of their country to build their own calendar to reflect their understandings of the environment and its cycles. For instance, in south-western Victoria seasonal charts have been developed for the Lake Condah, Budj Bim and Gariwerd (Grampians) areas in order to inform local land management practices (Jones *et al.* 2013).

Calendar plants

The literature of Indigenous Australian calendars contains numerous examples of what scholars have termed 'calendar plants', which to Aboriginal people indicate the change of a season (Davis 1989, 1997; Reid 1995; Clarke 2009a). For instance, in the temperate region of Yorke Peninsula in South Australia it was recorded that:

> During [a] portion of the year these natives resided in belts of ti-tree and sheoak country, living mainly upon kangaroo [Western Grey Kangaroo, *Macropus fuliginosus*], wallaby [Dama (Tammar) Wallaby, *Macropus eugenii*], Emu [*Dromaius novaehollandiae*], and similar game. When the Billy Buttons [*Craspedia glauca*] came into flower in the spring these natives read it as a signal from Mother Nature that the time had come for a 'walk-about' to the sea coast. For this also was the time when the butterfish [Mulloway, *Argyrosomus japonicus*] make their way into the shoal water along that part of the coastline for some months (Thom 1953, p. 2).

In the south-east of South Australia, the appearance of yellow flowers on a species of large Daisy Bush (possibly *Senecio* species) in mid-July was a sign that the greatly desired Black Swan (*Cygnus atratus*) eggs were available in swamps, such as in the lakes behind Teeluc north of Kingston (Clarke 2015b). While the records of calendar plants for south-eastern Australia are slim, it is likely that plants such as the billy button and various daisies, were widely used in this manner.

Other activities were also subjected to the onset of seasons that were indicated by calendar plants. In the Lower Murray region of South Australia, the Ngarrindjeri residents of Raukkan (Point McLeay) believed that it was not safe to swim in the nearby lake if what is locally called the 'dandelion' was still in flower, and those who did so risked contracting 'dandelion-fever', particularly if they were children (Clarke 1994). In 1969, Ngarrindjeri woman Annie (Fofon) Rankine explained the connection between the flowering season of this plant and the celestial movements of the Seven Sisters constellation and said:

> My father [Clarence Long, Milerum] used to tell us children of a special group of stars which is called the Seven Sisters, and before they were moving we weren't allowed to swim because the dandelions were in bloom then, and it was said that when the dandelions are out the water is still chill, and this is why our people are very strict and don't allow us to swim. When the flowers all died off and the stars moved over a bit further, this is when we were allowed to swim because in that time the dandelion flower which would cause a fever to anyone would not be out to make us sick (Rankine 1969 [cited Clarke 1994, p. 123]).

The 'dandelion' mentioned above was the Yam Daisy (Native Dandelion, *Microseris lanceolata*), which has become locally scarce since the country was transformed into a rural landscape (Clarke 2014a).

The sudden surface appearance of fungi, such as the Earthstar (*Geastrum* species), with the autumn rains in some temperate parts of Australia signified the change in season and

Yam Daisy (*Microseris lanceolata*), a staple food in south-eastern Australia and a 'calendar plant'. PA Clarke, Mount George, South Australia, 2009.

Earthstar (*Geastrum* species), a fungus species that appears in autumn at the same time as the autumn stars. PA Clarke, Aldgate, South Australia, 1983.

was associated with the arrival of certain seasonal constellations (Clarke 2015a). The Murray River people, between Wellington and Rufus River, may also have made this seasonal association, with *pidli* recorded as the Ngaiawung term for 'mushroom, a star' (Moorhouse 1846 [1935, p. 34]). An association of fallen stars with mushrooms occurs in other cultures outside of Australia (e.g. Beech 1986; Hamacher and Norris 2010a). There are also Australian records of fungi having cultural associations with Creation ancestors and spirit beings (Kalotas 1996).

Weather makers

During the Creation, the actions of ancestors created weather. For instance, the Wurunjerri (Woiworung) people on the northern side of Melbourne in Victoria believed that in their sky country, which they called *Tharangalk-bek*, was 'the place to which Bunjil [Wedge-tailed Eagle ancestor] ascended with all his people in a whirlwind' (Howitt 1904, p. 433). It was said that Bunjil 'finally went up to the sky-land with all his people (the legend says his 'sons') in a whirlwind, which Bellin-bellin (the Musk-crow [Australian Raven]) let out of his skin bag at his order' (Howitt 1904, pp. 491–2). When people on Earth saw whirlwinds, it caused fear.

At least some of the weather was believed to emanate from the Underworld. In south-western Victoria, a monstrous being called Muuruup was said to sometimes visit Earth 'in the form of lightning, knocking trees to pieces, setting fire to wuurns [shelters], and killing people by 'striking them on the back', although he was said to live 'deep under the ground in a place called Ummekulleen' (Dawson 1881, pp. 48–9). This Underworld origin for some of the weather is also suggested by the record of the Aboriginal name for Merrijig Creek in Gippsland, which was *Nunga-bruggu-lar* and said to mean 'Wind-hole. There is said to be a rock through which the wind blows' (AW Howitt [cited Smyth 1878, vol. 2, p. 190]).

It is Aboriginal tradition that extreme inclement weather conditions are associated with calamities caused by sorcery or through the breaking of religious protocols (Clarke 2009b). In the corpus of Aboriginal Creation beliefs for western Victoria, Gartuk (Gertuk) the Mopoke (Boobook Owl, *Ninox novaeseelandiae*) ancestor caught powerful storms in bags, unleashing them later upon others (Howitt 1889; Mathews 1904a). Groups living in south-western Victoria feared their western neighbours, who were the Buandik (Booandik, Bungandaetch) people of Mount Gambier in the south-east of South Australia, for their supposed ability in sending them lightning and heavy rain to cause them injury (Dawson 1881). This perception would partly have been based upon the direction from where the anticyclonic weather of winter originated. The motivation for sending storms to a neighbour's country included such things as preventing them from hunting and fishing, or for concealing a party of warriors on a revenge expedition (McCarthy 1953; Clarke 2009a).

Inclement weather was often interpreted as having been caused by sorcery. Aboriginal people living around Lake Boga in central northern Victoria believed many of their number had once died because of 'Yathunge jah [the] poison-fog' in low-lying areas, although the settlers who came later blamed smallpox (Stone 1911, p. 462). The Wurunjerri considered that small whirlwinds were created by the Wirrarap or 'medicine-man' as a form of 'evil magic' and if they caught a man they would cause pain, along with a chill and shortness of breath (Howitt 1904, p. 365). During the early years of British settlement Aboriginal people continued to interpret any unusual event according to their own pre-European cosmological beliefs. For instance, to the immediate west of the Murray River in South Australia it was reported that 'Old residents at Mount Barker state that in '49 [1849] the aborigines were dumb founded with the appearance of a heavy fall of snow, the cause of which they attributed to be the appearance in their midst of the white man' (Anonymous 1901, p. 2). Similarly, west of Ballarat in central Victoria, Aboriginal people told a colonist that the drying up of Lake Burrumbeet was due to the apocalyptic arrival of 'white people' (Kirkland 1845, p. 8).

Aboriginal people had prohibitions to prevent unfavourable weather from occurring. In south-eastern Australia, it was believed if certain species of frog and toad were killed that heavy rains would come (Fraser 1901; Krappe 1940). Many of such taboos or prohibitions would have been linked to an event during the Creation (see Chapter 1). Another version of this myth from Gippsland in eastern Victoria, recorded by missionary John Bulmer, again possibly referring to the Giant Burrowing Frog (*Heleioporus australiacus*), is shown below.

> Ages ago there was no water in what are now the lakes, rivers, and seas known to the Kurnai [Ganai] … for an immense frog had swallowed up the whole of it. This state of things, it appears, was a source of great discomfort to the animals generally, and especially to the fishes, so they held a consultation of the subject, and came to the conclusion that the only remedy was to make the frog laugh, and that if this could be accomplished there would soon be plenty of water. To give effect to this idea, every animal presented himself before the frog in the most ludicrous postures he could assume, and went through his funniest antics. For a time, however, they were unsuccessful, until the eel stood upon the tip of his tail, which so tickled the overgrown frog that he literally burst with laughing, and the water poured from him in such vast streams that there was presently a deluge, and all the Blacks would have been drowned, had not one of them, Loon by name, made a large canoe, in which he saved a great many (Bulmer [cited Curr 1886–87, vol. 3, pp. 547–8]).

After saving so many people, Loon was denied a wife so covered parts of his body with pipeclay and was transformed into a pelican.

The order established at the close of the Creation period was maintained by adherence to rituals that kept the seasonal cycle going. In the Lower Murray of South Australia, Howitt reported that:

> There is a spot at Lake Victoria [Lake Alexandrina], in the Narrinyeri [Ngarrindjeri] country, where when the water is, at long intervals, exceptionally low, it causes a tree-stump to become visible. This is in charge of a family, and it is the duty of one of the men to anoint it with grease and red ochre. The reason for this is that they believe that if it is not done the lake would dry up and the supply of fish be lessened. This duty is hereditary from father to son. (Howitt 1904, pp. 399–400).

The role of 'medicine men' as weather makers, who often received their powers through their totemic links, was to control the severity of the weather (Berndt 1947; Elkin 1977).

Apart from the Aboriginal use of magic to combat sorcery and to prevent harmful things from occurring, there were rituals that were believed to have positive influences, such as affecting short-term changes to the weather generated in the Skyworld. It was Aboriginal tradition that certain individuals were 'rainmakers', having the ritual power to alter the weather and bring rains to their country (Berndt 1947; Clarke 2009b; Elkin 1977; McCarthy 1953). For instance, Dawson recorded that:

> They believe that, in dry weather, if any influential person take water into his mouth and blow it towards the setting sun, saying, 'Come down, rain,' the wind will blow and the rain will pour for three days. When they wish for rain to make the grass grow at any particular place, they dig up the root of the convolvulus [Blushing Bindweed, *Convolvulus erubescens*], called 'tarruuk,' and throw it in the direction of the place, saying, 'Go and make the grass grow there!' (Dawson 1881, p. 98).

It is believed that *tarook* ('tarruuk') is a general term for starchy root, and therefore used to describe the roots of the Blushing Bindweed and the Small-leaved Clematis (*Clematis microphylla*) (Gott and Conran 1991).

Rainmakers used a variety of ritual tools. For instance, settler Alfred Charles Stone at Lake Boga in central Victoria recorded that:

> The Boonyenge mundar, or Rain bringing stone, is a round, smooth stone, resembling white loaf sugar. It was placed in water when rain was required, and when sufficient had fallen it was taken out, dried carefully, and placed away in the doctor's ('Barngnull') poison-bag ('Neilgnooye'). Should, however, the rain be required to stop very soon, the stone or charm was dried quickly by the fire (Stone 1911, pp. 465–6).

In the western Victorian Mallee region, rainmakers made a ball from their own hair, which was then soaked in water (Stanbridge 1861; Dawson 1881; Massola 1968). The rainmaker chanted and then sucked water from the ball for squirting from the mouth towards the west, from where the wet winter weather chiefly arose. Similarly, in eastern Victoria the Ganai (Kurnai) people ritually created rain by squirting water in the direction from which showers generally came, and both men and women would sing rain chants (Howitt 1904; McCarthy 1953). In this region, the *bungil* were a class of people who possessed certain strong ritual powers; a man called *bungil krowero* was a wind controller, the *bungil willang* controlled the rain, and others had power over particular mammals and fish (Bulmer 1855–1908 [1999]; Pepper and de Araugo 1985]).

Similar rituals were recorded in Central Australia, where special stones of quartz or flint were wrapped with human hair mixed with blood for ritually making rain, and if this proved unsuccessful it was reasoned that a sorcerer some distance away was using more powerful magic to hold back the weather (Crystal 1912). In many northern areas of Australia, objects with the property of refracting light into 'rainbows', such as gypsum and shell, were also used in rainmaking rituals (McCarthy 1953; Akerman 1994; Kimber 1997). The mimicking of the calls and movements of frogs were part of rainmaking rituals in north-west Queensland (Roth 1897).

There are also records in Aboriginal Australia for the use of a variety of rituals to bring on the warmer weather. For instance, in coastal south-east Queensland there was the practice in late winter of waving a flowering *Acacia* branch at the sky in order to remind the ancestors that it was time for it to become warmer so that spring flowers would come for the bees (Grant-Swan 1931). Here, it was reported that:

> One tribe gives us the rite of burying lighted branches in the earth, with the idea of warming it for the coming of summer. Somewhat in contradiction, another tribe threw fire-tipped spears into the air, to tell the sun that they needed his greater warmth (Grant-Swan 1931, p. 20).

In northern Queensland, it was recorded that specially chosen Aboriginal women would swing on the lianas 'with the object of helping the golden sun to swing himself up to his zenith – a golden swing across the sky' (Grant-Swan 1931, p. 20). It was apparently a common practice for many groups, such as those at Groote Eylandt in the Northern Territory, for using:

> … the method of making mock suns, consisting of round stones, stuck round with owls' feathers, and hung on casuarina trees to tell summer, who had possibly forgotten, it was time to come back. Throwing up into the air mock suns of various kinds was a common practice of many tribes (Grant-Swan 1931, p. 20).

It is likely that in south-eastern Australia similar rituals were observed with the aim of bringing on warmer weather.

Discussion

In Aboriginal Australia, celestial changes observed in the Skyworld were an analogue for seasons occurring on Earth. Through the use of calendars that linked together such phenomena as star movements, plant flowerings, growth of toadstools, animal migrations and weather changes, hunter-gatherers were able to position themselves in the landscape in order to maximise subsistence foraging and comfort. Calendars were also relevant to the ceremonial life, within which ancestors responsible for the reproduction of the environment on Earth were honoured.

The scholarly investigations into the Aboriginal Biocultural Knowledge associated with seasons are not just of antiquarian value. For academic ecologists and land managers, the study of Indigenous calendars offers a means to more holistically understand the rhythms of specific regions (Jones 2010). A group of climate change researchers from the social sciences have observed that:

> In Indigenous community perceptions, environmental mapping of patterns determined and influenced land management curatorship, burning, and food harvesting activities, including the semi-sedentary seasonal shifting of camp sites. This knowledge underpins long-term (1,000-10,000 years) land

management strategies but can also offer insights as to climate change implications because of the recording of shifts and variations in seasonal cues (Jones *et al.* 2013, p. 10).

Studies of Indigenous ethnoastronomy and ethnometeorology have valuable contributions to make to the understanding of Aboriginal use and perception of the Australian environment. Scientists can gain insights into past climate and astronomical events through the study of Aboriginal Biocultural Knowledge, although due to the difference between Indigenous and modern Western European worldviews there is plenty of scope for misinterpretations to be made. Therefore, care must be taken when evaluating the effectiveness of Indigenous astronomy as part of general astronomy education and public outreach projects that typically ignore a non-Western perspective on the heavens (Norris and Hamacher 2009).

Conclusion: The future of Aboriginal Biocultural Knowledge

This book is the first systematic attempt to construct an overarching framework of archived Aboriginal Biocultural Knowledge in south-eastern Australia. It has provided examples of the kind of knowledge which was once held by Indigenous peoples in temperate Australia, including plant distribution, animal behaviour, astronomy, water flow patterns, weather and climate cycles, fire regimes, food collection and preparation, pharmacology of medicinal sources and the properties of materials used in the material culture.

Aboriginal Biocultural Knowledge in south-eastern Australia has been under-researched and subsequently poorly understood by most scholars, although this is the case for many other parts of Australia with the exception of the northern tropics (Ens *et al.* 2015). The current work is an overview that was written for a broad readership, which includes people from Indigenous communities, regional councils, natural resource management agencies, heritage tourism industry, universities and the wider community. It is a timely engagement with national and international debates about the role of Indigenous traditional knowledge in cultural renewal, sharing of histories, biodiversity, cultural and natural resource management, education and wellbeing. It represents an opportune engagement with worldwide debates about the potential roles of Indigenous knowledge in the 21st century.

Only a small fraction of the Aboriginal Biocultural Knowledge that existed at the time of first European settlement in the late 18th century has been recorded. By the time scholars, such as Stanbridge, Howitt and the Berndts, were trying to record an Indigenous perspective of country in south-eastern Australia during the late 19th and early 20th centuries, their living sources were largely part of a 'memory culture' of older people who could remember accounts from their parents and grandparents about life before the Europeans arrived (Clarke 2016a, p. 278; Tonkinson 1993, p. xix). The same processes of dispossessing and alienating Aboriginal people from their land also occurred in other parts of Australia, albeit much later in remote regions. In modern times, researchers, such as ethnobiologists, have struggled against time to record knowledge from an elderly generation of people who had grown up with close relationships with the environment, through having camped widely on their country and foraged for food, medicine and artefact-making materials from the wild resources available to them (Clarke 2003b).

Aboriginal Biocultural Knowledge is of great interest for both non-Aboriginal people, who want to gain fresh insights into ecological processes, and to Aboriginal people who want their connection to country recognised by all. These differing interests are reflected in how it is documented. Anthropologist Francesca Merlan remarked in a foreword of a book published on the ethnobiology of the Wardaman people in the Northern Territory that: 'the material is presented, with its emphasis on western [scientific] classification systems, is different to the way such information comes to Aboriginal people themselves, or is presented by them to those who spend time in their country' (Merlan 1999, p. ix). Across northern Australia and the arid interior, the Indigenous community support that biologists and environmentalists have received during their fieldwork for publications has in large part come from the desire of knowledge-holders to teach future generations about their country (Blake *et al.* 1998; Goddard and Kalotas 1988; Jakubowski and Atkinson 2016; Karadada *et al.* 2011; Nambatu *et al.* 2009; Puruntatameri *et al.* 2001; Smith 2013; Smith *et al.* 1993; Smith and Kalotas 1985; Smith and Wightman 1990; Wightman *et al.* 1991, 1992a,b, 1994; Wightman and Smith 1989, Wiynjorrotj *et al.* 2005; Yunupingu *et al.* 1995).

The acknowledged existence of Aboriginal Biocultural Knowledge should have the power to transform the perceptions of the broader public. In the foreword of a Victorian Aboriginal plant use book, Gunditjmara woman Joan Vickery said: 'Every time I learn that someone has written an article or book about my people's dietary habits and lifestyle I shudder because some non-Aboriginal people still don't know there is a need to listen to what my people have to say, to be able to understand the importance of nature and all it has to offer' (Vickery 1992, p. vii). For a modern Indigenous identity in Australia, the acknowledgment of a pre-European heritage and an understanding of environmental resources are essential elements for establishing a vital connection to the land. An Aboriginal custodian of Gariwerd (Grampian Ranges area), Tim Chatfield, stated that:

> Gariwerd is part of my people. It is place of power and awe. Generations of my people were born here. This land nurtured them. They caught and ate kangaroo, emu and goanna; dug holes in the ground and cooked eel and yam daisies in the embers of their fires. On the rocks of Gariwerd they left their symbols of education, initiation, law and life (Chatfield 1999).

The act of recording of Indigenous Biocultural Knowledge by itself will not preserve it in its earlier forms, when the fundamental relationships that its associated cultural group has with the biota is being irreversibly altered through changing living patterns. Anthropologist Nicholas Peterson questioned whether recording ecological knowledge, mythology and other aspects of Australian Indigenous culture would help to maintain an Aboriginal culture in the modern world. He came to the conclusion that: 'doing so turns knowledge and belief into information, inevitably distanced from the contexts in which such knowledge and belief were a part of lived experience, and literally disembodies them' (Peterson 2017, p. 236). Through linking knowledge with the worldview of its holders, it is apparent that contemporary Indigenous people are capable of creating new Indigenous Biocultural Knowledge, which is relevant to their own worldviews and is derived from their unique relationships with the modern environment. It is in this context that historical records of knowledge, derived from foragers who had lived before European settlement, could inform contemporary Indigenous peoples who are looking back into the past as inspiration for developing a modern non-Western perspective of country.

During the early 1990s, a close link between environmental issues and Indigenous identity building activities emerged when Ngarrindjeri people of the Lower Murray region became heavily involved in the broader environmental movement to stop the building of a bridge across the Goolwa Channel from the mainland to Hindmarsh Island (Tonkinson 1997). The controversy was expressed through a series of public meetings and intense media coverage, which eventually led to a royal commission, federal inquiry and finally a federal court case. From a local Indigenous perspective, the bridge development in an already heavily modified physical environment threatened further disruption to the totemic traditions, as the fate of the wild birds and their associated totemic groups were interlinked (Clarke 2016c). Aboriginal people had drawn upon their own traditions to justify their fears concerning a rapidly changing environment. One of the positive outcomes for Indigenous people becoming involved in public debates over environmental issues is the broader recognition for their own heritage values. In the case of the Ngarrindjeri people, they made a formal agreement with the state government over their involvement in management of the Lower Murray environment, and this is the basis upon which a resolution to native title issues in the region may ultimately be formed (KNYA Taskforce 2012).

The Gunditjmara people of south-western Victoria have successfully used the existence of their traditions concerning the wetlands, which had been modified for controlling water levels and eel numbers, to gain control of the Budj Bim National Heritage Landscape, gazetted in 2004, which includes the Mount Eccles/Lake Condah area (Gunditjmara People and Wettenhall 2010; Jones 2011; Jones *et al.* 2013). In 2007, the Gunditjmara Peoples received a favourable native title determination from the federal court (Weir 2009). The Creation narratives for the Gunditjmara detailed the actions and tracks of ancestral beings who, before leaving the Earth, modified the physical environment for the benefit of the people they left behind (Bell and Johnston 2008). Deeply embedded in the origins of the landscape, Indigenous people use these accounts today to assert a raft of new transformations and responsibilities with respect to the country. Gunditjmara woman, Eileen Albert, stated that:

> In the Dreamtime, the ancestral creators gave the Gunditjmara people the resources to live a settled lifestyle. They diverted the waterways, and gave us the stones and rocks to help us to build the aquaculture systems. They gave us the wetlands where the reeds grew so that we could make the eel baskets, and gave us the food-enriched landscape for us to survive (Albert [Gunditjmara People and Wettenhall 2010, p. 7]).

In this setting, the Budj Bim environment and landscape retains meaning in the present, which is supported by the records of past Aboriginal Biocultural Knowledge.

Public awareness of contemporary Indigenous beliefs concerning such things as animal ancestors and spirits, as conveyors of information about country, gives Aboriginal people a more privileged position as knowledge-holders for their country, in spite of all the change that has occurred. Such traditions are often presented to environmental managers as evidence of Aboriginal connections with the past and as a reason for Aboriginal people gaining a major voice in the present (Clarke 2016a, 2016b, 2016c). Too often, authorities have incorrectly assumed that local Aboriginal cultural values for wild plants and animals would be closely aligned to those of scientists (Smith NM 2013). As Indigenous people, rather than colonisers or immigrants, they are able to embed their cultural identity, albeit largely derived from written sources, in the country in ways that are not available to any non-Indigenous cultural groups. In the future globally, an increased understanding and appreciation of the Indigenous Biocultural Knowledge – which is associated with cultural groups who continue to live as minorities in countries colonised by others – may assist with highlighting the imperative for Indigenous peoples to be given greater involvement in environmental affairs.

The ancestors of contemporary Aboriginal people had a long custodianship of Australia (Lourandos 1997; Mulvaney and Kamminga 1999; Smith 2013), which was followed by a relatively short but intense period of exotic plant and animal introductions, along with native species extinctions, at the hands of Europeans (Crosby 2004; Low 1999, 2002; Rolls 1984). With no true Australian wildernesses – as defined as areas with an ecology totally unaffected by humans – the management of the environment will continue to require human intervention, if the high level of biological diversity is to be maintained. Scientist Tim Flannery remarked that: 'the entire continent has been actively and extensively managed for 60 000 years by its Aboriginal occupants. To leave it untouched will be to create something new, and less diverse, than that which went before' (Flannery 1994, p. 379).

Aboriginal Biocultural Knowledge as examined here is an integrated source of knowledge about the environment that is dynamic, constantly incorporating new experiences and organising them within an Indigenous worldview that is holistic in the way that culture and nature are inseparable. The collection and analysis of written sources of information into an overview provides several benefits for any researcher or organisation examining Aboriginal Biocultural Knowledge. Across the world, this type of knowledge is often perceived as a precious asset in the possession of Indigenous peoples with close connections to a landscape (Devereux 2013; Nabhan 2009). Perhaps of greater value is the robustness of the processes behind this learning, constantly integrating new knowledge in order to derive deeper understandings about the total physical and cultural environment (Ens *et al.* 2014, 2016; Hill *et al.* 2011, 2012). It is this latter aspect of Aboriginal Biocultural Knowledge that will make it a useful tool when Indigenous and non-Indigenous environmental managers alike strive to combat the emergence of major cascading ecological problems, such as climate change, land clearance, water shortages, altered fire regimes and the invasion of exotic organisms.

References

Addis EB (1842) *Report of the Crown Lands Commissioner for the County of Grant.* Manuscript. Mitchell Library, Sydney.

Adeney W (1842–43) *Diary of William Adeney, 19 August, 1842 – 17 March, 1843, Covering his Voyage to Hobart Town on Board the Ship Jane Frances, his Subsequent Arrival in Port Phillip and Journeys in the Western district in Search of Land.* Ms. 8536. State Library of Victoria, Melbourne.

Agrawal A (2002) Indigenous knowledge and the politics of classification. *International Social Science Journal* **54**(173), 287–297. doi:10.1111/1468-2451.00382

Ahmed AK, Johnson KA (2000) Horticultural development of Australian native edible plants. *Australian Journal of Botany* **48**(4), 417–426. doi:10.1071/BT99042

Aicken M, Ryan C (Eds) (2010) *Indigenous Tourism.* Routledge, London.

Akerman K (1979) Honey in the life of the aboriginals of the Kimberleys. *Oceania* **49**(3), 169–178. doi:10.1002/j.1834-4461.1979.tb01387.x

Akerman K (1994) *Riji and Jakoli: Kimberley Pearlshell in Aboriginal Australia.* Monograph Series 4. Northern Territory Museum of Arts & Sciences, Darwin.

Akerman K (2005) Shoes of invisibility and invisible shoes: Australian hunters and gatherers and ideas on the origins of footwear. *Australian Aboriginal Studies* **2**, 55–64.

Allen H, Holdaway S (2009) The archaeology of Mungo and the Willandra Lakes: looking back, looking forward. *Archaeology in Oceania* **44**(2), 96–106. doi:10.1002/j.1834-4453.2009.tb00052.x

Allen J (1853) *Journal of an Experimental Trip by the 'Lady Augusta' on the River Murray.* C.G.E. Platts, Adelaide.

Altman JC, Cochrane M (2003) *Innovative Institutional Design for Sustainable Wildlife Management in the Indigenous-owned Savanna.* Centre for Aboriginal Economic Policy, Canberra.

Altman JC, Larsen L (2006) *Natural and Cultural Resource Management.* CAEPR Topical Issue 2006/07. Centre for Aboriginal Economic Policy Research, Australian National University, Canberra.

Altman JC, Jordan K, Kerins S, Buchanan G, Biddle N, Ens EJ, May K (2009) Indigenous interests in land and water. In *Northern Australia Land and Water Science Review 2009 Full Report.* pp. 1–56. CSIRO National Research Flagships Sustainable Agriculture, Northern Australian Land and Water Taskforce, Canberra.

Anderson C (1989) Aborigines and conservatism: the Daintree-Bloomfield road. *The Australian Journal of Social Issues* **24**(3), 214–227. doi:10.1002/j.1839-4655.1989.tb00866.x

Andrews AEJ (1981) *Hume and Hovell 1824.* Blubber Head Press, Hobart.

Andrews AEJ (ed.) (1986) *Stapylton with Major Mitchell's Australia Felix Expedition.* Blubberhead Press, Hobart.

Andrews TD, Buggey S (2008) Authenticity in Aboriginal cultural landscapes. *APT Bulletin* **39**(2/3), 63–71.

Angas GF (1847a) *South Australia Illustrated.* T. McLean, London.

Angas GF (1847b) *Savage Life and Scenes in Australia.* Smith, Elder and Co., London.

Angas GF (1877) The lonely Coorong. *Narracoorte Herald, South Australia* **17**, 4.

Annear R (1999) *Nothing But Gold: The Diggers of 1852.* Text Publishing, Melbourne.

Anonymous (1844a) Kangaroo Island. *Adelaide Observer,* South Australia. 28 September 1844, p. 5.

Anonymous (1844b) No title. *Geelong Advertiser and Squatters' Advocate,* Victoria. 23 May 1844, p. 2.

Anonymous (1845) Wonderful discovery of a new animal. *Geelong Advertiser and Squatters' Advocate,* Victoria. 2 July 1845, p. 2.

Anonymous (1847a) The bunyip. *Port Phillip Gazette and Settler's Journal,* Victoria. 9 January 1847, p. 2.

Anonymous (1847b) The bunyip. *Geelong Advertiser and Squatters' Advocate*, Victoria. 12 January 1847, p. 2.

Anonymous (1847c) The bunyip. *Melbourne Argus*, Victoria. 29 June 1847, p. 2.

Anonymous (1847d) No title. *Geelong Advertiser and Squatters' Advocate*, Victoria. 30 July 1847, p. 2.

Anonymous (1851) *Map of Bendigo Goldfields*. La Trobe University Library, Bendigo.

Anonymous (1857) A veritable bunyip. *Age*, Melbourne. 13 July 1857, p. 3.

Anonymous (1858) Curious animal. *Argus*, Melbourne. 2 September 1858, p. 4.

Anonymous (1861) The diseases of the Aborigines. *Age*, Melbourne. 30 September 1861, p. 6.

Anonymous (1863) Importing a governess. *Bendigo Advertiser*, Victoria. 15 January 1863, p. 1.

Anonymous (1864) Native funeral on the Murray. *Illustrated Sydney News*, Sydney. 16 July 1864, p. 9.

Anonymous (1865a) General news. *Hamilton Spectator and Grange District Advertiser*, South Melbourne. 3 May 1865, p. 3.

Anonymous (1865b) What is the bunyip? *Ballarat Star*, Victoria. 5 May 1865, p. 3.

Anonymous (1865c) The decaying race. *London Times*, England. 1865, p. 5.

Anonymous (1867a) Current topics. *Geelong Advertiser*, Victoria. 20 March 1867, p. 2.

Anonymous (1867b) Current topics. *Geelong Advertiser*, Victoria. 17 October 1867, p. 2.

Anonymous (1867c) Country news. *The Argus*, Victoria. 17 October 1867, p. 7.

Anonymous (1871a) A real bunyip. *Ovens and Murray Advertiser*, Beechworth. 23 March 1871, p. 3.

Anonymous (1871b) Editorial. *Avoca Mail*, Victoria. 16 September 1871, p. 2.

Anonymous (1876) Report of Eddington Bridge Fire. *Dunolly & Bet Bet Shire Express*, Central Goldfields. 22 February 1876, n.p.

Anonymous (1877) Editorial. *Bendigo Advertiser*, Victoria. 17 March 1877, p. 2.

Anonymous (1878) Town talk. *Geelong Advertiser*, Victoria. 20 September 1878, p. 2.

Anonymous (1882) Fernihurst. *Bendigo Advertiser*, Victoria. 22 February 1882, p. 3.

Anonymous (1883a) Editorial. *Bacchus Marsh Express*, Victoria. 16 June 1883, p. 2.

Anonymous (1883b) Twenty years station life in New South Wales. *The Graphic*, n.d, n.p.

Anonymous (1885a) The pelican. *Border Watch*, Mount Gambier. 4 February 1885, p. 4.

Anonymous (1885b) Heywood. *Hamilton Spectator*, Victoria. 25 April 1885, p. 3.

Anonymous (1886) The sea-cow. *Mount Alexander Mail*, Victoria. 21 September 1886, p. 4.

Anonymous (1887a) Port Fairy. *Hamilton Spectator*, Victoria. 31 May 1887, p. 3.

Anonymous (1887b) What the first settlers found. *South Australian Register*, Adelaide. 21 June 1887, pp. 3–4.

Anonymous (1888a) A mysterious animal. Does the bunyip really exist. *Morwell Advertiser and Weekly Chronicle*, Morwell. 19 May 1888, p. 2.

Anonymous (1888b) The history of Victoria. *Illustrated Australian News*, Melbourne. 1 August 1888, pp. 2–8.

Anonymous (1888c) Editorial. *Ballarat Star*, Victoria. 10 August 1888, p. 2.

Anonymous (1889) Our native plants. *Adelaide Observer*, South Australia. 2 February 1889, p. 9.

Anonymous (1893) Obtaining water in the Mallee. *Weekly Times*, Melbourne. 18 March 1893, p. 9.

Anonymous (1898) Olla Podrida. *Kerang Times*, Victoria. 24 May 1898, p. 2.

Anonymous (1901) A snow storm. Record falls in the hills. Snow-balling at Mount Barker. *Mount Barker Courier, Onkaparinga & Gumeracha Advertiser*, South Australia. 2 August 1901, p. 2.

Anonymous (1902) Dolphins on the lakes. *Advertiser*, Adelaide. 20 November 1902, p. 7.

Anonymous (1904a) Aboriginal mythology. *The Newsletter. An Australian Paper for Australian People*, Sydney. 2 January 1904, p. 16.

Anonymous (1904b) Weather forecasts. *The Brisbane Courier*, Queensland. 24 September 1904, p. 4.

Anonymous (1911) The killers. *Twofold Bay Magnet, and South Coast and Southern Monaro Advertiser*, New South Wales. 3 July 1912, p. 7.

Anonymous (1912) The killers. *Twofold Bay Magnet, and South Coast and Southern Monaro Advertiser*, New South Wales. 8 July 1912, p. 4.

Anonymous (1914a) Aboriginal methods of obtaining water. *Australasian*, Melbourne. 17 January 1914, p. 11.

Anonymous (1914b) General news. Porpoise in Lake Alexandrina. *Advertiser*, Adelaide. 2 May 1914, p. 20.

Anonymous (1916) Notice. *Twofold Bay Magnet, and South Coast and Southern Monaro Advertiser*, New South Wales. 10 July 1916, p. 3.

Anonymous (1917) Death of a district pioneer. *Mornington Standard*, Frankston. 27 January 1917, p. 3.

Anonymous (1924) Bunyips and legends. *The Mail*, Adelaide. 1 March 1924, p. 1.

Anonymous (1927a) Porpoises at Wellington. Strange scene in the Murray. *Register*, Adelaide. 16 August 1927, p. 10.

Anonymous (1927b) General news. Porpoise at Tailem Bend. *Register*, Adelaide. 30 August 1927, p. 11.

Anonymous (1934a) Relics of Aboriginal tribes in the South-East. *The Narracoorte Herald*, South Australia. 6 March 1934, p. 2.

Anonymous (1934b) Bunyips migrates. Finds home in Scottish waters. Highland zoologist recognises Loch Ness monster. *Kalgoorlie Miner*, Western Australia. 5 April 1934, p. 3.

Anonymous (1935) Native, older than S.A., dies. *Mail*, Adelaide. 9 February 1935, p. 1.

Arbury J (1998) Breeding brambles. *The Garden. Journal of the Royal Horticultural Society* **123**(8), 578–583.

Archer M, Beale B (2004) *Going Native. Living in the Australian Environment.* Hodder, Sydney.

Arnot J (1852) *An Emigrants Journal.* Ms 989/3. State Library of Victoria, Melbourne.

Arthur JM (1996) *Aboriginal English. A Cultural Study.* Oxford University Press, Melbourne.

A.S.T. (1888) The bunyip. *Alexandra and Yea Standard, Gobur. Thornton and Acheron Express*, Victoria. 28 December 1888, p. 6.

Atkinson W (2005) Yorta Yorta occupation and the search for common ground. *Proceedings of the Royal Society of Victoria* **17**, 1–21.

Attwood B (1994) *A Life Together, a Life Apart: A History of Relations between Europeans and Aborigines.* Melbourne University Press, Carlton, Victoria.

Bailey FM (1880) Medicinal plants of Queensland. *Proceedings of the Linnean Society of New South Wales* **5**(1), 1–29.

Baker RM (1988) Yanyuwa Canoe Making. *Records of the South Australian Museum* **22**(2), 173–188.

Baker RM (1999) *Land is Life: From Bush to Town. The Story of the Yanyuwa People.* Allen & Unwin, Sydney.

Balme JM, Beck W (1996) Earth mounds in southeastern Australia. *Australian Archaeology* **42**, 39–51. doi:10.1080/03122417.1996.11681571

Bamblett M (2014) The possum skin cloak – being warmed by culture. *Children Australia* **39**(3), 134–136. doi:10.1017/cha.2014.15

Bancroft J (1877–80) *Papers on Pituri and Duboisia.* Warwick and Sapsford, Brisbane.

Bannerman C (2006) Indigenous food and cookery books: redefining Aboriginal cuisine. *Journal of Australian Studies* **30**(87), 19–36. doi:10.1080/14443050609388048

Barrett C (1946) *The Bunyip.* Reed and Harris, Melbourne.

Barwick DE (1971a) Changes in the Aboriginal population of Victoria, 1863–1966. In *Aboriginal Man and Environment in Australia.* (Eds DJ Mulvaney and J Golson) pp. 288–315. Australian National University Press, Canberra.

Barwick DE (1971b) What it means to be an Aborigine. In *The Aborigine Today.* (Ed. B Leach) pp. 15–23. Paul Hamlyn, Sydney.

Barwick DE (1984) Mapping the past: an atlas of Victorian clans. *Aboriginal History* **8**(2), 100–131.

Basedow H (1907) Anthropological notes on the western coastal tribes of the Northern Territory of South Australia. *Transactions of the Royal Society of South Australia* **31**, 1–62.

Bassett M (1962) *The Hentys: An Australian Colonial Tapestry.* Oxford University Press, London.

Bates DM (1906) The West Australian Aborigines. Marriage laws and some customs. *Western Mail, Western Australia* **28**, 44.

Batey I (1916) *Reminiscences.* Manuscript (Ms 780). Royal Historical Society of Victoria, Melbourne.

Baylie WH (1843) On the Aborigines of the Goulburn district. *Port Phillip Magazine* **1**, 86–92.

Bayly IAE (1999) Review of how Indigenous people managed for water in desert regions of Australia. *Journal of the Royal Society of Western Australia* **82**, 17–25.

Beatty B (1969) *A Treasury of Australian Folk Tales and Traditions.* Ure Smith, Sydney.

B.E.C. (1944) Nature notes. Ghost bird. *Portland Guardian*, Victoria. 27 April 1944, p. 1.

Beck W, Balme JM (2003) Dry rainforests: a productive habitat for Australian hunter-gatherers. *Australian Aboriginal Studies* **2**, 4–20.

Becke L (1904) *A Memory of the Southern Seas.* J.B. Lippincott Company, Philadelphia.

Beech M (1986) On meteors and mushrooms. *The Journal of the Royal Astronomical Society of Canada* **81**(605), 27–29.

Bell J (2010) Language and linguistic knowledge: a cultural treasure. *Ngoonjook* **35**, 84–96.

Bell D, Johnston C (2008) Budj Bim: caring for spirit and the people. In *16th ICOMOS General Assembly and International Symposium: 'Finding The Spirit of Place – Between the Tangible and the Intangible', 29 Sept - 4 Oct 2008, Quebec, Canada* (www.openarchive.icomos.org/12/1/77-7xnf-104.pdf; http://openarchive.icomos.org/12/; accessed 1 June 2017).

Bellchambers JP (1931) *A Nature-Lovers Notebook.* Nature Lovers League, Adelaide.

Bennett G (1834) *Wanderings in New South Wales, Batavia, Pedir Coast, Singapore and China Being the Journal of a Naturalist in Those Countries During 1832, 1833 and 1834.* Richard Bentley, London.

Bennett G (1860) *Gatherings of a Naturalist in Australasia: Being Observations Principally on the Animal and Vegetable Productions of New South Wales, New Zealand, and some of the Austral Islands.* John Van Voorst, London.

Berkes F (2009) Indigenous ways of knowing and the study of environmental change. *Journal of the Royal Society of New Zealand* **39**(4), 151–156. doi:10.1080/03014220909510568

Berkes F, Berkes MK (2009) Ecological complexity, fuzzy logic, and holism in indigenous knowledge. *Futures* **41**(1), 6–12. doi:10.1016/j.futures.2008.07.003

Berkes F, Colding J, Folke C (2000) Rediscovery of Traditional Ecological Knowledge as adaptive management. *Ecological Applications* **10**(5), 1251–1262. doi:10.1890/1051-0761(2000)010[1251:ROTEKA]2.0.CO;2

Berkinshaw TD (1999) The Business of Bush Foods: Ecological and Socio-cultural Implications. Postgraduate thesis, University of Adelaide, Adelaide.

Berndt RM (1940a) Some aspects of Jaraldi culture, South Australia. *Oceania* **11**(2), 164–185. doi:10.1002/j.1834-4461.1940.tb00283.x

Berndt RM (1940b) A curlew and owl legend from the Narunga tribe, South Australia. *Oceania* **10**(4), 456–462. doi:10.1002/j.1834-4461.1940.tb00306.x

Berndt RM (1947) Wuradjeri magic and "clever men". *Oceania* **17**(4), 327–365.

Berndt RM (1951) Ceremonial exchange in western Arnhem Land. *Southwestern Journal of Anthropology* **7**, 156–176. doi:10.1086/soutjanth.7.2.3628621

Berndt CH (1964) The role of native doctors in Aboriginal Australia. In *Magic, Faith and Healing* (Ed. A Kiev) pp. 264–282. Free Press, New York.

Berndt CH (1981) Interpretations and 'facts' in Aboriginal Australia. In *Woman the Gatherer.* (Ed. F Dahlberg) pp. 153–203. Yale University Press, New Haven.

Berndt RM, Berndt CH (1970) *Man, Land and Myth in North Australia. The Gunwinggu People.* Ure Smith, Sydney.

Berndt RM, Berndt CH (1989) *The Speaking Land.* Penguin, Melbourne.

Berndt RM, Berndt CH (1999) *The World of the First Australians: Aboriginal Traditional Life: Past and Present.* Aboriginal Studies Press, Canberra.

Berndt RM, Berndt CH, Stanton JE (1993) *A World That Was. The Yaraldi of the Murray River and the Lakes, South Australia.* Melbourne University Press at the Miegunyah Press, Melbourne.

Berndt RM, Berndt CH, Stanton JE (1999) *Aboriginal Art. A Visual Perspective.* Revised edition. Methuen Australia, Sydney.

Berry HL, Butler JRA, Burgess CP, King UG, Tsey K, Cadet-James YL, Rigby CW, Raphael B (2010) Mind, body, spirit: co-benefits for mental health from climate change adaption and caring for country in remote Aboriginal Australian communities. *NSW Public Health Bulletin* **21**(6), 139–145. doi:10.1071/NB10030

Beveridge P (1861–1864) A few notes on the dialects, habits, customs and mythology of the Lower Murray Aborigines. *Transactions and Proceedings Royal Society of Victoria* **6**, 14–24.

Beveridge P (1883) Of the Aborigines inhabiting the Great Lacustrine and Riverine Depression of the Lower Murray, Lower Murrumbidgee, Lower Lachlan, and Lower Darling. *Journal and Proceedings of the Royal Society of New South Wales* **17**, 19–74.

Beveridge P (1889) *The Aborigines of Victoria and Riverina, as Seen by Peter Beveridge.* M.L. Hutchinson, Melbourne.

Bibimus (1850) Water in the Mallee scrub. *Geelong Advertiser*, Victoria. 7 June 1850, p. 2.

Biernoff DC (1978) Safe and dangerous places. In *Australian Aboriginal Concepts.* (Ed. LR Hiatt) pp. 93–106. Australian Institute of Aboriginal Studies, Canberra.

Billot CP (1979) *John Batman: The Story of John Batman and the Founding of Melbourne.* Hyland House, Melbourne.

Bindon PR (1997) Aboriginal people and granite domes. *Journal of the Royal Society of Western Australia* **80**, 173–179.

Bird MI, Hutley LB, Lawes MJ, Lloyd J, Luly JG, Ridd PV, Roberts RG, Ulm S, Wurster CM (2013) Humans, megafauna and environmental change in tropical Australia. *Journal of Quaternary Science* **28**(5), 439–452. doi:10.1002/jqs.2639

Blacklock F (2002) *Aboriginal Skin Cloaks.* National Quilt Register http://www.collectionsaustralia. net/nqr/fabri.php (accessed 1 June 2017).

Blake BJ (2003) *The Warrnambool Language: A Consolidated Account of the Aboriginal Language of the Warrnambool Area of the Western District of Victoria Based on Nineteenth-century Sources.* Pacific Linguistics, Australian National University, Canberra.

Blake NM, Wightman G, Williams L (1998) 'Iwaidja Ethnobotany. Aboriginal Plant Knowledge from Gurig National Park, Northern Australia'. Northern Territory Botanical Bulletin No. 23. Parks and Wildlife Commission of the Northern Territory, Darwin.

Bland W (1965) *Journey of Discovery to Port Phillip, New South Wales by Messrs. W.H. Hovell, and Hamilton Hume in 1824 and 1825.* Libraries Board of South Australia, Adelaide.

Blandowski W (1855) Personal observations made in an excursion towards the central parts of Victoria, including Mount Macedon, McIvor and Black Ranges. *Transactions of the Philosophical Society of Victoria* **1**, 50–74.

Blandowski W (1858) Recent discoveries in natural history on the Lower Murray. *Transactions of the Philosophical Institute of Victoria* **2**, 124–137.

Blows M (1975) Eaglehawk and crow: birds, myths and moieties in south-east Australia. In *Australian Aboriginal Mythology.* (Ed. LR Hiatt) pp. 24–45. Australian Institute of Aboriginal Studies, Canberra.

Blyth T (c.1850s) *Diary.* MS 2310. National Library of Australia, Canberra.

Board for the Protection of the Aborigines in the Colony of Victoria (1872) *Report of the Board for the Protection of the Aborigines in the Colony of Victoria.* John Ferres, Melbourne.

Bohensky EL, Butler JR, Davies J (2013) Integrating indigenous ecological knowledge and science in natural resource management: perspectives from Australia. *Ecology and Society* **18**(3), 1–20. doi:10.5751/ES-05846-180320

Boldrewood R (1889) Old time memories – pioneer's luck. *Australasian*, Melbourne. 2 November 1889, p. 52.

Bonney N (1987) *Carpenter Rocks and Beyond.* The author, Millicent.

Bonney N (1994) *Uses of Native Plants in the South East of South Australia by the Indigenous Peoples Before 1839.* Celebration South East volume 4. Southeast Book Promotions, Naracoorte, South Australia.

Bonney N (2004) *Common Native Plants of the Coorong Region. Identification. Propagation. Historical Uses.* Australian Plants Society (SA Region) Inc., Adelaide.

Bonwick J (1856) *William Buckley, The Wild White Man. And his Port Phillip Black Friends.* George Nichols, Melbourne.

Bonwick J (1973) *John Batman – the Founder of Victoria.* (Edited by CE Sayers.) Wren, Melbourne.

Bonyhady T (1991) *Burke and Wills. From Melbourne to Myth.* David Ell Press, Sydney.

Braddon ME (1892) Australian myth. *Belgravia: A London Magazine* **78**, 86–94.

Bradley WA (1838–68) *Daily Journals 1838–1868*. In private hands. [Cited in McBryde 1984a].

Brand-Miller JC, Holt SHA (1998) Australian Aboriginal plant foods: a consideration of their nutritional composition and health implications. *Nutrition Research Reviews* **11**, 5–23. doi:10.1079/NRR19980003

Brand-Miller JC, James KW, Maggiore PMA (1993) *Tables of Composition of Australian Aboriginal Foods*. Aboriginal Studies Press, Canberra.

Bride TF (Ed.) (1898) *Letters from Victorian Pioneers: Being a Series of Papers on the Early Occupation of the Colony, the Aborigines, etc., Addressed by Victorian Pioneers to … Charles Joseph La Trobe*. RS Brain, Melbourne.

Bride TF (Ed.) (1969) *Letters from Victorian Pioneers: Being a Series of Papers on the Early Occupation of the Colony, the Aborigines, etc., Addressed by Victorian Pioneers to … Charles Joseph La Trobe*. 2nd edn. RS Brain, Melbourne.

Bridges W (no date) *The Travels of Walter Bridges*. Ms. No. 4140265. Ballarat and District Municipal Library, Ballarat, Victoria.

Brodribb WA, Bennett LH (1883) *Recollections of an Australian Squatter and Account of a Journey to Gipps Land*. Queensbury Hill Press, Melbourne.

Brout C (1861) *Guide for Emigrants to the Australian Gold Mines*. Paris.

Brown AR (1918) Notes on the social organisation of Australian tribes. *Journal of the Anthropological Institute of Great Britain* **48**, 222–253. doi:10.2307/2843422

Brown PL (1952) *Clyde Company Papers: Vol. 2: 1836–40*. Oxford University Press, London.

Brumm A (2010) 'The falling sky': symbolic and cosmological associations of the Mt William greenstone axe quarry, central Victoria. *Cambridge Archaeological Journal* **20**(2), 179–196. doi:10.1017/S0959774310000223

Builth H (1998) Lake Condah revisited: archaeological constructions of a cultural landscape. *Australian Archaeology* **47**, 68.

Builth H (2000) *The Connection Between the Gunditjmara Aboriginal People and Their Environment: The Case for Complex Hunter-gatherers in Australia*. People and Physical Environment Research, Sydney.

Builth H (2002) The Archaeology and Socioeconomy of the Gunditjmara: A Landscape Analysis from Southwest Victoria, Australia. PhD thesis. Flinders University of South Australia, Adelaide.

Builth H (2005) What we can learn from Lake Condah about sustainable living. *Local-Global: Identity, Security, Community* **1**, 16–22.

Builth H (2014) *Ancient Aboriginal Aquaculture Rediscovered: The Archaeology of an Australian Cultural Landscape*. Lambert Academic Publishing, Saarbrücken, Germany.

Builth H, Kershaw AP, White C, Roach A, Hartney L, McKenzie M, Jacobsen G (2008) Environmental and cultural change on the Mt Eccles lava-flow landscapes of southwest Victoria, Australia. *The Holocene* **18**(3), 413–424. doi:10.1177/0959683607087931

Bulmer J (1855–1908 [1999]) *John Bulmer's Recollections of Victorian Aboriginal Life, 1855–1908*. No. 3. (Compiled by A Campbell and edited by R Vanderwal.) Museum Victoria, Melbourne.

Bulmer J (1887) Some account of the Aborigines of the Lower Murray, Wimmera, Gippsland and Maneroo. *Transactions and Proceedings of the Royal Geographical Society of Australasia. Victorian Branch* **5**(1), 15–43.

Bunce D (1859) *Travels with Dr. Leichhardt in Australia*. Steam Press, Melbourne.

Burchett C (1840s) *Letters*. MS 014506, Box 33–4. Royal Historical Society of Victoria, Melbourne.

Burrows N (2003) Using and sharing Indigenous knowledge. In *Australia Burning: Fire Ecology, Policy and Management Issues*. (Eds G Cary, D Lindenmayer and S Dovers) pp. 205–210. CSIRO Publishing, Collingwood, Victoria.

Butler R, Hinch T (Eds) (2007) *Tourism and Indigenous Peoples: Issues and Implications*. Butterworth-Heinemann, Oxford.

Butler DW, Fensham RJ, Murphy BP, Haberle SG, Bury SJ, Bowman DM (2014) Aborigine-managed forest, savanna and grassland: biome switching in montane eastern Australia. *Journal of Biogeography* **41**(8), 1492–1505. doi:10.1111/jbi.12306

Butlin NG (1993) *Economics and the Dreamtime: A Hypothetical History*. Cambridge University Press, Cambridge.

Buultjens J, Gale D, White NE (2010) Synergies between Australian indigenous tourism and ecotourism: possibilities and problems for future development. *Journal of Sustainable Tourism* **18**(4), 497–513. doi:10.1080/09669581003653518

Byrne CJ (1848) *Twelve Years' Wanderings in the British Colonies from 1835 to 1847.* R Bentley, London.

Byrt P (2011) *I Succeeded Once: The Aboriginal Protectorate on the Mornington Peninsula, 1839–1840.* ANU E Press, Canberra.

Cahir D (2001) The Wathawurrung People's Encounters with Outside Forces 1797–1849: A History of Conciliation and Conflict. MA thesis. University of Ballarat, Mt Helen, Vic.

Cahir D (2010) 'Are you off to the Diggings?': Aboriginal guiding to and on the goldfields. *The La Trobe Journal* **85**, 22–36.

Cahir F (2005) Dallong – possum skin rugs: a study of an inter-cultural trade item in Victoria. *Provenance: The Journal of Public Record Office Victoria* **4**(5), 1–14.

Cahir F (2006) Black gold: a history of the role of Aboriginal people on the goldfields of Victoria, 1850–70. PhD thesis. University of Ballarat, Ballarat.

Cahir F (2012) Murnong: much more than a food. *The Artefact: The Journal of the Archaeological and Anthropological Society of Victoria* **35**, 29–39.

Cahir F (2013) *Black Gold: Aboriginal People on the Goldfields of Victoria, 1850–1870.* ANU Press, Canberra.

Cahir F (2014) Finding Indigenous history in the RHSV collections. *The Victorian Historical Journal* **85**(1), 12–30.

Cahir F, Clark ID (2013) The historic importance of the dingo in Aboriginal society in Victoria (Australia): a reconsideration of the archival record. *Anthrozoos* **26**(2), 185–198. doi:10.2752/1753 03713X13636846944088

Cahir F, McMaster S, Clark I, Kerin R, Wright W (2016) Winda Lingo Parugoneit or Why Set the Bush [On] Fire? Fire and Victorian Aboriginal People on the Colonial Frontier. *Australian Historical Studies* **47**(2), 225–240. doi:10.1080/1031461X.2016.1156137

Cameron ALP (1885) Notes on some tribes of New South Wales. *Journal of the Anthropological Institute of Great Britain and Ireland* **14**, 344–370. doi:10.2307/2841627

Cameron-Bonney L (1990) *Out of the Dreaming.* Kingston, South East Kingston Leader.

Campbell AH (1965) Elementary food production by the Australian Aborigines. *Mankind* **6**(5), 206–211.

Campbell J (2002) *Invisible Invaders. Smallpox and Other Diseases in Aboriginal Australia, 1780–1880.* Melbourne University Press, Melbourne.

Campbell TD, Cleland JB, Hossfeld PS (1946) Aborigines of the lower south east of South Australia. *Records of the South Australian Museum* **8**, 445–502.

Campbell TG (1926) Insect foods of the Aborigines. *Australian Museum Magazine* **2**(12), 407–410.

Campbell WS (1932) The use and abuse of stimulants in the early days of settlement in New South Wales. *Journal & Proceedings of the Royal Australian Historical Society* **18**(2), 74–99.

Cane S (1987) Australian Aboriginal subsistence in the Western Desert. *Human Ecology* **15**(4), 391–434. doi:10.1007/BF00887998

Cann JH, De Deckker P, Murray-Wallace CV (1991) Coastal Aboriginal shell middens and their palaeoenvironmental significance, Robe Range, South Australia. *Transactions of the Royal Society of South Australia* **115**(4), 161–175.

Cannon M (1982) *The Aborigines of Port Phillip, 1835–1839. Vol. 2A.* Victorian Government Printing Office, Melbourne.

Cannon M (1983) *Aborigines and Protectors. The Aborigines of Port Phillip, 1835–1839. Vol. 2B.* Victorian Government Printing Office, Melbourne.

Cannon M (1991) *Historical Records of Victoria: Volume 6.* Victorian Government Printing Office, Melbourne.

Cannon M, MacFarlane I (1991) *The Crown, the Land and the Squatter, 1835–1840.* Melbourne University Press, Melbourne.

Carmichael DL, Hubert J, Reeves B, Schanche A (Eds) (2013) *Sacred Sites, Sacred Places.* Routledge, London.

Carr-Boyd W (1892) A cure for snake-bite. *The Argus*, Victoria. 7 April 1892, p. 9.

Carroll E, Patenaude N, Alexander A, Steel D, Harcourt R, Childerhouse S, Smith S, Bannister J, Constantine R, Baker CS (2011) Population structure and individual movement of southern right whales around New Zealand and Australia. *Marine Ecology Progress Series* **432**, 257–268. doi:10.3354/meps09145

Cary JJ (1904) Canoes of Geelong Aborigines. *Geelong Naturalist* **1**, 30–36.

Castlemaine Association of Pioneers and Old Residents (1972) *Records of the Castlemaine Pioneers.* Rigby, Melbourne.

Cawte J (1974) *Medicine is the Law.* University Press of Hawaii, Honolulu.

Cawte J (1996) *Healers of Arnhem Land.* University of New South Wales Press, Sydney.

Cayley NW (2011) *What Bird is That?* Revised by TR Lindsey. Australia's Heritage Publishing, Sydney.

Chatfield T (1999) Foreword. In *The People of Gariwerd. The Grampians Aboriginal Heritage.* (G Wettenhall). Aboriginal Affairs, Melbourne.

Cherikoff V, Isaacs J (1989) *The Bush Food Handbook. How to Gather, Grow, Process and Cook Australian Wild Foods.* Ti Tree Press, Sydney.

Cherikoff V, Johnson L (2000) 'Marketing the Australian Native Food Industry'. Publication no. 00/061. Rural Industries Research and Development Corporation, Barton, Australian Capital Territory.

Christie M (2007) Knowledge management and natural resource management. In *Investing in Indigenous Natural Resource Management.* (Eds MK Luckert, B Campbell and JT Gorman) pp. 86–90. Charles Darwin University Press, Darwin

Clark ID (1982) The Ethnocide of the Tjapwurong: The Nexus Between Conquest and Non-being. BA (Hons) thesis. Monash University, Melbourne.

Clark ID (1990a) *Aboriginal Languages and Clans: An Historical Atlas of Western and Central Victoria, 1800–1900.* Department of Geography and Environmental Science, Monash University, Melbourne.

Clark ID (1990b) In quest of the tribes: G.A. Robinson's unabridged report of his 1841 expedition among Western Victorian Aboriginal tribes; Kenyon's 'condensation' reconsidered. *Memoirs of the Museum of Victoria* **1**(1), 97–130[Anthropology and History].

Clark ID (2000) *The Papers of George Augustus Robinson, Chief Protector, Port Phillip Aboriginal Protectorate: Volume two: Aboriginal Vocabularies: South East Australia, 1839–1852.* Heritage Matters, Clarendon.

Clark ID (2001a) *The Papers of George Augustus Robinson, Chief Protector, Port Phillip Aboriginal Protectorate, Volume Two: 1 October 1840 – 31 August 1841.* Heritage Matters, Clarendon.

Clark ID (2001b) *The Papers of George Augustus Robinson, Chief Protector, Port Phillip Aboriginal Protectorate. Vol. 4: Annual and Occasional Reports 1841–1849.* Heritage Matters, Clarendon.

Clark ID (2002) The ebb and flow of tourism at Lal Lal Falls, Victoria: a tourism history of a sacred Aboriginal site. *Australian Aboriginal Studies* **2**, 45–53.

Clark ID (2003) *That's My Country Belonging to Me: Aboriginal Land Tenure and Dispossession in Nineteenth Century Western Victoria.* Heritage Matters, Melbourne.

Clark ID (2007a) The abode of malevolent spirits and creatures – caves in Victorian Aboriginal social organization. *Helictite* **40**(1), 3–10.

Clark ID (2007b) In quest of Nargun and Nyols: a history of indigenous tourism at the Buchan Caves Reserve. *Australasian Cave and Karst Management Association Inc Journal* **69**, 30–38.

Clark ID (2008) The northern Wathawurrung and Andrew Porteous, 1860–1877. *Aboriginal History* **32**, 97–108.

Clark ID (2010) Colonial tourism in Victoria, Australia, in the 1840s: George Augustus Robinson as a nascent tourist. *International Journal of Tourism Research* **12**(5), 561–573.

Clark ID (2013) *'Prettily situated' at Mungallook: A History of the Goulburn River Aboriginal Protectorate Station at Murchison, Victoria, 1840–1853.* Ballarat Heritage Services, Ballarat.

Clark ID (Ed.) (2014a) *The Papers of George Augustus Robinson, Chief Protector, Port Phillip Aboriginal Protectorate, 1 January 1839–30 September 1852*, Createspace, USA [single volume]. 1st edn.

Clark ID (Ed.) (2014b) *The Journals of George Augustus Robinson, Chief Protector, Port Phillip Aboriginal Protectorate, 1 January 1839–30 September 1852*, Createspace, USA [single volume]. 1st edn.

Clark ID (2016a) Appendix 11. The Challicum bunyip. In *'We are all of one blood' – A History of the Djabwurrung Aboriginal people of Western Victoria, 1836–1901. Vol. 3 Anthology of Sources*. pp. 42–51. CreateSpace Publishing, USA.

Clark ID (2016b) *The Last Matron of Coranderrk: Natalie Robarts's Diary of the Final Years of Coranderrk Aboriginal Station, 1909–1924*. Createspace Publishing, USA.

Clark ID (2017) Aboriginal people and frontier violence: the letters of Richard Hanmer Bunbury to his father, 1841–1847. *La Trobeana Journal of the C.J. La Trobe Society* **16**(1), 25–40.

Clark ID, Cahir DA (2003) Aboriginal people, gold, and tourism: the benefits of inclusiveness for goldfields tourism in regional Victoria. *Tourism. Culture and Communication* **4**, 123–136.

Clark ID, Cahir F (2008) The comfort of strangers: hospitality on the Victorian goldfields, 1850–1860. *Journal of Hospitality and Tourism Management* **15**(1), 2–7. doi:10.1375/jhtm.15.2

Clark ID, Cahir DA (2011) Understanding 'Ngamadjidj': Aboriginal perceptions of Europeans in nineteenth century western Victoria. *Journal of Australian Colonial History* **13**, 105–124.

Clark ID, Cahir F (Eds) (2016) *The Children of the Port Phillip Aboriginal Protectorate: An Anthology of Their Reminiscences*. Australian Scholarly Publishing, Melbourne.

Clark ID, Heydon TG (2002) *Dictionary of Aboriginal Placenames of Victoria*. Victorian Aboriginal Corporation for Languages, Melbourne.

Clark ID, Heydon TG (2004) *A Bend in the Yarra: a History of the Merri Creek Protectorate Station and Merri Creek Aboriginal School 1841–1851*. Aboriginal Studies Press, Canberra.

Clarke A (1991) 'Lake Condah Project, Aboriginal Archaeology – Resource Inventory'. Victorian Archaeological Survey Occasional Report 36. Department of Conservation and Environment, Melbourne.

Clarke A (1994) Romancing the stones. The cultural construction of archaeological landscape in the western district of Victoria. *Archaeology in Oceania* **29**, 1–15.

Clarke M (2012) 'Australian Native Food Industry Stocktake'. 12/066 RIRDC publication. Rural Industries Research and Development Corporation, Canberra.

Clarke PA (1985a) The importance of roots and tubers as a food source for southern South Australian Aborigines. *Journal of the Anthropological Society of South Australia* **23**(6), 2–12.

Clarke PA (1985b) Fruits and seeds as food for southern South Australian Aborigines. *Journal of the Anthropological Society of South Australia* **23**(9), 9–22.

Clarke PA (1986a) Aboriginal use of plant exudates, foliage and fungi as food and water sources in southern South Australia. *Journal of the Anthropological Society of South Australia* **24**(3), 3–18.

Clarke PA (1986b) The study of ethnobotany in southern South Australia. *Australian Aboriginal Studies* **2**, 40–47.

Clarke PA (1987) Aboriginal uses of plants as medicines, narcotics and poisons in southern South Australia. *Journal of the Anthropological Society of South Australia* **25**(5), 3–23.

Clarke PA (1988) Aboriginal use of subterranean plant parts in southern South Australia. *Records of the South Australian Museum* **22**(1), 63–76.

Clarke PA (1989) The computerization of the South Australian Museum's Anthropology Registers. *Conference of Museum Anthropologists Bulletin* **21**, 2–7.

Clarke PA (1990) *Fieldnotes, Lower Murray*. The author, Adelaide.

Clarke PA (1991) Adelaide as an Aboriginal landscape. *Aboriginal History* **15**(1), 54–72.

Clarke PA (1994) Contact, Conflict and Regeneration. Aboriginal Cultural Geography of the Lower Murray, South Australia. Unpublished Ph.D. dissertation. Departments of Geography and Anthropology, University of Adelaide, Adelaide.

Clarke PA (1995a) Computerisation of the Anthropology Division's artefact registers at the South Australian Museum. *Conference of Museum Anthropologists Bulletin* **26**, 48–55.

Clarke PA (1995b) Myth as history: the Ngurunderi mythology of the Lower Murray, South Australia. *Records of the South Australian Museum* **28**(2), 143–157.

Clarke PA (1996) Early European interaction with Aboriginal hunters and gatherers on Kangaroo Island, South Australia. *Aboriginal History* **20**(1), 51–81.

Clarke PA (1997) The Aboriginal cosmic landscape of southern South Australia. *Records of the South Australian Museum* **29**(2), 125–145.

Clarke PA (1998a) The Aboriginal presence on Kangaroo Island, South Australia. In *Aboriginal Portraits of 19th Century South Australia*. (Eds J Simpson and L Hercus) pp. 14–48. Aboriginal History Monograph. Australian National University, Canberra.

Clarke PA (1998b) Early Aboriginal plant foods in southern South Australia. *Proceedings of the Nutrition Society of Australia* **22**, 16–20.

Clarke PA (1999a) Waiyungari and his role in the mythology of the Lower Murray, South Australia. *Records of the South Australian Museum* **32**(1), 51–67.

Clarke PA (1999b) Spirit beings of the Lower Murray, South Australia. *Records of the South Australian Museum* **31**(2), 149–163.

Clarke PA (2001) The significance of whales to the Aboriginal people of southern South Australia. *Records of the South Australian Museum* **34**(1), 19–35.

Clarke PA (2002) Early Aboriginal fishing technology in the Lower Murray, South Australia. *Records of the South Australian Museum* **35**(2), 147–167.

Clarke PA (2003a) *Where the Ancestors Walked. Australia as an Aboriginal Landscape.* Allen & Unwin, Sydney.

Clarke PA (2003b) Australian ethnobotany: an overview. *Australian Aboriginal Studies* **2**, 21–38.

Clarke PA (2005a) Aboriginal 'fire-stick' burning practices on the Adelaide Plains. In *Adelaide. Nature of a City. The Ecology of a Dynamic City from 1836 to 2036.* (Eds CB Daniels and CJ Tait) pp. 424, 428–429. BioCity: Centre for Urban Habitats, Adelaide.

Clarke PA (2005b) Aboriginal relationships with grass. In *Grasslands Conservation and Production: Both Sides of the Fence. Proceedings of the Fourth Stipa National Conference on the Management of Native Grasses and Pastures, Burra, SA. 11th – 13th October 2005*, 1–5. Stipa, Wellington, New South Wales.

Clarke PA (2007a) *Aboriginal People and Their Plants.* Rosenberg Publishing, Dural Delivery Centre, New South Wales.

Clarke PA (2007b) Indigenous spirit and ghost folklore of 'settled' Australia. *Folklore* **118**(2), 141–161. doi:10.1080/00155870701337346

Clarke PA (2008a) *Aboriginal Plant Collectors. Botanists and Aboriginal People in the Nineteenth Century.* Rosenberg Publishing, Dural Delivery Centre, New South Wales.

Clarke PA (2008b) Aboriginal healing practices and Australian bush medicine. *Journal of the Anthropological Society of South Australia* **33**, 3–38.

Clarke PA (2009a) An overview of Australian Aboriginal ethnoastronomy. *Archaeoastronomy. The Journal of Astronomy in Culture* **21**, 39–58.

Clarke PA (2009b) Australian Aboriginal ethnometeorology and seasonal calendars. *History and Anthropology* **20**(2), 79–106. doi:10.1080/02757200902867677

Clarke PA (2009c) Aboriginal culture and the riverine environment. In *The Natural History of the Riverland and Murraylands*. (Ed. JT Jennings) pp. 142–161. Royal Society of South Australia, Adelaide.

Clarke PA (2012) *Australian Plants as Aboriginal Tools.* Rosenberg Publishing, Dural Delivery Centre, New South Wales.

Clarke PA (2013a) The Aboriginal ethnobotany of the Adelaide region, South Australia. *Transactions of the Royal Society of South Australia* **137**(1), 97–126. doi:10.1080/3721426.2013.10887175

Clarke PA (2013b) The use and abuse of Aboriginal ecological knowledge. In *The Aboriginal Story of Burke and Wills. Forgotten Narratives.* (Eds ID Clark and F Cahir) pp. 61–79. CSIRO Publishing, Melbourne.

Clarke PA (2013c) Review of 'Roving Mariners: Australian Aboriginal Whalers and Sealers in the Southern Oceans 1790–1870' by Lynette Russell. *Aboriginal History* **37**(2), 185–187.

Clarke PA (2014a) *Discovering Aboriginal Plant Use. Journeys of an Australian Anthropologist.* Rosenberg Publishing, Dural Delivery Centre, New South Wales.

Clarke PA (2014b) Australian Aboriginal astronomy and cosmology. In *Handbook of Archaeoastronomy and Ethnoastronomy*. (Ed. CLN Ruggles) pp. 2223–2230. Springer, New York.

Clarke PA (2014c) The ethnobotany of the Skyworld. Part 1: the flora and the aesthetics of the heavens in Aboriginal Australia. *Journal of Astronomical History and Heritage* **17**(3), 307–335.

Clarke PA (2014d) Review of 'Science & Sustainability. Learning from Indigenous Wisdom' by Joy Hendry. *Times Higher Education* December issue, **48**.

Clarke PA (2015a) The ethnobotany of the Skyworld. Part 2: plants' connections with the heavens of Aboriginal Australia. *Journal of Astronomical History and Heritage* **18**(1), 23–37.

Clarke PA (2015b) The Aboriginal ethnobotany of the South East of South Australia region. Part 1: seasonal life and material culture. *Transactions of the Royal Society of South Australia* **139**(2), 216–246. doi:10.1080/03721426.2015.1073415

Clarke PA (2015c) The Aboriginal ethnobotany of the South East of South Australia region. Part 2: foods, medicines and narcotics. *Transactions of the Royal Society of South Australia* **139**(2), 247–272. doi:10.1080/03721426.2015.1074339

Clarke PA (2015d) The Aboriginal ethnobotany of the South East of South Australia region. Part 3: mythology and language. *Transactions of the Royal Society of South Australia* **139**(2), 273–305. doi: 10.1080/03721426.2015.1074340

Clarke PA (2016a) Birds as totemic beings and creators in the Lower Murray, South Australia. *Journal of Ethnobiology* **36**(2), 277–293. doi:10.2993/0278-0771-36.2.277

Clarke PA (2016b) Early indigenous practices of bird foraging in the Lower Murray, South Australia. *Transactions of the Royal Society of South Australia* **140**, 1–22.

Clarke PA (2016c) Birds and the spirit world of the Lower Murray, South Australia. *Journal of Ethnobiology* **36**(4), 746–764. doi:10.2993/0278-0771-36.4.746

Cleland JB (1939) Some aspects of the ecology of the Aboriginal inhabitants of Tasmania and Southern Australia. *Papers and Proceedings of the Royal Society of Tasmania*, 1–18.

Clode D (2011) *Killers in Eden. The Story of a Rare Partnership Between Men and Killer Whales*. Museum Victoria, Melbourne.

Clowes EM (1911) *On the Wallaby: Through Victoria*. Heinemann, London.

Clutterbuck S (c.1855) *Diary 1849–1854*. MS 011230. Royal Historical Society of Victoria, Melbourne.

Cobcroft M (1983) Medical aspects of the Second Fleet. In *Australia's Quest for Colonial Health. Some Influences on Early Health and Medicine in Australia*. (Eds JH Pearn and C O'Carrigan) pp. 13–33. Department of Child Health, Royal Children's Hospital, Brisbane.

Cocker M, Tipling D (2013) *Birds and People*. Jonathan Cape, London.

Collins D (1798–1802) *An Account of the English Colony in New South Wales: With Remarks on the Dispositions, Customs, Manners, &c. of the Native Inhabitants of that Country*. Two volumes. T. Cadell and W. Davies, London.

Collins P (2006) *Burn: The Epic Story of Bushfire in Australia*. Allen & Unwin, Sydney.

Condon HT (1941) The stone plover (*Burhinus magnirostris*). *South Australian Naturalist* **21**(1), 8–12.

Conrick J (1923) The story of John Conrick, pioneer told by himself. *The News*. 28 July 1923, p. 4.

Costello P (1841–1860) *Patrick Costello; Narrative of His Life as a Port Phillip Pioneer*. MS 776. Royal Historical Society of Victoria, Melbourne.

Coutts PJF (1981) 'Readings in Victorian Prehistory. Volume 2. The Victorian Aboriginals 1800 to 1860'. Victoria Archaeological Survey, Ministry for Conservation, Melbourne.

Coutts PJF, Frank RK, Hughes P (1978) 'Aboriginal Engineers of the Western District, Victoria'. Records of the Victorian Archaeological Survey No. 7. Ministry of Conservation, Melbourne.

Craig BF (1969) *Central Australian and Western Desert Regions: An Annotated Bibliography*. Australian Institute of Aboriginal Studies, Canberra.

Craw C (2012) Tasting territory. Imagining place in Australian native food packing. *Locale: The Australasian-Pacific Journal of Regional Food Studies* **2**, 1–25.

Crawfurd J (1868) On the vegetable and animal food of the natives of Australia in reference to social position, with a comparison between the Australians and some other races of man. *Transactions of the Ethnological Society of London* **6**, 112–122. doi:10.2307/3014250

Cribb AB, Cribb JW (1982) *Useful Wild Plants in Australia*. Revised edition. Fontana/Collins, Sydney.

Cribb AB, Cribb JW (1987) *Wild Food in Australia*. Fontana/Collins, Sydney.

Critchett J (1990) *A 'distant field of murder': Western District Frontiers, 1834–1848*. Melbourne University Press, Melbourne.

Crook WP (1983) *An Account of the Settlement at Sullivan Bay, Port Phillip, 1803*. Colony Press, Melbourne.

Crook DA, Macdonald JI, Morrongiello JR, Belcher CA, Lovett D, Walker A, Nicol SJ (2014) Environmental cues and extended estuarine residence in seaward migrating eels (Anguilla australis). *Freshwater Biology* **59**(8), 1710–1720. doi:10.1111/fwb.12376

Crosby AW (2004) *Ecological Imperialism. The Biological Expansion of Europe, 900–1900*. Second edition. Cambridge University Press, Cambridge.

Cruz MG, Sullivan AL, Gould JS, Sims NC, Bannister AJ, Hollis JJ, Hurley RJ (2012) Anatomy of a catastrophic wildfire: The Black Saturday Kilmore East fire in Victoria, Australia. *Forest Ecology and Management* **284**, 269–285. doi:10.1016/j.foreco.2012.02.035

Crystal (1912) Aboriginal rain-makers. *Sydney Mail*, New South Wales. 4 December 1912, p. 27.

Curoc (1861) Protection to native industry by a blackfellow. *Ballarat Star*, Victoria. 16 July 1861, pp. 7–8.

Curr EM (1883) *Recollections of Squatting in Victoria: then called the Port Phillip District, (from 1841 to 1851)*. George Robertson: Melbourne, 1883. [Republished 1965 Melbourne University Press: Melbourne]

Curr EM (1886–87) *The Australian Race. Its Origin, Languages, Customs, Place of Landing in Australia and the Routes by which it Spread Itself over that Continent*. Four volumes. Trubner, London.

Daniels C, Nelson E, Roy J, Dixon P, Ens E, Towler G (2012) Commitment to our country. In *People on Country, Vital Landscapes, Indigenous Futures*. (Eds J Altman and S Kerins) pp. 174–189. Federation Press, Sydney.

Dannock J (c.1855) *Autobiography*. Mss. M1862. National Library of Australia, Canberra.

Darlot JM (1834) *James Monckton Darlot Reminiscences* 1834–1868. Box 21/1. State Library of Victoria, Melbourne.

David B (2002) *Landscapes, Rock-art and the Dreaming: An Archaeology of Preunderstanding*. Continuum, London.

Davidson DS (1932) *The Chronology of Australian Watercraft*. Thomas Avery & Sons, New Plymouth, U.S.A.

Davidson DS (1938) An ethnic map of Australia. *Proceedings of the American Philosophical Society* **79**(4), 649–679.

Davidson J (1898) Language of the Pinejunga people. In *Corartwalla. A History of Penola, the Land and its People*. (Ed. C Hanna) pp. 328–333. Magill Publications, Adelaide.

Davies E (1881) *The Story of an Earnest Life. A Woman's Adventures in Australia, and in Two Voyages Around the World*. Central Book Concern, Cincinnati.

Davies J (2007) *Walking Together, Working Together: Aboriginal Research Partnerships*. Desert Knowledge Cooperative Research Centre, Alice Springs.

Davies J, Holcombe S (2009) Desert knowledge: integrating knowledge and development in arid and semi-arid drylands. *GeoJournal* **74**(5), 363–375. doi:10.1007/s10708-009-9279-4

Davis S (1989) *Man of All Seasons. An Aboriginal Perspective of the Natural Environment*. Angus & Robertson, Sydney.

Davis S (1997) Documenting an Aboriginal seasonal calendar. In *Windows on Meteorology. Australian Perspective*. (Ed. EK Webb) pp. 29–33. CSIRO Publishing, Melbourne.

Dawson J (1881) *Australian Aborigines*. Robertson, Melbourne.

De Castella H, Thornton-Smith CB (1987) *Australian Squatters*. Melbourne University Press, Melbourne.

De Chabrillan C (1998) *The French Consul's Wife: Memoirs of Celeste De Chabrillan in Goldrush Australia. 1877*. Translated by P Clancy and J Allen. Miegunyah Press, Melbourne.

De Guchteneire P, Krukkert I, von Liebenstein G (Eds) (2002) *Best Practices on Indigenous Knowledge*. NUFFIC, Amsterdam.

DETE (2001) *Ngarrindjeri Dreaming Stories, with Paintings by Jacob Stengle*. Department of Education, Training and Employment Publishing, Adelaide.

Devereux E (2013) Traditional ecological knowledge: learning from the landscape – Emma Devereux. *Women* 1(2).

Dixon RMW, Ramson WS, Thomas M (1992) *Australian Aboriginal Words in English. Their Origin and Meaning*. Oxford University Press Australia, Melbourne.

Dollin AE, Dollin LJ, Sakagami SF (1997) Australian stingless bees of the genus *Trigona* (Hymenoptera: Apidae). *Invertebrate Taxonomy* **11**, 861–896. doi:10.1071/IT96020

Donaldson I, Donaldson T (Eds) (1985) *Seeing the First Australians*. Allen & Unwin, Sydney.

Draper N (2015) Islands of the dead? Prehistoric occupation of Kangaroo Island and other southern offshore islands and watercraft use by Aboriginal Australians. *Quaternary International* **385**, 229–242. doi:10.1016/j.quaint.2015.01.008

Dredge J (1845a) *Brief Notices of the Aborigines of New South Wales, including Port Phillip, in Reference to Their Past History and Present Condition*. James Harrison, Geelong.

Dredge J (1845b) *Letterbook 20 April 1839–3 January 1845*. MS 11625. State Library of Victoria, Melbourne.

Dugdale AE (2008) Where do Queensland's Indigenous People live? *The Medical Journal of Australia* **188**(10), 614.

Dumont d'Urville J, Rosenman H (1988) *An Account in Two Volumes of Two Voyages to the South Seas by Captain (later Rear-Admiral) Jules S-C Dumont d'Urville of the French Navy to Australia, New Zealand, Oceania, 1826–1829 in the Corvette Astrolabe and to the Straits of Magellan, Chile, Oceania, South East Asia, Australia, Antarctica, New Zealand, and Torres Strait, 1837–1840 in the Corvettes Astrolabe and Zelee*. University of Hawaii Press, Honolulu.

Earl GW (1846) On the Aboriginal tribes of the northern coast of Australia. *Royal Geographic Society of London* **16**, 239–251.

Education Department of South Australia (1991) *The Kai Kai Nature Trail: A Resource Guide for Aboriginal Studies*. Aboriginal Studies R-12. Education Department of South Australia, Adelaide.

Edwards R (1972) *Aboriginal Bark Canoes of the Murray Valley*. Rigby, Adelaide.

Eickelkamp U (2004) Egos and ogres: aspects of psychosexual development and cannibalistic demons in Central Australia. *Oceania* **74**(3), 161–189. doi:10.1002/j.1834-4461.2004.tb02849.x

E.K.V (1884) Aboriginal star lore. *The Queenslander*, Brisbane. 6 September 1884, p. 387.

Elder J, Taylor M (Ed.) (2013) *To a Land of Plenty: John Elder's Voyage from Scotland to Port Phillip and his Account of the Colony, 1849–51*. McComas Taylor, Canberra.

Elfin (1886) Old English customs and folk-lore. *South Australian Register*, Adelaide. 2 December 1886, p. 3.

Elkin AP (1932) The secret life of the Australian Aborigines. *Oceania* **3**(2), 119–138. doi:10.1002/j.1834-4461.1932.tb00061.x

Elkin AP (1964) *The Australian Aborigines. How to Understand Them*. Fourth edition. Angus and Robertson, Sydney.

Elkin AP (1977) *Aboriginal Men of High Degree*. Second edition. St Lucia, Queensland, University of Queensland Press.

Ellen R (1982) *Environment, Subsistence and System: The Ecology of Small-scale Social Formations*. Cambridge University Press, Cambridge.

Elliot RW, Jones DL, Blake T (1993) *Encyclopaedia of Australian Plants Suitable for Cultivation: Volume 6 (K-M)*. Lothian Press, Port Melbourne, Victoria.

Ely B (1980) *Murray/Murundi*. Experimental Art Foundation, Adelaide.

Ens EJ (2012a) *Indigenous Women Rangers Talking: Sharing Ideas and Information about Women Ranger's Work. Technical Report*. ANU Centre for Aboriginal Economic Policy Research, Canberra.

Ens EJ (2012b) Conducting two-way ecological research. In *People on Country, Vital Landscapes, Indigenous Futures*. (Eds J Altman and S Kerins) pp. 45–64. Federation Press, Sydney.

Ens EJ, Finlayson M, Preuss K, Jackson S, Holcombe S (2012) Australian approaches for managing 'country' using indigenous and non-indigenous knowledge. *Ecological Management & Restoration* **13**(1), 100–107. doi:10.1111/j.1442-8903.2011.00634.x

Ens EJ, Burns E, Russell-Smith J, Sparrow B, Wardle GM (2014) The cultural imperative: broadening the vision of long-term monitoring to enhance environmental policy and management outcomes. In *Biodiversity and Environmental Change: Monitoring, Challenges and Directions*. (Eds D Lindenmayer, E Burns, N Thurgate, AJ Lowe) pp. 83–107. CSIRO Publishing, Melbourne.

Ens EJ, Pert P, Clarke PA, Budden M, Clubb L, Doran B, Douras C, Gaikwad J, Gott B, Leonard S, Locke J, Packer J, Turpin G, Wason S (2015) Indigenous biocultural knowledge in ecosystem science and management: review and insight from Australia. *Biological Conservation* **181**, 133–149. doi:10.1016/j.biocon.2014.11.008

Ens EJ, Scott ML, Yugul Mangi Rangers, Moritz C, Pirzl R (2016) Putting Indigenous conservation policy into practice delivers biodiversity and cultural benefits. *Biodiversity and Conservation* **25**(14), 2889–2906. doi:10.1007/s10531-016-1207-6

Ens EJ, Walsh F, Clarke PA (2017) Aboriginal people's past and current influence on Australia's vegetation. In *Australian Vegetation*. Third edition. (Ed. D Keith) pp. 89–112. Cambridge University Press, Cambridge.

Erskine JE (1852) *A Short Account of the Late Discoveries of Gold in Australia: With Notes of a Visit to the God District*. T. & W. Boone, London.

Evans N (2010) *Dying Words: Endangered Languages and What They Have to Tell Us*. Wiley, West Sussex.

Eyre EJ (1832–39 [1984]) *Autobiographical narrative of residence and exploration in Australia 1832–1839*. Published from the original manuscript. Caliban Books, London.

Eyre EJ (1845) *Journals of Expeditions of Discovery*. 2 volumes. Boone, London.

Faull J (1983) *The Cornish in Australia, Australian Ethnic Heritage Series*. AE Press, Melbourne.

Fawcett SG (1955) 'Upper Hume Catchment: Ecological Report'. Unpublished report. Soil Conservation Authority, Melbourne.

Fels MH (2011) 'I Succeeded Once' The Aboriginal Protectorate on the Mornington Peninsula, 1839–1840. Aboriginal History Monograph 22. Australian National University Press, Canberra.

Ferguson CD (1888) *The Experiences of a Forty-niner During Thirty-Four Years' Residence in California and Australia*. Williams Publishing Company, Cleveland, Ohio, U.S.A.

Fernihurst District History Committee (1992) *Reflections from the Kinypaniel*. Fernihurst District History Committee, Fernihurst, Victoria.

Fison L, Howitt AW (1880) *Kamilaroi and Kurnai*. George Robertson, Melbourne.

Flannery TF (1994) *The Future Eaters. An Ecological History of Australasian Lands and People*. Reed, Melbourne.

Flinders M (1814) *A Voyage to Terra Australis; Undertaken for the Purpose of Completing the Discovery of that Vast Country, and Prosecuted in the Years 1801, 1802, and 1803 in His Majesty's Ship The Investigator*. G and W Nicol, London.

Flood J (1980) *The Moth Hunters: Aboriginal Prehistory of the Australian Alps*. Australian Institute of Aboriginal Studies, Canberra.

Foster R (1990) Two early reports on the Aborigines of South Australia. *Journal of the Anthropological Society of South Australia* **28**(1&2), 38–63.

Foster R, Monaghan P, Mühlhäusler P (2003) *Early Forms of Aboriginal English in South Australia, 1840s-1920s*. Pacific Linguistics, Australian National University, Canberra.

Frankel D (1982) Earth rings at Sunbury, Victoria. *Archaeology in Oceania* **17**(2), 89–97. doi:10.1002/j.1834-4453.1982.tb00043.x

Fraser A (1901) Hot and cold weather, and who caused it. *Science of Man* **4**(2), 29.

Fraser J (1892) *The Aborigines of New South Wales*. Government Printers, Sydney.

Fraser JG (1930) *Myths of the Origin of Fire*. Macmillan, London.

Fred S (1893) Some southern beasts. Under which head are included the Aborigines of this country. *Adelaide Observer*, South Australia. 2 December 1893, p. 33.

Froggatt WW (1903) Insects used as food by the Australian natives. *Science of Man* **6**(1), 11–13.

Fuller RS, Hamacher DW, Norris RP (2013) Astronomical orietnations of bora ceremonial grounds in southeast Australia. *Australian Archaeology* **77**(1), 30–37.

Gammage W (2011) *The Biggest Estate on Earth: How Aborigines Made Australia*. Allen & Unwin, Sydney.

Gara T (1985) Aboriginal techniques for obtaining water in South Australia. *Journal of the Anthropological Society of South Australia* **23**(2), 6–11.

Gaughwin D, Fullagar R (1995) Victorian offshore islands in a mainland coastal economy. *Australian Archaeology* **40**, 38–50. doi:10.1080/03122417.1995.11681546

Gerritsen R (2000) *The Traditional Settlement Pattern in South West Victoria Reconsidered*. Intellectual Property Publications, Canberra.

Gerritsen R (2001) Aboriginal fish hooks in southern Australia: evidence, arguments and implications. *Australian Archaeology* **52**, 18–28. doi:10.1080/03122417.2001.11681699

Gerstaecker F (1853) *Narrative of a Journey Round the World: Comprising a Winter-passage Across the Andes to Chili, with a Visit to the Gold Regions of California and Australia, the South Sea Islands, Java, &c.* Harper & Brothers, New York.

Gilbert G (1842) Unregistered Correspondence Regarding the Dismissal of W. Le Souëf from the Office of Assistant Protector of Aborigines, North Eastern District. See papers relating to the sale of skins by George Bertram in Unit 1, Folder No. 1. Public Record Office of Victoria. VPRS 4398/PO.

Gilligan I (2008) Clothing and climate in Aboriginal Australia. *Current Anthropology* **49**(3), 487–495. doi:10.1086/588199

Gilmore M (1934) *Old Days: Old Ways*. Angus & Robertson, Sydney.

Gisborne HF (1839) *Report to C.J La Trobe of His Visit to Portland Bay, December 1839*. PROV VPRS 6766 Outward Correspondence Vol.

Glen AS, Dickman CR (2006) Home range, denning behaviour and microhabitat use of the carnivorous marsupial *Dasyurus maculatus* in eastern Australia. *Journal of Zoology* **268**(4), 347–354. doi:10.1111/j.1469-7998.2006.00064.x

Goddard C, Kalotas A (Eds) (1988) *Punu. Yankunytjatjara Plant Use*. Angus & Robertson, Sydney.

Godden L, Cowell S (2016) Conservation planning and Indigenous governance in Australia's Indigenous Protected Areas. *Restoration Ecology* **24**(5), 692–697. doi:10.1111/rec.12394

Godelier M (1977) *Perspective in Marxist Anthropology*. Cambridge University Press, Cambridge.

Godfrey FR, Drought ML (1926) *Extracts from Old Journals*. Tytherleigh Press, Melbourne.

Gorenflo LJ, Romaine S, Mittermeier RA, Walker-Painemilla K (2012) Co-occurrence of linguistic and biological diversity in biodiversity hotspots and high biodiversity wilderness areas. *Proceedings of the National Academy of Sciences of the United States of America* **109**, 8032–8037. doi:10.1073/pnas.1117511109

Gosford R (2010) The Bush Stone Curlew as a Harbinger of Death ... and More. *The Northern Myth* 27 September 2010 (http://blogs.crikey.com.au/northern/2010/09/27/bird-of-the-week-the-bush-stone-curlew-as-a-harbinger-of-death-and-more/, accessed 23 February 2016).

Gott B (1982a) Ecology of root use by the Aborigines of Southern Australia. *Archaeology in Oceania* **17**, 59–67. doi:10.1002/j.1834-4453.1982.tb00039.x

Gott B (1982b) Kunzea pomifera – Dawson's "Nurt". *Artefact* **7**(1–2), 3–17.

Gott B (1983) Murnong – 'Microseris scapigera'. A study of a staple food of Victorian Aborigines. *Australian Aboriginal Studies* **2**, 2–17.

Gott B (1985) Plants mentioned in Dawson's *Australian Aborigines*. *Artefact* **10**, 3–14.

Gott B (2005) Aboriginal fire management in south-eastern Australia: aims and frequency. *Journal of Biogeography* **32**, 1203–1208. doi:10.1111/j.1365-2699.2004.01233.x

Gott B (2008) Indigenous use of plants in south-eastern Australia. *Telopea* **12**(2), 215–226. doi:10.7751/telopea20085811

Gott B (no date) *SALANG. Electronic Resource, Aboriginal Studies Electronic Data Archive.* (Retrieved from http://www.language-archives.org/archive/aseda.aiatsis.gov.au, accessed 10 March 2017).

Gott B, Conran J (1991) *Victorian Koorie Plants: Some Plants Used by Victorian Koories for Food, Fibre, Medicines and Implements*. Yangennanock Women's Group, Aboriginal Keeping Place, Hamilton, Victoria.

Gould RA (1969) *Yiwara: Foragers of the Australian Desert*. Collins, London.

Graham C, Hart D (1997) 'Prospects for the Australian Native Bushfood Industry'. Publication no.97/22. Rural Industries Research and Development Corporation, Canberra.

Grant J (1803) *Narrative of a Voyage of Discovery: Performed in His Majesty's Vessel the Lady Nelson....* Printed by C Roworth for T. Egerton, London.

Grant-Swan E (1931) The coaxers. Bringing summer. Quaint rites. Aboriginal ceremonies. *Sunday Mail*, Brisbane. 13 September 1931, p. 20 (republished as 'Hastening summer: quaint Aboriginal rites' by E. Grant Swan, *Sydney Mail*, New South Wales. 4 November 1936, p. 2).

Grassie (1876) Notes by the way. Mustering "the big paddock". *Border Watch*. Mount Gambier, South Australia. 9 December 1876, p. 4.

Grassie (1878) The traveller. *Border Watch*. Mount Gambier, South Australia 29 June 1878, p. 4.

Gray H (c.1850s) *Letters*. MS 6200. National Library of Australia, Canberra.

Gray A, Smith LR (1983) The size of the Aboriginal population. *Australian Aboriginal Studies* **1**, 2–9.

Great Britain Parliament House of Commons & Great Britain Colonial Office (1844) *Aborigines (Australian Colonies): Return to an Address of the House of Commons, dated 5 August 1844 for, Copies or Extracts from the Despatches of the Governors of the Australian Colonies, with Reports of the Protectors of Aborigines, and any other Correspondence to Illustrate the Condition of the Aboriginal Population*. House of Commons, London.

Green D, Raygorodetsky G (2010) Indigenous knowledge of a changing climate. *Climatic Change* **100**, 239–242. doi:10.1007/s10584-010-9804-y

Green D, Alexander L, McInnes K, Church J, Nicholls N, White N (2010a) An assessment of climate change impacts and adaptation for the Torres Strait Islands, Australia. *Climatic Change* **102**(3–4), 405–433. doi:10.1007/s10584-009-9756-2

Green D, Billy J, Tapim A (2010b) Indigenous Australians' knowledge of weather and climate. *Climatic Change* **100**(2), 337–354. doi:10.1007/s10584-010-9803-z

Greenway CC (1901) The constellation Pleiades – Mei-mei. *Science of Man* **4**(11), 190–191.

Gregory JW (1904) The antiquity of man in Victoria. *Proceedings of the Royal Society of Victoria*, Vol. XVII (New Series). Part 1, 121–141.

Griffin JG (1923) Australian Aboriginal astronomy. *The Journal of the Royal Astronomical Society of Canada*. *Royal Astronomical Society of Canada* **17**, 156–163.

Griffith C (1845) *The Present State and Prospects of the Port Phillip District of New South Wales*. William Curry, Junior and Company, Dublin.

Griffith CJ (c.1845–c.1859) *Reminiscences and Diaries, ca. 1845 – ca. 1859*. Ms. 10875. State Library of Victoria, Melbourne.

Griffiths PM (1988) *Three Times Blest – A History of Buninyong and District 1837–1901*. Buninyong and District Historical Society, Buninyong, Victoria.

Gross A (1924) *The Winning of Fire, Youloin Keear: A Legend of the Aborigines of South-western Victoria, Rendered into Rhymed English Verse*. Alan Gross, Melbourne.

Gunditjmara People, Wettenhall G (2010) *The People of Budj Bim: Engineers of Aquaculture, Builders of Stone House Settlements and Warriors Defending Country*. EmPress Publishing, Ballarat, Victoria.

Gunn RC (1847) The bunyip. *Port Phillip Patriot and Morning Advertiser*, Victoria. 20 March 1847, p. 2.

Haddon AC (1913) The outrigger canoes of Torres Strait and North Queensland. In *Essays and Studies Presented to Wm. Ridgeway*. (Ed. EC Quiggen) pp. 609–634. Cambridge University, Cambridge.

Hagen R (2001) Ethnographic information and anthropological interpretations in a native title claim: the Yorta Yorta experience. *Aboriginal History* **25**, 216–227.

Hagger J (1979) *Australian Colonial Medicine*. Rigby, Adelaide.

Hahn DM (1838–39 [1964]) Extracts from the 'Reminiscences of Captain Dirk Meinertz Hahn, 1838–1839.' Translated by FJH Blaess and LA Triebel. *South Australiana* **3**(2), 97–134.

Hale HM, Tindale NB (1929) Further notes on Aboriginal rock carvings in South Australia. *South Australian Naturalist* **10**(2), 30–34.

Hallam SJ (1975) *Fire and Hearth: A Study of Aboriginal Usage and European Usurpation in South-western Australia*. Australian Institute of Aboriginal Studies, Canberra.

Halls Gap and Grampians Historical Society (2006) *Victoria's Wonderland*. Author, Halls Gap, Victoria.

Hamacher DW (2011) On the Astronomical Knowledge and Traditions of Aboriginal Australians. PhD thesis. Macquarie University, Sydney.

Hamacher DW, Norris RP (2010a) Meteors in Australian Aboriginal Dreamings. *WGN, Journal of the International Meteor Organization* **38**(3), 87–98.

Hamacher DW, Norris RP (2010b) Comets in Australian Aboriginal astronomy. *Journal of Astronomical History and Heritage* **14**(1), 31–40.

Hamacher DW, Norris RP (2011) Eclipses in Australian Aboriginal astronomy. *Journal of Astronomical History and Heritage* **14**(2), 103–114.

Hamacher DW, Fuller RS, Norris RP (2012) Orientations of linear stone arrangements in New South Wales. *Australian Archaeology* **75**, 46–54. doi:10.1080/03122417.2012.11681949

Hamilton A (1972) Aboriginal man's best friend? *The Australian Journal of Anthropology* **8**(4), 287–295. doi:10.1111/j.1835-9310.1972.tb00449.x

Hardwick C (1870) Snake-bite. *Hamilton Spectator*, Victoria. 31 December 1870, p. 3.

Hardy B (1969) *West of the Darling*. Jacaranda, Milton, Queensland.

Harvey A (1939) *Field Notebook*. Fry Collection, South Australian Museum Archives, Adelaide.

Hateley RF (2010) The Victorian bush: its 'original and natural' condition. 1st edn. Polybractea Press, South Melbourne.

Hawdon J (1952) *The Journal of a Journey from New South Wales to Adelaide, the Capital of South Australia, Performed in 1838*. Georgian House, Melbourne.

Haydon GH (1846) *Five Years Experience in Australia Felix, Comprising a Short Account of its Early Settlement and its Present Position, with Many Particulars Interesting to Intending Emigrants*. Hamilton, Adams, and Co., London.

Haygarth H (1861) *Recollections of Bush Life in Australia: During a Residence of Eight Years in the Interior*. John Murray, London.

Haynes RD (1992) Aboriginal Astronomy. *Australian Journal of Astronomy* **4**(3), 127–140.

Haynes RD (2009) Dreaming the stars. *Earth Song Journal: Perspectives in Ecology. Spirituality and Education* **11**(5), 5–12.

Haynes RF, Haynes RD, Malin D, McGee R (1996) *Explorers of the Southern Sky. A History of Australian Astronomy*. Cambridge, Cambridge University Press.

Head L (1989) Using palaeoecology to date Aboriginal fishtraps at Lake Condah, Victoria. *Archaeology in Oceania* **24**(3), 110–115. doi:10.1002/j.1834-4453.1989.tb00220.x

Healy T, Cropper P (2006) *The Yowie. In Search of Australia's Bigfoot*. Strange Nation, Sydney.

Hegarty MP, Hegarty EE, Wills RBH (2001) 'Food Safety of Australia Plant Bushfoods'. Publication no. 01/028. Rural Industries Research and Development Corporation, Canberra.

Hele A (2001) *Muntries Production*. Australian Native Produce Industries Pty Ltd, Primary Industries and Resources SA, Adelaide.

Hele AE (2003) 'Researchers' Extension Program for the Native Foods Industry'. Publication no. 03/013. Rural Industries Research and Development Corporation, Canberra.

Helman P (2009) 'Droughts in the Murray Darling Basin Since European Settlement'. Griffith Centre for Coastal Management Research Report No. 100. Griffith University and Murray-Darling Basin Authority, Canberra.

Hemming SJ, Rigney D (2010) Decentring the new protectors: transforming Aboriginal heritage in South Australia. *International Journal of Heritage Studies* **16**(1–2), 90–106. doi:10.1080/13527250903441804

Hemming SJ, Jones PG, Clarke PA (2000) *Ngurunderi: An Aboriginal Dreaming*. South Australian Museum, Adelaide.

Hemming SJ, Rigney D, Berg S (2010) Researching on Ngarrindjeri ruwe/ruwar: methodologies for positive transformation. *Australian Aboriginal Studies* **2**, 92–106.

Herbert F (1999) *Cultivation of Bushfoods: Preliminary Investment Analysis of the Commercial Cultivation of Six Bushfoods*. Marketing Economics and Rural Adjustment Service, Perth.

Hercus LA (1971) Eaglehawk and crow: a Madimadi version. *Mankind* **8**, 137–140.

Hercus LA (1980) "How we danced the Mudlunga": memories of 1901 and 1902. *Aboriginal History* **4**, 4–31.

Hercus LA (1986) 'Victorian Languages: A Late Survey'. Pacific Linguistics, Series B – No. 77. Department of Linguistics, Research School of Pacific Studies, Australian National University, Canberra.

Hiatt LR (Ed.) (1975) *Australian Aboriginal Mythology. Essays in Honour of W.E.H. Stanner*. Australian Institute of Aboriginal Studies, Canberra.

Hiatt LR (1996) *Arguments About Aborigines. Australia and the Evolution of Social Anthropology*. Cambridge University Press, Cambridge.

Hill RD, Hill FD (1875) *What we Saw in Australia*. Macmillan & Co., London.

Hill R, Turpin G, Canendo W, Standley P, Crayn D, Warne S, Keith K, Addicott E, Zich F (2011) Indigenous-driven tropical ecology. *Australasian Plant Conservation* **19**(4), 24–25.

Hill R, Grant C, George M, Robinson CJ, Jackson S, Abel N (2012) A typology of Indigenous engagement in Australian environmental management: implications for knowledge integration and social-ecological system sustainability. *Ecology and Society* **17**(1), 23. doi:10.5751/ES-04587-170123

Hiscock P (2014) Creators or destroyers? The burning questions of human impact in ancient Aboriginal Australia. *Humanities Australia* **5**, 40–52.

Hobson J, Lowe K, Poetsch S, Walsh M (2010) *Re-awakening Languages: Theory and Practice in the Revitalisation of Australia's Indigenous Languages*. Sydney University Press, Sydney.

Hodgson CP (1846) *Reminiscences of Australia, with Hints on the Squatter's Life*. W.N. Wright, London.

Holden R, Holden N (2001) *Bunyips. Australia's Folklore of Fear*. National Library of Australia, Canberra.

Horsburgh J (1849) James Horsburgh, General Report of the Goulburn River Aboriginal Station, 6 January 1849. Public Record Office of Victoria, Inward Registered and Unregistered Correspondence. VPRS 44/P0, Unit 484.

Horton D (Ed.) (1994) *The Encyclopaedia of Aboriginal Australia: Aboriginal and Torres Strait Islander History, Society and Culture*. Australian Institute of Aboriginal Studies, Canberra.

Horton D (2000) *The Pure State of Nature. Sacred Cows, Destructive Myths and the Environment*. Allen and Unwin, Sydney.

Howitt AW (no date a) *Papers*. Notes on the Kulin – information from William Barak c.1882 in Papers. MS 9356. State Library of Victoria, Melbourne.

Howitt AW (no date b) *Papers*. National Museum of Victoria, Melbourne.

Howitt AW (1830–1908) *Papers*. MS 9356. La Trobe Library Collection, State Library of Victoria, Melbourne.

Howitt AW (1884) On some Australian beliefs. *Journal of the Anthropological Institute of Great Britain and Ireland* **13**, 185–198. doi:10.2307/2841724

Howitt AW (1886) On the migrations of the Kurnai ancestors. *Journal of the Anthropological Institute of Great Britain and Ireland* **15**, 409–422. doi:10.2307/2841818

Howitt AW (1887) On Australian medicine men: or, doctors and wizards of some Australian tribes. *The Journal of the Anthropological Institute of Great Britain and Ireland* **16**, 23–59.

Howitt AW (1889) Further notes on the Australian class system. *Journal of the Anthropological Institute of Great Britain and Ireland* **18**, 31–70. doi:10.2307/2842513

Howitt AW (1904) *Native Tribes of South-East Australia*. Macmillan, London.

Howitt W (1855 [1972]) *Land, Labour and Gold*. Second edition. Lowden, Kilmore.

Hughes L (Ed.) (2003) *A Young Australian Pioneer: Henry Mundy*. Next Century Books, Leighton Buzzard, U.K.

Hyde P (2013) A curious cure – a bizarre moment in medical history. *Signals* **104**, 59–61.

Integrity (Cutter), Robbins C, Oxley J (1804–1805) *Log books 1804–1805*. MS 13. National Library of Australia, Canberra.

Irvine FR (1970) Evidence of change in the vegetable diet of Australian Aborigines. In *Diprotodon to Detribalisation*. (Eds AR Pilling and RA Waterman) pp. 278–284. Michigan State University Press, East Lansing.

Isaacs J (1980) *Australian Dreaming: 40,000 Years of Aboriginal History*. Lansdowne Press, Sydney.

Isaacs J (1987) *Bush Food: Aboriginal Food and Herbal Medicine*. Weldons, Sydney.

Iwaniszewski S (2014) Australian Aboriginal astronomy and cosmology. In *Handbook of Archaeoastronomy and Ethnoastronomy*. (Ed. CLN Ruggles) pp. 3–14. Springer, New York.

Iwaniuk AN, Lefebvre L, Wylie DR (2009) The comparative approach and brain-behaviour relationships: a tool for understanding tool use. *Canadian Journal of Experimental Psychology/Revue Canadienne de Psychologie Expérimentale* **63**(2), 150–159.

Jackson WD (1968) Fire, air, water and earth   an elemental ecology of Tasmania. *Proceedings of the Ecological Society of Australia* **3**, 9–16.

Jackson S, Morrison J (2007) Indigenous perspectives in water management, reforms and implementation. In *Managing Water for Australia: The Social and Institutional Challenges*. (Eds K Hussey and S Dovers) chapter 3. CSIRO Publishing, Melbourne.

Jakubowski P, Atkinson F (2016) *Pormpuraaw Cultural Uses for Plants*. Pormpuraaw Art and Cultural Centre, Pormpuraaw, Queensland.

Janca A, Bullen C (2003) The Aboriginal concept of time and its mental health implications. *Australasian Psychiatry* **11**(s1), S40–S44. doi:10.1046/j.1038-5282.2003.02009.x

Janke T (2009) *Writing Up Indigenous Research: Authorship, Copyright and Indigenous Knowledge Systems*. Terri Janke, Terri Janke and Company, Sydney.

J.D. (1931) Aboriginal ghosts. *Age*, Melbourne. 17 January 1931, p. 17.

Jenkins J (1850) *Voyage of the U.S Exploring Squadron, Commanded by Captain Charles Wilkies of the United States Navy in 1838–42*. James M Alden, Auburn.

Johns RE (c.1870s) *Papers*. MSF 10075. State Library of Victoria, Melbourne.

Johnson D (1998) *Night Skies of Aboriginal Australia. A Noctuary*. Oceania Monograph no. 47. University of Sydney, Sydney.

Johnson D (2005) The Southern Night Sky. In *Macquarie Atlas of Indigenous Australia. Culture and Society Through Space and Time*. (Eds B Arthur and F Morphy) pp. 108–113. Macquarie Library, Macquarie University, Sydney.

Johnson D (2011) Interpretations of the Pleiades in Australian Aboriginal astronomies. In *Oxford IX International Symposium on Archaeoastronomy Proceedings IAU Symposium No. 278, 2011*. (Ed. CLN Ruggles) pp. 291–297. International Astronomical Union, Oxford, UK.

Johnson P (c.1855) *Papers*. 7627. National Library of Australia, Canberra.

Jones DS (Ed.) (2010) 'Lake Condah and Budj Bim National Heritage Landscape Report'. School of Architecture, Landscape Architecture and Urban Design, University of Adelaide, Adelaide.

Jones DS (2011) The water harvesting landscape of Budj Bim and Lake Condah: whither world heritage recognition. *2011 International Conference of the Association of Architecture Schools of Australasia*. pp. 131–142. Deakin University, Geelong, Victoria.

Jones DS, Clarke PA (in press) Australian Aboriginal culture and food-landscape relationships: possibilities of Indigenous knowledge for the future Australian landscape. In *Routledge Companion to Landscape and Food*. (Eds J Zeunert and T Waterman). Routledge, London.

Jones DS, Low Choy D, Clarke PA, Hale R (2013) Watching clouds over country: reconsidering Australian Indigenous perspectives about environmental change and climate change. In *UPE10 2012: NEXT CITY: Planning for a New Energy and Climate Future: Proceedings of the 10th International Urban Planning and Environment Association Symposium*. (Eds N Gurran, P Phibbs and S Thompson) pp. 148–163. ICMS, Sydney.

Jones DS, Low Choy D, Clarke PA, Serrao-Neumann S, Hales R, Koschade O (2016) The challenge of being heard: understanding Wadawurrung climate change vulnerability and adaptive capacity. In *Conflict and Change in Australia's Peri-urban Landscapes. Urban Planning and Environment*. (Eds M Kennedy, A Butt and M Amati) pp. 260–279. Routledge, London.

Jones J (2014) Lighting the fire and the return of the boomerang cultural renaissance in the south-east. *Artlink* **34**(2), 35–38.

Jones P (Ed.) (1981) *Historical Records of Victoria. Volume. 1, Beginnings of Permanent Government*. Victorian Government Printing Office, Melbourne.

Jones PG (2007) *Ochre and Rust. Artefacts and Encounters on Australian Frontiers*. Wakefield Press, Adelaide.

Jones RM (1969) Firestick farming. *Australian Natural History* **16**, 224–228.

Jones RM (1978) Why did the Tasmanians stop eating fish? In *Explorations in Ethnoarchaeology*. (Ed. RA Gould) pp. 11–47. School of American Research Advanced Seminar Series. University of New Mexico Press, Albuquerque.

Jordan JW (2012) The engineering of Budj Bim and the evolution of a societal structure in Aboriginal Australia. *Australian Journal of Multi-Disciplinary Engineering* **9**(1), 63–68. doi:10.7158/N12-H07.2012.9.1

Kalotas AC (1996) Aboriginal knowledge and use of fungi. In *Fungi of Australia. Volume 1B. Introduction – Fungi in the Environment.* (Ed. AE Orchard) pp. 268–295. ABRS/CSIRO, Canberra.

Kamminga J (1982) 'Over the edge': functional analysis of Australian stone tools. *Occasional Papers in Anthropology* **12**. Anthropology Museum, University of Queensland, St Lucia, Queensland.

Karadada J, Karadada L, Goonack W, Mangolamara G, Bunjuck W, Karadada L, Djanghara B, Mangolamara S, Oobagooma J, Charles A, Williams D, Karadada R, Saunders T, Wightman G (2011) 'Uunguu. Plants and Animals. Aboriginal Biological Knowledge from Wunambal Gaambera Country in the North-west Kimberley, Australia'. Northern Territory Botanical Bulletin No. 35. Wunambal Gaambera Aboriginal Corporations and Department of Natural Resources, Environment, the Arts and Sport. Kalumburu, Western Australia.

Keeler C, Couzens V (2010) *Meerreeng-an Here is My Country: the Story of Aboriginal Victoria Told Through Art*. Koorie Heritage Trust Inc., Melbourne.

Keen I (Ed.) (1994) *Being Black. Aboriginal Cultures in 'Settled' Australia*. Aboriginal Studies Press, Canberra.

Kenny R (2007) *The Lamb Enters the Dreaming: Nathanael Pepper and the Ruptured World*. Scribe, Melbourne.

Kenyon AS (1914) The story of the Mallee. *The Victorian Historical Magazine* **4**(2), 57–74.

Kerr T (2004) As if bunyips mattered ... cross-cultural mytho-poetic beasts in Australian subaltern planning. *Journal of Australian Studies* **80**, 15–27.

Kimber RG (1997) Cry of the plover, song of the desert rain. In *Windows on Meteorology. Australian Perspective*. (Ed. EK Webb) pp. 7–13. CSIRO Publishing, Melbourne.

King B (1997) Acacia – research, field trials and databases. *Australian Bushfoods Magazine* **4**, 10–11, 14.

King B (1998) Muntari – much more than a ground cover. *Australian Bushfoods Magazine* **6**, 10–11.

King PP (1826) *Narrative of a Survey of the Intertropical and Western Coasts of Australia Performed Between 1818 and 1822, 2 Volumes*. John Murray, London.

Kirby J (1895) *Old Times in the Bush of Australia: Trials and Experiences of Early Bush Life in Victoria During the Forties*. George Robertson, Melbourne.

Kirkland K (1845) *Life in the Bush. By a Lady. Chambers's Miscellany of Useful and Entertaining Tracts*. William and Robert Chambers, London.

KNYA Taskforce (2012) 'Kungun Ngarrindjeri Yunnan Agreement, Listening to Ngarrindjeri People Talking'. Department of Environment, Water and Natural Resources, and Ngarrindjeri Regional Authority, Adelaide and Murray Bridge.

Koeler H, Mühlhäusler P (2006) *Hermann Koeler's Adelaide: Observations on the Language and Culture of South Australia by the First German Visitor*. Australian Humanities Press, Unley, South Australia.

Koenig W (1935) *The History of the Winchelsea Shire*. Winchelsea Shire Council, Winchelsea, Victoria.

Kolig E (1981) *The Silent Revolution: The Effects of Modernization on Australian Aboriginal Religion*. Institute for the Study of Human Issues, Philadelphia.

Kolig E (1988) Mission not accomplished: Christianity in the Kimberleys. In *Aboriginal Australians and Christian Missions: Ethnographic and Historical Studies* (Eds T Swain and DB Rose) pp. 376–390. Australian Association for the Study of Religions, Bedford Park, South Australia.

Krappe AH (1940) The lunar frog. *Folklore* **51**(3), 161–171. doi:10.1080/0015587X.1940.9718229

Krefft G (1862) The vertebrated animals of the Lower Murray and Darling. *Argus*, Melbourne. 5 November 1862, p. 7.

Krefft G (1862–65) On the manners and customs of the Aborigines of the River Murray and Darling. *Transactions of the Philosophical Society of New South Wales* **1862–65**, 357–374.

Laird SA (2010) *Biodiversity and Traditional Knowledge: Equitable Partnerships in Practice*. Routledge, London.

Lakic M, Wrench R (1994) *Through their Eyes: An Historical Record of Aboriginal People of Victoria as Documented by the Officials of the Port Phillip Protectorate, 1839–1841*. Museum of Victoria, Melbourne.

Lang GS (1865) *The Aborigines of Australia: in their Original Condition and in their Relations with the White Men*. Revised edtion. Wilson and Mackinnon, Melbourne.

Lang JD (1847) *Phillipsland, or, The Country Hitherto Designated Port Phillip: its Present Condition and Prospects, as a Highly Eligible Field for Emigration*. Longman, Brown, Green, and Longmans, London.

Langhorne G (1837) *Reminiscences of James Buckley who Lived for Thirty Years Among the Wallawarro or Watourong Tribes at Geelong Port Phillip, Communicated by Him to George Langhorne*. MS 13483. State Library of Victoria, Melbourne.

Langhorne G (1838) DLADD 90 ITEM 02 Mission Report for month ended 31st January 1838 SLNSW.

Langton M, Ma Rhea Z (2005) Traditional Indigenous biodiversity-related knowledge. *Australian Academic and Research Libraries* **36**(2), 45–69. doi:10.1080/00048623.2005.10721248

Lassak EV, McCarthy T (1983) *Australian Medicinal Plants*. Methuen, Melbourne.

Latz P (1995) *Bushfires and Bushtucker. Aboriginal Plant Use in Central Australia*. Institute of Aboriginal Development, Alice Springs.

Laudine C (2009) *Aboriginal Environmental Knowledge. Rational Reverence*. Ashgate, Farnham, UK.

Law SP (1986) The regulation of menstrual cycle and its relationship to the moon. *Acta Obstetricia et Gynecologica Scandinavica* **65**(1), 45–48. doi:10.3109/00016348609158228

Lawrence R (1968) 'Aboriginal Habitat and Economy'. Geography Occasional Paper no. 6. Australian National University, Canberra.

Le Souëf A (c.1890) *Personal Recollections of Early Victoria*. ML A2762. Mitchell Library, Sydney.

Le Souëf WHD (1897) *Wild Life in Australia*. Whitcombe & Tombs, Melbourne.

Leatherbee A (Ed.) (1984) *Knocking About: Being Some Adventures of Augustus Baker Pierce in Australia*. Shoe String Press, Wangaratta.

Leckie G (1932) Black magic "cures" and curses of the Aborigines. *The Argus*, Victoria. 27 August 1932, p. 7.

Lee I, Grant J (1915) *The Logbooks of the 'Lady Nelson': with the Journal of Her First Commander, Lieutenant James Grant, R.N. 1772–1833*. Grafton and company, London.

Lefebvre L, Gaxiola A, Dawson S, Timmermans S, Rosza L, Kabai P (1998) Feeding innovations and forebrain size in Australasian birds. *Behaviour* **135**(8), 1077–1097. doi:10.1163/156853998792913492

Leigh WH (1839) *Reconnoitering Voyages and Travels with Adventures in the New Colonies of South Australia*. Smith, Elder & Co., London.

Lepofsky D (2009) The past, present, and future of traditional resource and environmental management. *Journal of Ethnobiology* **29**, 161–166. doi:10.2993/0278-0771-29.2.161

Lertzman K (2009) The paradigm of management, management systems, and resource stewardship. *Journal of Ethnobiology* **29**, 339–358. doi:10.2993/0278-0771-29.2.339

Levin RB (2010) Meet the muntries. A "super fruit" of the Australian bush. *Thefoodpaper* (http://www.gayot.com/lifestyle/health/features/muntries-health-benefits.html, accessed 25 April 2017).

Lingard J (1846) *A Narrative of the Journey to and from New South Wales: Including a Seven years' Residence in that Country*. s.n, [Chapel-en-le-Frith].

Link CA (2012) Challenges to Flavour: Influences on the Cultural Identity of Cuisines in the Australian Foodscape. PhD thesis. University of Western Sydney, Sydney.

Linn R (1988) *A Diverse Land: A History of the Lower Murray, Lakes and Coorong.* Meningie Historical Society, Meningie.

Lintermans M (2007) 'Fishes of the Murray-Darling Basin: An Introductory Guide'. MDBC Publication No. 10/07. Murray-Darling Basin Authority, Canberra.

Lloyd GT (1862) *Thirty-three Years in Tasmania and Victoria.* Houlston & Wright, London.

Locke (1866) The native language. *Argus*, Melbourne. 24 August 1866, p. 3.

Longmire A (1985) *Nine Creeks to Albacutya – a History of the Shire of Dimboola.* Shire of Dimboola in conjunction with Hargreen, Melbourne.

Lourandos H (1977) Aboriginal spatial organisation and population: south western Victoria reconsidered. *Archaeology and Physical Anthropology in Oceania* **12**(3), 202–225.

Lourandos H (1980) Change or stability?: hydraulics, hunter-gatherers and population in temperate Australia. *World Archaeology* **11**(3), 245–264. doi:10.1080/00438243.1980.9979765

Lourandos H (1997) *Continent of Hunter-gatherers. New Perspectives in Australian Prehistory.* Cambridge University Press, Melbourne.

Low T (1988) *Wild Food Plants of Australia.* Angus & Robertson, Sydney.

Low T (1989) *Bush Tucker. Australia's Wild Food Harvest.* Angus & Robertson, Sydney.

Low T (1990) Witchetty grubs. *Australian Natural History* **23**(4), 284–285.

Low T (1991) *Wild Food Plants of Australia.* Angus & Robertson, Sydney.

Low T (1999) *Feral Future. The Untold Story if Australia's Exotic Invaders.* Viking Penguin Books, Melbourne.

Low T (2002) *The New Nature. Winners and Losers in Wild Australia.* Viking Penguin Books, Melbourne.

Low Choy D, Clarke PA, Jones DS, Serrao-Neumann S, Hales R, Koschade O (2013) 'Indigenous Climate Change Adaptation: Understanding Coastal Urban and Peri-urban Indigenous People's vulnerability and adaptive capacity to Climate Change'. A report for the National Climate Change Adaptation Research Facility. Griffith University, Brisbane.

Low Choy D, Clarke PA, Serrao-Neumann S, Hales R, Koschade O, Jones DS (2016) Coastal Urban and Peri-Urban Indigenous People's Adaptive Capacity to Climate Change. In *Balanced Urban Development: Options and Strategies for Liveable Cities.* (Eds B Maheshwari, VP Singh and B Thoradeniya) pp. 441–461. Springer International Publishing.

Luebbers R (1978) Meals and Menus: A Study of Change in Prehistoric Coastal Settlements in South Australia. Postgraduate thesis. Australian National University, Canberra.

Lynch AJJ, Thackway R, Specht A, Beggs PJ, Brisbane S, Burns EL, Byrne M, Capon SJ, Casanova MT, Clarke PA, Davies JM, Dovers S, Dwyer RG, Ens E, Fisher DO, Flanigan M, Garnier E, Guru SM, Kilminster K, Locke J, MacNally R, McMahon KM, Mitchell PJ, Pierson JC, Rodgers EM, Russell-Smith J, Udy J, Waycott M (2015) Transdisciplinary synthesis for ecosystem science, policy and management: The Australian experience. *The Science of the Total Environment* **534**, 173–184. doi:10.1016/j.scitotenv.2015.04.100

Lyons K (1988) Prehistoric Aboriginal relationships with the forests of the Riverine Plain in South-Eastern Australia. In *Australia's Ever Changing Forests: Proceedings of the First National Conference on Australian Forest History*, 9–11 May, 1988, Canberra. (Eds K Frawley and N Semple) pp. 169–177. Department of Geography and Oceanography, Australian Defence Force Academy, Campbell, ACT.

Macdonald D (1917) Nature notes and queries. *The Argus*, Melbourne. 20 July 1917, p. 5.

MacDonald S (c.1850s) *The Member for Mt Ida.* MS 7033, Folder 17, Box 3. National Library of Australia, Canberra.

Macdougall JH (1912) Wild life and the Aborigines. *Western Mail*, Perth. 6 July 1912, p. 35.

Mackay R (1916) *Recollections of Early Gippsland Goldfields: The Reminiscences of Richard Mackay; Gold Buyer, Businessman and Adventurer of the Mountain Goldfields 1860s–1916.* Mackay, Langwarrin.

MacPherson J (1925) The gum-tree and wattle in Australian Aboriginal medical practice. *The Australian Nurses' Journal* **23**(12), 588–596.

MacPherson P (1881) Astronomy of the Australian Aborigines. *Journal and Proceedings of the Royal Society of New South Wales* **15**, 71–80.

Maddock K (1970) Myths of the acquisition of fire in northern and eastern Australia. In *Australian Aboriginal Anthropology. Modern Studies in the Social Anthropology of the Australian Aborigines*. (Ed. RM Berndt) pp. 174–199. Australian Institute of Aboriginal Studies and University of Western Australia Press, Perth.

Maddock K (1982) *The Australian Aborigines. A Portrait of Their Society*. Penguin Books. Melbourne.

Maegraith BG (1932) The Astronomy of the Aranda and Luritja Tribes. *Transactions of the Royal Society of South Australia* **56**, 19–26.

Maffi L (2001) *On Biocultural Diversity: Linking Language, Knowledge and the Environment*. Smithsonian Institution Press, Washington.

Maffi L (2007) Biocultural diversity and sustainability. In *The Sage Handbook of Environment and Society*. (Eds J Pretty, A Ball, T Benton, J Guivant, DR Lee, D Orr, M Pfeffer, H Ward) pp. 267–275. Sage Publications, London.

Magarey AT (1895a) Aborigines' water quest in arid Australia. *Proceedings of the Australasian Association for the Advancement of Science* 647–658.

Magarey AT (1895b) Aboriginal water quest. *Advocate*, Victoria. 10 August 1895, p. 8.

Maiden JH (1889) *The Useful Native Plants of Australia*. Trubner, London.

Mallyon WK (1910) Notes and queries. An Aboriginal custom. *Observer*, Adelaide. 31 December 1910, p. 44.

Maningrida Arts and Culture Annandale Galleries (2007) *Spirit in Variation: The Art of Maningrida: Bark Paintings, Ceremonial Poles, Yawk Yawks, Mimihs, Woven Forms, Video Installation "The Power of Rarrk"*. Annandale Galleries in association with Maningrida Arts and Culture, Annandale, New South Wales.

Martin T (1909) Aboriginal surgery. *The Argus*, Melbourne. 23 August 1909, p. 8.

Massola A (1956) Australian fish hooks and their distribution. *National Museum of Victoria Memoirs* **22**, 1–16.

Massola A (1957) The Challicum bun-yip. *Victorian Naturalist* **74**, 76–83.

Massola A (1962) The Aborigines of the Victorian High Plains. *Proceedings of the Royal Society of Victoria* **75**, 319–325.

Massola A (1968) *Bunjil's Cave. Myths, Legends and Superstitions of the Aborigines of South-East Australia*. Lansdowne Press, Melbourne.

Massola A (1969) *Journey to Aboriginal Victoria*. Rigby Limited, Adelaide.

Massola A (1971) *The Aborigines of South-Eastern Australia as They Were*. Heinemann, Melbourne.

Mathew J (1899) *Eaglehawk and Crow. A Study of the Australian Aborigines Including an Inquiry into Their Origin and a Survey of Australian Languages*. Melville, Mullen and Slade, Melbourne.

Mathew J (1918) Australian Aborigines. *Footscray Chronicle*, Melbourne. 27 July 1918, p. 2.

Mathew J (1925) Aboriginal sketch. Gleanings in Aboriginal magic. *Australasian*, Melbourne. 7 March 1925, p. 66.

Mathew J (no date) *Papers*. MS 950. Australian Institute of Aboriginal and Torres Strait Islander Studies, Canberra.

Mathews D (1899) Native Tribes of the Upper Murray. *Proceedings of the Royal Geographical Society of Australasia, South Australian Branch* **4**, 43–52.

Mathews RH (1894) Some stone implements used by the Aborigines of New South Wales. *Journal and Proceedings of the Royal Society of New South Wales* **28**, 301–305.

Mathews RH (1897) The Burbung or initiation ceremonies of the Murrumbidgee tribes. *Journal and Proceedings of the Royal Society of New South Wales* **31**, 111–153.

Mathews RH (1898) The Victorian Aborigines: their initiation ceremonies and divisional systems. *American Anthropologist* **11**(11), 325–343. doi:10.1525/aa.1898.11.11.02a00000

Mathews RH (1903) Language of the Bungandity tribe, South Australia. *Journal and Proceedings of the Royal Society of New South Wales* **37**, 59–74.

Mathews RH (1904a) Ethnological Notes on the Aboriginal Tribes of New South Wales and Victoria. *Journal of the Royal Society of New South Wales* **38**, 203–381.

Mathews RH (1904b) The native tribes of Victoria: their languages and customs. *Proceedings of the American Philosophical Society* **43**(175), 54–70.

Mathews RH (1905) *Ethnological Notes on the Aboriginal Tribes of New South Wales and Victoria*. FW White, Sydney.

Mathews RH (1907a) Language of the Birdhawal tribe, in Gippsland, Victoria. *Proceedings of the American Philosophical Society* **46**(187), 346–359.

Mathews RH (1907b) *Aboriginal Navigation and Other Notes*. pp. 215–216. Government Printer, Sydney.

Mathews RH (no date) *Notebooks*. Manuscript. Australian Institute of Aboriginal and Torres Strait Islander Studies, Canberra.

Matthews EG (1976) *Insect Ecology*. University of Queensland Press, St Lucia, Queensland.

McBryde I (1984a) Exchange in south eastern Australia: an ethnohistorical perspective. *Aboriginal History* **8**(2), 132–153.

McBryde I (1984b) Kulin greenstone quarries: the social contexts of production and distribution for the Mt William site. *World Archaeology* **16**(2), 267–285. doi:10.1080/00438243.1984.9979932

McCarthy FD (1940) Aboriginal stone arrangements in Australia. *Australian Museum Magazine*, Sydney. 2 September 1940, pp. 184–189.

McCarthy FD (1953) Aboriginal rain-makers. *Weather* **8**, 72–77. doi:10.1002/j.1477-8696.1953.tb01603.x

McCarthy FD (1938–40) 'Trade' in Aboriginal Australia and 'trade' relationships with Torres Strait, New Guinea and Malaya. *Oceania* **9**, 405–38; **10**, 80–104, 171–95.

McCartney J (1859) *Diaries* MS 12929, Box 166. Royal Historical Society of Victoria, Melbourne.

McCorquodale J (1987) *Aborigines and the Law: A Digest*. Aboriginal Studies Press, Canberra.

McCourt T, Mincham H (1987) *The Coorong and Lakes of the Lower Murray*. National Trust, Beachport, South Australia.

McCrae G (1934) *Georgiana's Journal: Melbourne a Hundred Years Ago*. Angus & Robertson Limited, Sydney.

McDonald D (1887) *Gum Boughs and Wattle Bloom, Gathered on Australian Hills and Plains*. Cassell, London.

McDonald H (2003) The fats of life. *Australian Aboriginal Studies* **2**, 53–61.

McDougall AC (1901) Manners, customs and legends of the Coombangree tribe. *Science of Man* **4**(8), 46–47, 63.

McGivern M (1983) *Big camp Wahgunyah: History of the Rutherglen district*. Spectrum, Melbourne.

McIntyre-Tamwoy S, Buhrich A (2012) The Cultural Assets and Climate Change Literature Review and Research Synthesis. Unpublished report to the NSW Office of Environment and Heritage. Cairns Institute, James Cook University, Cairns.

McIntyre-Tamwoy S, Fuary M, Buhrich A (2013) Understanding climate, adapting to change: indigenous cultural values and climate change impacts in North Queensland. *Local Environment* **18**(1), 91–109. doi:10.1080/13549839.2012.716415

McKinlay J (1863) *McKinlay's Diary of his Journey Across the Continent of Australia*. E & D Syme, Melbourne.

McLaren G (1996) *Beyond Leichhardt. Bushcraft and the Exploration of Australia*. Fremantle Arts Centre, Fremantle.

McNiven IJ, Crouch J, Richards T, Dolby N, Jacobsen G (2012) Dating Aboriginal stone-walled fishtraps at Lake Condah, southeast Australia. *Journal of Archaeological Science* **39**(2), 268–286. doi:10.1016/j.jas.2011.09.007

McNiven I, Crouch J, Richards T, Sniderman K, Dolby N, Mirring G (2015) Phased redevelopment of an ancient Gunditjmara fish trap over the past 800 years: Muldoons Trap Complex, Lake Condah, southwestern Victoria. *Australian Archaeology* **81**(1), 44–58. doi:10.1080/03122417.2015.11682064

McPhee C (1996) The botanist at Como: Mueller and the Armytage family. *Victorian Naturalist* **113**(4), 227–228.

Meckel BL, Port Phillip Pioneers' Group (1991) *Pioneer Profiles. Vol. 2: The Stories, in Brief, of Some of Victoria's Early Settlers*. Betty Meckel, Mentone.

Meehl GA, Arblaster JM (1998) The Asian-Australian monsoon and El Nino-Southern Oscillation in the NCAR climate system model. *Journal of Climate* **11**(6), 1356–1385. doi:10.1175/1520-0442(1998)011<1356:TAAMAE>2.0.CO;2

Memmott P (2007) *Gunyah, Goondie + Wurley: The Aboriginal Architecture of Australia*. University of Queensland Press, St Lucia, Qld.

Menkhorst P, Knight F (2001) *A Field Guide to the Mammals of Australia*. Oxford University Press, Melbourne.

Merlan F (1999) Foreword. In 'Wardaman Ethnobiology: Aboriginal Plant and Animal Knowledge from the Flora River and South-west Katherine Region, North Australia'. (E Raymond, J Blutja, L Gin.gina, M Raymond, O Raymond, L Raymond, J Brown, Q Morgan, D Jackson, N Smith and G Wightman) pp. ix–x. Centre for Indigenous Natural and Cultural Resource Management, Occasional Paper No. 2. Northern Territory Botanical Bulletin No. 25. Parks and Wildlife Commission of the Northern Territory, Darwin.

Meyer HAE (1843) *Vocabulary of the Language Spoken by the Aborigines of South Australia*. Allen, Adelaide.

Meyer HAE (1846) Manners and Customs of the Aborigines of the Encounter Bay Tribe, South Australia. Reprinted 1879 in *The Native Tribes of South Australia*. (Ed. JD Woods) pp. 183–206. ES Wigg & Son, Adelaide.

Miller JB, James KW, Maggiore PMA (1993) *Tables of Composition of Australian Aboriginal Foods*. Aboriginal Studies Press, Canberra.

Milligan AW (1894) Native birds and their languages. *Gippsland Farmers' Journal*, Traralgon. 9 November 1894, p. 3.

Milne E (1910) Twofold Bay. The gateway to Monaro and eastern Gippsland. *Twofold Bay Magnet, and South Coast and Southern Monaro Advertiser*, New South Wales. 17 January 1910, p. 3.

Mitchell TL (1839) *Three Expeditions into the Interior of Eastern Australia: With Descriptions of the Recently Explored Region of Australia Felix and of the Present Colony of New South Wales*. Two volumes. T. & W. Boone, London.

Mohamad S (2010) The Ethnobotany of the Semelai Community at Tasek Bera, Pahang, Malaysia: An Ethnographic Approach for Re-settlement. PhD thesis. University of Adelaide, Adelaide.

Moorhouse M (1846) *A Vocabulary and Outline of the Grammatical Structure of the Murray River Language Spoken by the Natives of South Australia from Wellington on the Murray, as Far as Rufus*. Republished 1935 in *The Autochthones of Australia*. (Ed. TA Parkhouse) pp. 1–47. The editor, Adelaide.

Morgan J (1852) *The Life and Adventures of William Buckley: Thirty-two Years a Wanderer Amongst the Aborigines of the Unexplored Country Round Port Phillip*. Republished in 1980. Australian National University Press, Canberra.

Morieson J (2003) The astronomy of the Boorong. *Journal of Australian Indigenous Issues* 2(4), 19–28.

Moritz C, Ens EJ, Potter S, Catullo RA (2013) The Australian monsoonal tropics: an opportunity to protect unique biodiversity and secure benefits for Aboriginal communities. *Pacific Conservation Biology* 19(4), 343–355. doi:10.1071/PC130343

Morphy H (1995) Landscape and the Reproduction of the Ancestral Past. In *The Anthropology of the Landscape: Perspectives on Place and Space*. (Eds E Hirsch and M O'Hanlon) pp. 184–209. Clarendon Press, Oxford.

Morphy H (1998) *Aboriginal Art*. Phaidon Press, London.

Morphy H (1999) Australian Aboriginal Concepts of Time. In *The Story of Time*. (Ed. K Lippincott) pp. 264–267. Merrell Holberton Publishers, London.

Morrison E (c.1971) *The Loddon Aborigines: Tales of Old Jim Crow*. Author, Yandoit, Victoria.

Mossman S, Banister T (1853) *Australia Visited and Revisited: A Narrative of Recent Travels and Old Experiences in Victoria and New South Wales*. Addey and Co., London.

Mountford CP (1927) Aboriginal stone structures in South Australia. *Transactions and Proceedings of the Royal Society of South Australia* 51, 169–172.

Mountford CP (1929) A unique example of Aboriginal rock carving at Panaramitee North. *Transactions of the Royal Society of South Australia* 53, 245–248.

Mountford CP (1958) *The Tiwi: Their Art, Myth and Ceremony*. Phoenix House, London.

Mühlhäusler P (2003) *Language of Environment, Environment of Language: a Course in Ecolinguistics*. Battlebridge, London.

Mühlhäusler P, Fill A (Eds) (2001) *The Ecolinguistics Reader: Language, Ecology, and Environment*. Continuum, London.

Muller S (2012) Two ways: bringing Indigenous and non-indigenous knowledges together. In *Country, Native Title and Ecology*. (Ed. J Weir) pp 59–79. Australian National University e-press and Aboriginal History Incorporated (Monograph 24), Canberra.

Mulvaney DJ, Kamminga J (1999) *Prehistory of Australia*. Allen & Unwin, Sydney.

Mulvaney J (1994) The Namoi bunyip. *Australian Aboriginal Studies* **1**, 36–38.

Murray RD (1843) *A Summer at Port Phillip*. W Tait, Edinburgh.

Museum Board (1887) Notes Upon Additions to the Museum of the South Australian Public Library, Museum, and Art Gallery by "An Amateur Naturalist". Reprinted from *The South Australian Register* and *Adelaide Observer*. W.K. Thomas & Co. Printers, Adelaide.

Musharbash Y (2016) Evening play: acquainting toddlers with dangers and fear at Yuendumu, Northern Territory. In *Social Learning and Innovation in Contemporary Hunter-Gatherers*. (Eds H Terashima and BS Hewlett) pp. 171–177. Springer, Japan.

Nabhan G (2009) Ethnoecology: bridging disciplines, cultures, and species. *Journal of Ethnobiology* **29**, 3–7. doi:10.2993/0278-0771-29.1.3

Nambatu NJ, Nudjulu PP, Nama LJ, Munar KJ, Kungul DA, Munar LR, Tchinburrurr KB, Jongmin MJ, Kungul YB, McTaggart PM, Crocombe M, Wightman G (2009) 'Marri Ngarr and Magati ke Plants and Animals – Aboriginal Knowledge of Flora and Fauna from Moyle River and Neninh Areas, North Australia'. Northern Territory Botanical Bulletin no. 32. Northern Territory Department of Natural Resources and Environment, Darwin.

Nash D (1992) Hot and Cold Over Clockwise. In *The Language Game. Papers in Memory of Donald C. Laycock, Pacific Linguistics Series C-110*. (Eds TE Dutton, MD Ross and DT Tryon) pp. 291–297. Australian National University, Canberra.

Nash D (2004) *Aboriginal Plant Use in South-eastern Australia*. Australian National Botanic Gardens, Canberra.

Netzel M, Netzel G, Tian Q, Schwartz S, Konczak I (2007) Native Australian fruits – a novel source of antioxidants for food. *Innovative Food Science & Emerging Technologies* **8**, 339–346. doi:10.1016/j.ifset.2007.03.007

New South Wales (1896) *Historical Records of New South Wales, Volume 4, Hunter and King, 1801, 1802, 1803*. C Potter, Government Printer, Sydney.

Newland S (1921) Annual Address. *Journal of the Royal Geographical Society of South Australia* **3**.

Ngaanyatjarra Pitjantjatjara Yankunytjatjara Women's Council (2003) *Ngangkari Work-Anangu Way. Traditional Healers of Central Australia*. NPYW Council, Alice Springs.

Nicholls CJ (2014) 'Dreamings' and place – Aboriginal monsters and their meanings'. *The Conversation* 30 April 2014 (http://theconversation.com/dreamings-and-place-aboriginal-monsters-and-their-meanings-25606, accessed 20 February 2017).

Nicholls P, Wrede RW (2012) *A Wonderful Change: The Story of Robert Wrede Including his Journal 1837–41*. LuLu.com: Raleigh.

Nicholson PH (1981) Fire and the Australian Aborigine. In *Fire and the Australian Biota*. (Eds AM Gill, RH Groves and IR Noble) pp. 55–76. Australian Academy of Science, Canberra.

Noble JC, Kimber RG (1997) On the ethno-ecology of mallee root-water. *Aboriginal History* **21**, 170–202.

Norman J (1951) Aboriginal legends and the bunyip. *The Advertiser*, Adelaide. 24 February 1951, p. 6.

Norris RP (2007) Searching for the astronomy of Aboriginal Australians. *Conference Proceedings from the VIIIth Oxford International Conference on Archaeoastronomy and Astronomy in Culture* (http://www.atnf.csiro.au/people/rnorris/papers/n214.pdf, accessed 25 April 2017).

Norris RP (2016) Dawes Review 5: Australian Aboriginal astronomy and navigation. *arXiv preprint arXiv:1607.02215*.

Norris RP, Hamacher DW (2009) The astronomy of Aboriginal Australia. In *The Role of Astronomy in Society and Culture Proceedings IAU Symposium. Vol. 260*. (Eds D Valls-Gabaud and A Boksenberg) pp. 10–17. International Astronomical Union.

Norris RP, Hamacher DW (2014) Australian Aboriginal Astronomy – An Overview. In *Handbook of Archaeoastronomy and Ethnoastronomy*. (Ed. CLN Ruggles) pp. 2215–2222. Springer, New York.

Norris RP, Norris C, Hamacher DW, Abrahams R (2012) Wurdi Youang: an Australian Aboriginal stone arrangement with possible solar indications. *Rock Art Research* **30**, 55–65.

Oates A (1977) Plant Food Utilization by Victorian Aborigines. MSc preliminary thesis. La Trobe University, Melbourne.

Oates WJ, Oates LF (1970) *A Revised Linguistic Survey of Australia*. Australian Institute of Aboriginal Studies, Canberra.

Oedipus (1870) Scientific gossip. *Leader*, Melbourne 3 December 1870, p. 5.

Orchard AE, Wilson AJG (1999) Utilisation of the Australian flora. In *Australian Biological Resources Study. Flora of Australia. Volume 1*. (Ed. AE Orchard) pp. 437–466. CSIRO, Melbourne.

Orchard K, Ross H, Young E (2003) Institutions and processes for resource and environmental management in the Indigenous domain. In *Managing Australia's Environment*. (Eds S Dovers and S River) pp. 413–441. Federation Press, Sydney.

Oriolus (1941) Nature notes. Birds and the bunyip. *Age*, Melbourne. 17 May 1941, p. 12.

Osborne Mr (1950) Mr. Osborne's memories. *Recorder*, Port Pirie. 31 July 1950, p. 3.

Owens D, Hayden B (1997) Prehistoric rites of passage: a comparative study of transegalitarian hunter-gatherers. *Journal of Anthropological Anthropology* **16**, 121–161. doi:10.1006/jaar.1997.0307

Oxley J (1820) *Journals of Two Expeditions into the Interior of New South Wales, Undertaken by Order of the British Government in the Years 1817–18*. John Murray, London.

Parmington A, Griffin D, McConachy F, Rakoczy S (2012) Partnerships and Indigenous cultural values recording within Victoria: the Merri Creek cultural values project. *Excavations, Surveys and Heritage Management in Victoria* **1**, 57–66.

Pascoe B (2014) *Dark Emu. Black Seeds: Agriculture or Accident*. Magabala Books, Broome, Western Australia.

Pate JS, Dixon KW (1982) *Tuberous, Cormous and Bulbous Plants: Biology of An Adaptive Strategy in Western Australia*. University of Western Australia Press, Nedlands, Western Australia.

Pateshall N (1803) *A Short Account of a Voyage Round the Globe, Performed in Fifteen Months, by H.M. Ship Calcutta by N. Pateshall, Lieut. of the Same Ship* [ca. 1803–1804]. MS 13479. State Library of Victoria, Melbourne.

Paton JG (1891) *John G. Paton: Missionary to the New Hebrides An Autobiography*. Hodder & Stoughton, London.

Pearn JH (1990) *Medicine and Botany. An Australian Cadaster: Australian Flora Named After Those Whose Lives Have Served Medicine and Health*. Amphion Press, Brisbane.

Pearn JH (2001) *A Doctor in the Garden. Nomen Medici in Botanicis: Australian Flora and the World of Medicine*. Amphion Press, Brisbane.

Pearn JH, O'Carrigan C (Eds) (1983) *Australia's Quest for Colonial Health. Some Influences on Early Health and Medicine in Australia*. Department of Child Health, Royal Children's Hospital, Brisbane.

Peck CW (1933) *Australian Legends: Tales Handed Down from the Remotest Times by the Autochthonous Inhabitants of our Land*. Republished 2014. Netlancers Inc.

Peel L (Ed.) (1996) *The Henty Journals: A Record of Farming, Whaling and Shipping at Portland Bay, 1834–1839*. Miegunyah Press, Melbourne.

Penney R (as 'Cuique') (1842) Letter to the editor. *South Australian Magazine*, Adelaide. June–July 1842, 389–394.

Pepper P, de Araugo T (1985) *The Kurnai of Gippsland. What Did Happen to the Aborigines of Victoria. Volume 1*. Hyland House, Melbourne.

Perry F, de Quetteville Robin A (1984) *Australian Sketches: The Journals and Letters of Frances Perry*. Queensbury Hill Press, Melbourne.

Pert PL, Ens EJ, Locke J, Clarke PA, Packer JM, Turpin G (2015) An online spatial database of Australian Indigenous Biocultural Knowledge for contemporary natural and cultural resource management. *The Science of the Total Environment* accessed 25 April 2017. doi:10.1016/j.scitotenv.2015.01.073

Peterson N (1978) The traditional pattern of subsistence to 1975. In *The Nutrition of Aborigines in Relation to the Ecosystem of Central Australia*. (Eds BS Hetzel and HJ Frith) pp. 25–35. Commonwealth Scientific and Industrial Research Organisation, Melbourne.

Peterson N (2017) Is there a role for anthropology in cultural reproduction? Maps, mining, and the "cultural failure" in Central Australia. In *Entangled Territorialities. Negotiating Indigenous Lands in Australia and Canada*. (Eds F Dussart and S Poirier) pp. 235–252. University of Toronto Press, Toronto.

Peterson N, Lampert R (1985) A Central Australian ochre mine. *Records of the Australian Museum* **37**(1), 1–9. doi:10.3853/j.0067-1975.37.1985.333

Peterson N, Rigsby B (Eds) (1998) *Customary Marine Tenure in Australia*. Oceania Monograph no. 48. University of Sydney, Sydney.

Petheram L, Zander KK, Campbell BM, High C, Stacey N (2010) 'Strange changes': Indigenous perspectives of climate change and adaptation in NE Arnhem Land (Australia). *Global Environmental Change* **20**(4), 681–692. doi:10.1016/j.gloenvcha.2010.05.002

Petri H (1952) *Der Australische Medizenmann*. Republished as *The Australian Medicine Man*, translated by I Campbell and edited by K Akerman, 2014. Hesperian Press, Perth.

Plomley B, Henley KA (1990) *The Sealers of Bass Strait and the Cape Barren Island Community*. Blubber Head Press, Hobart.

Polehampton A (1862) *Kangaroo Land*. Richard Bentley, London.

Porter L (2004) Planning's Colonial Culture: An Investigation of the Contested Process of Producing Place in (Post) Colonial Victoria. Ph.D. thesis. University of Melbourne, Faculty of Architecture, Building and Planning, Melbourne.

Pounder DJ (1985) A new perspective on kadaitja killings. *Oceania* **56**(1), 77–82. doi:10.1002/j.1834-4461.1985.tb02109.x

Powell JM (1990) Early impressions of the vegetation of the Sydney region: exploration and plant use by the First Fleet officers. In *History of Systematic Botany in Australasia. Proceedings of a symposium held at the University of Melbourne*. (Ed. PS Short) pp. 87–96. Australian Systematic Botany Society, Melbourne.

Prangnell J, Ross A, Coghill B (2010) Power relations and community involvement in landscape-based cultural heritage management practice: an Australian case study. *International Journal of Heritage Studies* **16**(1–2), 140–155. doi:10.1080/13527250903441838

Presland G (2005) The Natural History of Melbourne – A Reconstruction. PhD thesis. History and Philosophy of Science/SAGES, University of Melbourne, Melbourne.

Pretty GL (1977) The cultural chronology of the Roonka Flat. In *Stone Tools as Cultural Markers: Change, Evolution and Complexity*. (Ed. RVS Wright) pp. 288–331. Australian Institute of Aboriginal Studies, Canberra and Humanities Press, Atlantic Highlands, N.J.

Pridden W (1843) *Australia: Its History and Present Condition, Containing an Account Both of the Bush and of the Colonies, with Their Respective Inhabitants*. James Burns, London.

Prout JS (1852) *An Illustrated Handbook of the Voyage to Australia*. Duff: London.

Puruntatameri J, Puruntatameri R, Pangiraminni A, Burak L, Tipuamantymirri C, Tipakalippa M, Puruntatameri J, Puruntatameri P, Pupangamirri JP, Kerinaiua R, Tipiloura D, Orsto M, Kantilla B, Kurrupuwu M, Puruntatameri PF, Puruntatameri TD, Puruntatameri L, Kantilla K, Wilson J, Cusack J, Jackson D, Wightman G (2001) *Tiwi Plants and Animals. Aboriginal Flora and Fauna Knowledge from Bathurst and Melville Islands, Northern Australia*. Parks and Wildlife Commission of the Northern Territory, and Tiwi Land Council, Darwin.

Quinlan LM (1967) *Here My Home: the Life and Times of Captain John Stuart Hepburn, 1803–1860, Master Mariner, Overlander and Founder of Smeaton Hill, Victoria*. Oxford University Press, Melbourne and New York.

Radcliffe-Brown AR (1958) A comparative method in social anthropology. In *Method in Social Anthropology: Selected Essays*. pp. 108–129. University of Chicago Press, Chicago.

Ramson WS (Ed.) (1988) *The Australian National Dictionary: A Dictionary of Australianisms on Historical Principles*. Oxford University Press, Oxford.

Rankine HJ (1991) A talk by Henry Rankine. *Journal of the Anthropological Society of South Australia* **29**(2), 108–127.

Ratzel F (1896) *The History of Mankind*. Translated from the second German edition by AJ Butler. Macmillan, London.

Read GF (no date) *Papers*. MS 8912. State Library of Victoria, Melbourne.

Reid A (1995) *Banksias and Bilbies: Seasons of Australia*. Gould League, Melbourne.

Renfrew C (1979) *Problems in European Prehistory*. Edinburgh University Press, Edinburgh.

Renwick C (2000) *Geebungs and Snake Whistles: Koori People and Plants of Wreck Bay*. Aboriginal Studies Press, Canberra.

Reynolds H (1987) *Frontier: Aborigines, Settlers and Land*. Allen & Unwin, Sydney.

Reynolds H (1990) *With the White People*. Penguin Books, Melbourne.

Rich M (Ed.) (1990) *Eugene Von Guerard in Ballarat: Journal of an Australian Gold Digger by Eugene Von Guerard*. Ballarat Fine Art Gallery, Ballarat, Victoria.

Richards T (2011) A late nineteenth-century map of an Australian Aboriginal fishery at Lake Condah. *Australian Aboriginal Studies* **2**, 64–87.

Richards T, Jordan J (1999) 'Aboriginal Archaeological Investigations in the Barwon Drainage Basin'. Occasional report 50, Aboriginal Affairs Victoria, Victorian Department of Human Services, Melbourne.

Rimmer MA, Battaglene SL, Dostine PL (1983) Observations on the distribution of *Bankia australis* calman (Mollusca: Teredinidae) in the Patonga Creek mangrove swamp, New South Wales. *Australian Journal of Marine and Freshwater Research* **34**(2), 355–357. doi:10.1071/MF9830355

Rinkevich S, Greenwood K, Leonetti C (2011) *Traditional Ecological Knowledge for Application by Service Scientists*. Native American Program, US Fish and Wildlife Service, Arlington.

RIRDC (1998) 'R & D Plan for the Bushfood Industry 1998–2002'. Publication no. 98/111. Rural Industries Research and Development Corporation, Canberra.

Roberts A (2003) Knowledge, Power and Voice: An Investigation of Indigenous South Australian Perspectives of Archaeology. PhD thesis. Department of Archaeology, Flinders University, Adelaide.

Robins J (1996) *Wild Lime. Cooking from the Bushfood Garden*. Allen & Unwin, Sydney.

Roff D (1983) *Bushtucker. Geo* **5**(3), 70–87.

Roheim G (1925) *Australian Totemism. A Pyscho-analytic Study in Anthropology*. Allen & Unwin, London.

Rolls E (1984) *They All Ran Wild. The Animals and Plants that Plague Australia*. Angus & Robertson, Sydney.

Rose DB (Ed.) (1995) *Country in Flames. Proceedings of the 1994 Symposium on Biodiversity and Fire in North Australia*. Biodiversity Unit, Department of the Environment, Sport and Territories and the North Australia Research Unit, Canberra and Darwin.

Rose DB (1996) *Nourishing Terrain. Australian Aboriginal Views of Landscape and Wilderness*. Australian Heritage Commission, Canberra.

Rose D, Clarke A (1997) *Tracking Knowledge in Northern Australian Landscapes: Studies in Indigenous and Settler Ecological Knowledge Systems*. Northern Australian Research Unit, Australian National University, Darwin.

Rose DB, Bell D, Crook DA (2016) Restoring habitat and cultural practice in Australia's oldest and largest traditional aquaculture system. *Reviews in Fish Biology and Fisheries* **26**(3), 589–600. doi:10.1007/s11160-016-9426-1

Ross A (1871) Extraordinary recovery from an Aboriginal from a spear wound. *Otago Witness* 987, (7 January 1871) p. 17.

Ross H, Grant C, Robinson CJ, Izurieta A, Smyth D, Rist P (2009) Co-management and Indigenous protected areas in Australia: achievements and ways forward. *Australasian Journal of Environmental Management* **16**, 242–252. doi:10.1080/14486563.2009.9725240

Ross JH (1996) The legacy of Mueller's collections. *Victorian Naturalist* **113**(4), 146–150.

Ross L (1915) *From Rossville to the Victorian Goldfields in 1852*. Sydney and Melbourne Publishing Co., Sydney.

Roth WE (1897) *Ethnological Studies Among the North-West-Central Queensland Aborigines*. Queensland Government Printer, Brisbane.

Roth WE (1901) *North Queensland Ethnography. Food. Its Search, Capture, and Preparation. Bulletin 3*. Government Printer, Brisbane.

Rowe G (1854) *Correspondence. Letters*. MS 3116. National Library of Australia, Canberra.

Rowley I (1973) The comparative ecology of Australian corvids. II. Social organization and behaviour. *Wildlife Research* **18**(1), 25–65. doi:10.1071/CWR9730025

Ruggles CLN (2014) Calendars and astronomy. In *Handbook of Archaeoastronomy and Ethnoastronomy*. (Ed. CLN Ruggles) pp. 15–30. Springer, New York.

Russell A (1840) *A Tour Through the Australian Colonies in 1839: With Notes and Incidents of a Voyage Round the Globe, Calling at New Zealand and South America*. David Robertson, Glasgow.

Russell HC (1876) Meteorological periodicities. *The Sydney Morning Herald*, New South Wales. 18 October 1876, pp. 7–8.

Russell L (2012) *Roving Mariners. Australian Aboriginal Whalers and Sealers in the Southern Oceans, 1790–1870*. State University of New York, Albany, United States.

Ryan J (2012) The six seasons: shifting Australian nature writing towards ecological time and embodied temporality. *Transformations* **21**, 12.

Ryder M, Latham Y (2005) *Cultivation of Native Food Plants in South-eastern Australia*. 04/178 RIRDC publication. Rural Industries Research and Development Corporation, Canberra.

Santich B (2011) Nineteenth-Century Experimentation and the Role of Indigenous Foods in Australian Food Culture. *Australian Humanities Review* No 51.

Sauer CO (1925 [1963]) The Morphology of Landscape. Reprinted in *Land and Life: Selections from the Writings of Carl Ortwin Sauer*. (Ed. J Leighly). University of California, Berkeley and Los Angeles.

Say M (2005) Black Thursday [Painting]: William Strutt's 'Itinerant picture'. *La Trobe Journal* **75**, 27–34.

Sayers A, National Gallery of Australia (1994) *Aboriginal Artists of the Nineteenth Century*. Oxford UP in association with the National Gallery of Australia, Melbourne.

Schultz CJ, Apps DJ, Johnson TE, Bastian SEP (2009) Testing consumer acceptability of new crops: an integrated sensory and marketing approach using the Australian berry muntries. *Food Australia* **61**, 335–341.

Seddon G (Ed.) (1989) *The Ballad of Bunjil Bottle; AW Howitt's Exploration of the Mitchell River by Canoe in 1875*. Centre for Gippsland Studies, Gippsland Institute of Advanced Education, Churchill, Victoria.

Selwyn A (1859) The Dandenong Goldfield. *Argus*, Melbourne. 17 February 1859, p. 4.

Shaw E (1949) *Early Days Among the Aborigines: The Story of Yelta and Coranderrk Missions*. The Author, Melbourne.

Shellard F (c.1860) *Reminiscences of an Old Digger*. MS 1890. National Library of Australia, Canberra.

Shillinglaw J (1972) *Historical Records of Port Phillip: The First Annals of the Colony of Victoria*. Heinemann, Melbourne.

Silverman EK (1997) Politics, gender, and time in Melanesia and Aboriginal Australia. *Ethnology* **36**(2), 101–121. doi:10.2307/3774078

Sinclair J (c.1860) *Memoirs*. MS 5052. National Library of Australia, Canberra.

Slater P (1970) *A Field Guide to Australian Birds. Non-passerines*. Rigby, Adelaide.

Slater P (1974) *A Field Guide to Australian Birds. Passerines*. Rigby, Adelaide.

Smith C (1880) *The Booandik Tribe of South Australian Aborigines*. South Australian Government Printer, Adelaide.

Smith JG (2002) *Reminiscences of the Ballarat Goldfield*. Pick Point, Ballarat, Victoria.

Smith M (1996) *Bunyips and Bigfoots. The Search of Australia's Mystery Animals*. Millennium Books, Sydney.

Smith M (2013) *The Archaeology of Australia's Deserts*. Cambridge World Archaeology, Cambridge.

Smith M, Kalotas AC (1985) Bardi plants: an annotated list of plants and their use by the Bardi Aborigines of Dampierland, in north-western Australia. *Records of the Western Australian Museum* **12**(3), 317–359.

Smith NM (2013) Contested Discourses: Aboriginal Attitudes Towards Non-invasive Plants and Engagement in Weed Management in Cape York, Northern Australia. PhD thesis, Charles Darwin University, Casuarina, Northern Territory.

Smith NM, Wightman GM (1990) 'Ethnobotanical Notes from Belyuen, Northern Territory, Australia'. Northern Territory Botanical Bulletin No. 10. Conservation Commission of the Northern Territory, Darwin.

Smith NM, Wididburu B, Harrington RN, Wightman GM (1993) 'Ngarinyman Ethnobotany. Aboriginal Plant Use from the Victoria River Area, Northern Australia'. Northern Territory Botanical Bulletin No. 16. Conservation Commission of the Northern Territory, Darwin.

Smith WR (1909) The Aborigines of Australia. Reprinted from *Official Year Book of the Commonwealth No.3*. McCarron, Bird and Co., Melbourne.

Smith WR (1930) *Myths and Legends of the Australian Aboriginals*. Harrap, Sydney.

Smyth D (1993) *A Voice in all Places. Aboriginal and Torres Strait Islander Interests in Australia's Coastal Zone*. Revised edition. Resource Assessment Commission, Canberra.

Smyth D (2011) 'Indigenous Land and Sea Management – a Case Study'. Report Prepared for the Australian Government Department of Sustainability, Environment, Water, Population and Communities. State of the Environment 2011 Committee. Canberra.

Smyth D, Szabo S, George M (2004) *Case Studies in Indigenous Engagement in Natural Resource Management in Australia*. Department of Environment and Heritage, Canberra.

Smyth RB (1878) *The Aborigines of Victoria*. 2 volumes. Victorian Government Printer, Melbourne.

Smyth RB (1886) Lecture on Aboriginal customs. *Mount Alexander Mail*, Victoria. 9 August 1886, p. 2.

Smyth RB (no date) *Papers*. MS. 8781. State Library of Victoria, Melbourne.

Snow CP (2012) *The Two Cultures*. Cambridge University Press, Cambridge.

Soerjohardjo W (2012) Remembering Yarrie: An Indigenous Australian and the 1852 Gundagai Flood. *Public History Review* 19, 120–129. doi:10.5130/phrj.v19i0.3096

Stanbridge WE (1857) On the astronomy and mythology of the Aborigines of Victoria. *Philosophical Institute of Victoria. Transactions* 12, 137–140.

Stanbridge WE (1861) Some particulars on the general characteristics, astronomy, and mythology of the tribes in the central part of Victoria, Southern Australia. *Transactions of the Ethnological Society of London* 1, 286–304. doi:10.2307/3014201

Stanbury P, Clegg J (1990) *A Field Guide to Aboriginal Rock Engravings with Special Reference to Those Around Sydney*. Sydney University Press, Sydney.

Stannard M (1873) *Memoirs of a Professional Lady Nurse*. Simpkin, Marshall, London.

Stanner WEH (1965) Aboriginal territorial organization: estate, range, domain and regime. *Oceania* 36(1), 1–26. doi:10.1002/j.1834-4461.1965.tb00275.x

Stanner WEH (1933–34) Ceremonial economics of the Mulluk Mulluk and Madngella tribes of the Daly River, North Australia. A preliminary paper. *Oceania* 4(2), 156–175; 4(4), 458–471.

Stead RE (1987) Towards a classification of Australian Aboriginal stone arrangements: an investigation of methodological problems with a gazetteer of selected sites. Master's thesis. Australian National University, Canberra.

Stephens M (Ed.) (2014) *The Journal of William Thomas Assistant Protector of the Aborigines of Port Phillip & Guardian of the Aborigines of Victoria 1839–1867*, 4 Vols. Victorian Aboriginal Corporation for Languages, Melbourne.

Stern BJ, Howitt AW, Fison L (1930) Selections from the letters of Lorimer Fison and AW Howitt to Lewis Henry Morgan. *American Anthropologist* 32(2), 257–279. doi:10.1525/aa.1930.32.2.02a00020

Stevens FH, Boort Historical Society (1969) *Smoke From the Hill: A Story of the Boort District 1836–1968*. Boort Historical Society, Bendigo, Victoria.

Stewart D (c.1870–c.1883) Papers. Reproduced in *Two Notable South Australians*, 1977 (Eds T McCourt and H Mincham). Beachport Branch of the National Trust, Beachport, South Australia.

Stokes C (1986) Possums in Australian Aboriginal economy. BA Hons. Thesis, La Trobe University, Bundoora, Victoria.

Stone AC (1911) The Aborigines of Lake Boga, Victoria. *Proceedings of the Royal Society of Victoria* **23**(2), 433–468.

Strahan R (Ed.) (1983) *The Australian Museum Complete Book of Australian Mammals*. Angus and Robertson, Sydney.

Strehlow TGH (1970) Geography and the totemic landscape in Central Australia: a functional study. In *Australian Aboriginal Anthropology. Modern Studies in the Social Anthropology of the Australian Aborigines*. (Ed. RM Berndt) pp. 92–140. Australian Institute of Aboriginal Studies and University of Western Australia Press, Perth.

Strutt W (c.1850s) *Off to Australia*. MS 15886. National Library of Australia, Canberra.

Strutt W (1864) *Black Thursday, February 6th. 1851* [art original]. State Library of Victoria, Melbourne.

Sturt C (1834) *Two Expeditions into the Interior of Southern Australia, During the Years 1828, 1829, 1830, and 1831: with Observations on the Soil, Climate and General Resources of the Colony of New South Wales*. Smith, Elder & Co., London.

Sturt C (1838) *An Account of a Journey to South Australia*. Sullivans Cove, Adelaide. Republished in 1990.

Stynes B (1997) Opportunities for contributing to the development of Aboriginal food plants. *Tropical Grasslands* **31**, 311–314.

Sugden G (c.1840–1870) *Reminiscences*. MS 000301. (Box 115-6). Royal Historical Society of Victoria, Melbourne.

Suggit D (2008) A Clever People: Indigenous healing traditions and Australian mental health futures. Short Thesis, Australian National University, Canberra.

Sutton P (1988) Dreamings. In *Dreamings: The Art of Aboriginal Australia*. (Ed. P Sutton) pp. 13–32. Penguin Books, Melbourne.

Sutton P (1994) Myth as history, history as myth. In *Being Black. Aboriginal Cultures in 'Settled' Australia* (Ed. I Keen) pp. 251–268. Aboriginal Studies Press, Canberra.

Sutton P (1998a) *Native Title and the Descent of Rights*. National Native Title Tribunal, Perth.

Sutton P (1998b) Icons of Country: Topographic Representations in Classical Aboriginal Traditions. In *The History of Cartography, Vol. 2.3: Cartography in the Traditional African, American, Arctic, Australian, and Pacific Societies*. (Eds D Woodward and CM Lewis) pp. 351–386. University Press Chicago, Chicago.

Sutton P (2003) *Native Title in Australia. An Ethnographic Perspective*. Cambridge University Press, Cambridge.

Swain T (1991) The Mother Earth conspiracy: an Australian episode. *Numen* **38**(1), 3–26.

Swain T (2000) *A Place for Strangers: Towards a History of Australian Being*. Cambridge University Press, Cambridge, MA.

Taçon PSC, South B, Hooper SB (2003) Depicting cross-cultural interaction: figurative designs in wood, earth and stone from south-east Australia. *Archaeology in Oceania* **38**(2), 89–101. doi:10.1002/j.1834-4453.2003.tb00532.x

Tan AC, Konczak I, Ramzan I, Sze DMY (2011) Antioxidant and cytoprotective activities of native Australian fruit polyphenols. *Food Research International* **44**(7), 2034–2040. doi:10.1016/j.foodres.2010.10.023

Taplin G (1859–79) *Journals*. Mortlock Library, Adelaide.

Taplin G (1874) The Narrinyeri. In *The Native Tribes of South Australia*, 1879 (Ed. JD Woods) pp. 1–156. ES Wigg & Son, Adelaide.

Taplin G (Ed.) (1879) *The Folklore, Manners, Customs, and Languages of the South Australian Aborigines Gathered from Inquiries Made by Authority of South Australian Government*. South Australian Government Printer, Adelaide.

Taylor J (1997) The contemporary demography of Indigenous Australians. *Journal of the Australian Population Association* **14**(1), 77–114. doi:10.1007/BF03029488

Taylor L (1996) *Seeing the Inside. Bark Painting in Western Arnhem Land*. Oxford University Press, Oxford.

Teichelmann CG (1841) *Aborigines of South Australia*. Committee of the South Australian Wesleyan Methodist Auxiliary Society, Adelaide.

Teichelmann CG, Schürmann CW (1840) *Outlines of a Grammar, Vocabulary, and Phraseology of the Aboriginal Language of South Australia, Spoken by the Natives in and for Some Distance Around Adelaide*. Thomas and Company, Adelaide.

Tellurian (1934) Science notes. Seals in inland waters. *Australasian*, Melbourne. 27 January 1934, p. 42.

TenHouten WD (1999) Text and temporality: patterned-cyclical and ordinary-linear forms of time-consciousness, Inferred from a corpus of Australian Aboriginal and Euro-Australian life-historical interviews. *Symbolic Interaction* **22**(2), 121–137. doi:10.1525/si.1999.22.2.121

Thom G (1953) A pioneer's story of southern Yorke Peninsula. Some native legends. *The Pioneer*. Yorketown, South Australia. 11 September 1953, p. 2.

Thomas NW (1905) Australian canoes and rafts. *Journal of the Anthropological Institute* **35**, 56–80.

Thomas W (1839) *William Thomas Journal, November 1839 – January 1840*. M-UNCAT MSS SET 214 Item 1, State Library of New South Wales, Sydney.

Thomas W (1840) *Aboriginal Protectorate Weekly, Monthly, Quarterly and Annual Reports and Journals*. VPRS 4410/P0. Unit 3, Folder No. 67, Periodical Report for the period February to August 1840). Public Record Office of Victoria, Melbourne.

Thomas W (1844) *Aboriginal Protectorate Weekly, Monthly, Quarterly and Annual Reports and Journals*. VPRS 4410/P0, Unit 3, Folder 82, William Thomas quarterly report, 31 November 1844, for the period 1 September 1844 – 30 November 1844. Public Record Office of Victoria, Melbourne.

Thomas W (1858) Papers of William Thomas. MSS 214, vol. 24, Number 11. 'Aborigines: Superior Races'. Mitchell Library, Sydney.

Thomas W (no date) 'Brief Account of the Aborigines of Australia Felix' and 'Account of the Aborigines'. Letters Nos 13 and 14. In *Letters from Victorian Pioneers, 1898*. (Eds TF Bride and CE Sayers) pp. 65–83, 84–100. Heinemann, Melbourne.

Thomas MH, Thomas EK (Ed.) (1925) *The Diary and Letters of Mary Thomas (1836–1866): Being a Record of the Early Days of South Australia*. Third edition. WK Thomas & Co., Adelaide.

Thomson DF (1939) The seasonal factor in human culture illustrated from the life of a contemporary nomadic group. *Proceedings of the Prehistoric Society* **5**(02), 209–221[New Series]. doi:10.1017/S0079497X00020545

Thomson DF (1949) *Economic Structure and the Ceremonial Exchange Cycle in Arnhem Land*. Macmillan, Melbourne.

Tibbett K (2004) Risk and economic reciprocity: an analysis of three regional Aboriginal food-sharing systems in late Holocene Australia. *Australian Archaeology* **58**(1), 7–10. doi:10.1080/03122417.2004.11681774

Timms BV, Rankin C (2016) The geomorphology of gnammas (weathering pits) of northwestern Eyre Peninsula, South Australia: typology, influence of haloclasty and origins. *Transactions of the Royal Society of South Australia* **140**, 28–45. doi:10.1080/03721426.2015.1115459

Tindale NB (1926) Natives of Groote Eylandt and of the West Coast of the Gulf of Carpentaria. Part 2. *Records of the South Australian Museum* **3**(2), 103–134.

Tindale NB (1930–52) *Murray River Notes*. AA338/1/31/1. South Australian Museum Archives, Adelaide.

Tindale NB (1931–34) *Journal of Researches in the South East of South Australia. Vol. I*. Manuscript. AA338/1/33/1. South Australian Museum Archives, Adelaide.

Tindale NB (c.1931–c.1991) *Place Names: N.B. Tindale Ms SE of S Australia*. AA338/7/1/44. South Australian Museum Archives, Adelaide.

Tindale NB (1934) Vanished tribal life of Coorong blacks. Tragedy of supplanted race. Country's changed aspect. *The Advertiser*, Adelaide. 7 April 1934, p. 9.

Tindale NB (1934–37) *Journal of Researches in the South East of South Australia. Vol. II*. AA338/1/33/2. South Australian Museum Archives, Adelaide.

Tindale NB (1936a) Exploring the Coorong. Fascinating glimpses of its past. *The Advertiser*, Adelaide. 12 May 1936, p. 18.

Tindale NB (1936b) Wonders of the Coorong. *The Advertiser*, Adelaide. 14 May 1936, pp. 18 & 23.

Tindale NB (1936c) Notes on the natives of the southern portion of Yorke Peninsula, South Australia. *Transactions of the Royal Society of South Australia* **60**, 55–70.

Tindale NB (1936d) *Map for Emu and Brolga Text*. AA338/16/19/1–2. South Australian Museum Archives, Adelaide.

Tindale NB (1937) Native songs of the South East of South Australia. *Transactions of the Royal Society of South Australia* **61**, 107–120.

Tindale NB (1938–56) *Journal of Researches in the South East of South Australia. Vol. III.* AA338/1/33/3. South Australian Museum Archives, Adelaide.

Tindale NB (1938a) Prupe and Koromarange: a legend of the Tanganekald, Coorong, South Australia. *Transactions of the Royal Society of South Australia* **62**, 18–23.

Tindale NB (1938b) Ghost moths of the family Hepialidae. *South Australian Naturalist* **19**(1), 1–6.

Tindale NB (1939) Eagle and crow myths of the Maraura tribe, Lower Darling River, N.S.W. *Records of the South Australian Museum* **6**, 243–261.

Tindale NB (1941) Native songs of the South East of South Australia. Part II. *Transactions of the Royal Society of South Australia* **65**(2), 233–243.

Tindale NB (1951) Comments on supposed representations of giant bird tracks at Pimba. *Records of the South Australian Museum* **9**(4), 381–382.

Tindale NB (1953) On some Australian Cossidae including the moth of the witjuti (witchety) grub. *Transactions of the Royal Society of South Australia* **76**, 56–65.

Tindale NB (1964) *Murray River Notes*. AA338/1/31/2. South Australian Museum Archives, Adelaide.

Tindale NB (1966) Insects as food for the Australian Aborigines. *Australian Natural History* **15**, 179–183.

Tindale NB (1974) *Aboriginal Tribes of Australia: Their Terrain, Environmental Controls, Distribution, Limits, and Proper Names*. Australian National University Press, Canberra.

Tindale NB (1976) Some ecological bases for Australian tribal boundaries. In *Tribes and Boundaries in Australia* (Ed. N Peterson) pp. 12–29. Australian Institute of Aboriginal Studies, Canberra.

Tindale NB (1977) Adaptive significance of the Panara or grass seed culture of Australia. In *Stone Tools as Cultural Markers*. (Ed. RVS Wright) pp. 345–349. Australian Institute of Aboriginal Studies, Canberra.

Tindale NB (1981) Desert Aborigines and the southern coastal peoples: some comparisons. In *Ecological Biogeography of Australia*. (Ed. A Keast) pp. 1855–1884. Junk, The Hague.

Tindale NB (1987a) The wanderings of Tjirbruki: a tale of the Kaurna people of Adelaide. *Records of the South Australian Museum* **20**(1), 5–13.

Tindale NB (1987b) *Yaralde Tribe Place Names, June 1987*. AA338/8/17. South Australian Museum Archives, Adelaide.

Tindale NB (2005) Celestial lore of some Aboriginal tribes. In *Songs from the Sky: Indigenous Astronomical and Cosmological Traditions of the World*. (Eds von Del Chamberlain, JB Carlson and MJ Young). Ocarina Books, West Sussex, United Kingdom [*Archaeoastronomy* **12–13**, 358–379].

Tindale NB (no date) *Milerum* manuscript, folder 1, draft A. *South Australian Museum Archives, Adelaide.*

Tindale NB, Maegraith BG (1931) Traces of an extinct Aboriginal population on Kangaroo Island. *Records of the South Australian Museum* **4**, 275–289.

Tindale NB, Pretty GL (1980) The surviving record. In *Preserving Indigenous Cultures: A New Role for Museums*. (Eds R Edwards and J Stewart) pp. 43–52, 228. Australian Government, Canberra.

Tipping M (1987) *Convicts Unbound: The Story of the Calcutta Convicts and Their Settlement in Australia*. Viking O'Neil, Melbourne.

Tonkinson R (1978) Semen versus spirit-child in a Western Desert culture. In *Australian Aboriginal Concepts* (Ed. LR Hiatt) pp. 81–92. Australian Institute of Aboriginal Studies, Canberra.

Tonkinson R (1993) Foreword. In *A World That Was. The Yaraldi of the Murray River and the Lakes, South Australia*. (RM Berndt, CH Berndt and JE Stanton) pp. xvii–xxxi. Melbourne University Press at the Miegunyah Press, Melbourne.

Tonkinson R (1997) Anthropology and Aboriginal tradition: the Hindmarsh Island bridge affair and the politics of interpretation. *Oceania* **68**, 1–26. doi:10.1002/j.1834-4461.1997.tb02639.x

Trezise P, Roughsey D (1978) *The Quinkins*. Collins, Sydney.

Trezise P, Roughsey D (1982) *Turramulli the Giant Quinkin*. Collins, Sydney.

Tuckey JH (1805) *An Account of a Voyage to Establish a Colony at Port Philip in Bass's Strait, on the South Coast of New South Wales, in His Majesty's Ship Calcutta, in the Years 1802–3–4*. Printed for Longman, Hurst, Rees and Orme, London.

Tudhope AW, Chilcott CP, McCulloch MT, Cook ER, Chappell J, Ellam RM, Lea DW, Lough JM, Shimmield GB (2001) Variability in the El Niño-Southern Oscillation through a glacial-interglacial cycle. *Science* **291**(5508), 1511–1517. doi:10.1126/science.1057969

Turney CS, Flannery TF, Roberts RG, Reid C, Fifield LK, Higham TF, Jacobs Z, Kemp N, Colhoun EA, Kalin RM, Ogle N (2008) Late-surviving megafauna in Tasmania, Australia, implicate human involvement in their extinction. *Proceedings of the National Academy of Sciences of the United States of America* **105**(34), 12150–12153. doi:10.1073/pnas.0801360105

Unaipon D (1924–25) *Legendary Tales of the Australian Aborigines*. (Edited and introduced by S Muecke and A Shoemaker, 2001.) The Miegunyah Press at Melbourne University Press, Melbourne.

Vaarzon-Morel P, Edwards G (2012) Incorporating Aboriginal people's perceptions of introduced animals in resource management: insights from the feral camel project. *Ecological Management and Restoration* **13**, 65–71.

Van Der Kaars S, Miller GH, Turney CS, Cook EJ, Nürnberg D, Schönfeld J, Kershaw AP, Lehman SJ (2017) Humans rather than climate the primary cause of Pleistocene megafaunal extinction in Australia. *Nature Communications* **8**(14142), 1–7.

Vickery J (1992) Foreword. In *Koorie Plants. Koorie People. Traditional Aboriginal Food, Fibre and Healing Plants of Victoria* (N Zola and B Gott) p. vii. Koorie Heritage Trust, Melbourne.

Victoria (1861) First Report of the Central Board appointed to watch over the interests of the Aborigines in the Colony of Victoria. *Victoria Legislative Council Votes & Proceedings*. Melbourne.

Victorian Parliament Legislative Council, Select Committee on the Aborigines (1859) *Report of the Select Committee of the Legislative Council on the Aborigines: Together with the Proceedings of Committee, Minutes of Evidence, and Appendices*. Government Printer, Melbourne.

Victoria Royal Commission on the Aborigines (1877) *Report of the Commissioners Appointed to Inquire into the Present Condition of the Aborigines of this Colony: and to Advise as to the Best Means of Caring …*. John Ferres, Government Printer, Melbourne.

Von Mueller F (1825–96) *Regardfully Yours. Selected Correspondence of Ferdinand von Mueller*. (Eds RW Home, AM Lucas, S Maroske, DM Sinkora, JH Voigt and M Wells). 1998–2006. 3 volumes. Peter Lang, New York.

Von Mueller F (1856) Account of the *gunyang*: a new indigenous fruit of Victoria. *Hooker's Journal of Botany and Kew Garden Miscellany* **8**, 336–338.

Von Mueller F (1858) On a general introduction of useful plants into Victoria. *Transactions of the Philosophical Institute of Victoria* **2**(2), 93–109.

Von Mueller F (1876) *Select Plants Readily Eligible for Industrial Culture or Naturalisation in Victoria*. McCarron, Bird and Co., Melbourne.

Von Mueller F (1888) *Select Extra-Tropical Plants, Readily Eligible for Industrial Culture or Naturalisation, with Indications of Their Native Countries and Some of Their Uses*. Revised edition. RS Brain, Government Printer, Sydney.

Vox (1938) Goolwa regatta. *Chronicle*, Adelaide. 29 December 1938, p. 62.

Walker S (1993) *Glenlyon Connections*. S Walker, Stanthorpe, Queensland.

Walker T (1965) *A Month in the Bush of Australia. A Journal of one of a party of gentlemen who recently travelled from Sydney to Port Phillip – with some remarks on the present state of the farming establishments and society in the settled parts of the Argyle Country*. Libraries Board of South Australia, Adelaide [First published J Cross, London, 1838].

Walkinshaw L (1973) *Cranes of the World*. Winchester Press, New York.

Ward R (1958) *The Australian Legend*. Oxford University Press, Melbourne.

Waterfield JH (1914) Extracts from the diary of the Rev Wm Waterfield, first Congregational minister at Port Phillip, 1838–43. *The Victorian Historical Magazine* **3**(3), 105–132.

Wathen GH (1855) *The Golden Colony: Or Victoria in 1854 with Remarks on the Geology of the Australian Gold Fields*. Longman, Brown, Green, and Longmans, London.

Webb JB (2003) *The Botanical Endeavour. Journey Towards a Flora of Australia*. Surrey Beatty & Sons, Chipping Norton, New South Wales.

Webster J (1908) *Reminiscences of an Old Settler in Australia and New Zealand*. Witcombe & Tombs, Christchurch.

Wedge JH (1980) Narrative of William Buckley. Reprinted in *The Life and Adventures of William Buckley* (Ed. J Morgan). Facsimile of 1852 Hobart edition, 165–171. ANU Press, Canberra.

Weir JK (2007) The traditional owner experience along the Murray River. In *Fresh Water: New Perspectives on Water in Australia* (Eds E Potter, A Mackinnon, S McKenzie and J McKay) pp. 44–58. Melbourne University Press, Carlton.

Weir JK (2009) *The Gunditjmara Land Justice Story*. Native Title Research Unit, Australian Institute of Aboriginal and Torres Strait Islanders Studies, Canberra.

Westgarth W (1848) *Australia Felix, or, A Historical and Descriptive Account of the Settlement of Port Phillip, New South Wales: Including Full Particulars of the Manners and Condition of the Aboriginal Natives, with Observations on Emigration, on the System of Transportation, and on Colonial Policy*. Oliver & Boyd, Edinburgh.

Wettenhall G (1999) *The People of Gariwerd: The Grampians' Aboriginal Heritage*. Aboriginal Affairs Victoria, Melbourne.

Wheelwright HW (1979) *Bush Wanderings of a Naturalist by an Old Bushman*. Oxford University Press: Melbourne. [First published in 1861, Routledge, Warne & Routledge, London] [Written under the pseudonym of 'An Old Bushman'].

White M (2009) *Prioritising Rock-holes of Aboriginal and Ecological Significance in the Gawler Ranges*. Knowledge and Information Division, Department of Water, Land and Biodiversity Conservation, South Australia. Adelaide.

Whitehead PJ, Bowman DMJS, Preece N, Fraser F, Cooke P (2003) Customary use of fire by indigenous peoples in northern Australia: its contemporary role in savanna management. *International Journal of Wildland Fire* **12**(4), 415–425. doi:10.1071/WF03027

Widdowson J (1971) The bogeyman: some preliminary observations on frightening figures. *Folklore* **82**(2), 99–115. doi:10.1080/0015587X.1971.9716716

Wiencke SW (1984) *When the Wattles Bloom Again: The Life and Times of William Barak, Last Chief of the Yarra Yarra Tribe*. Author, Victoria.

Wightman GM, Smith NM (1989) 'Ethnobotany, Vegetation and Floristics of Milingimbi, Northern Australia'. Northern Territory Botanical Bulletin No. 6. Conservation Commission of the Northern Territory, Darwin.

Wightman GM, Jackson DM, Williams LLV (1991) 'Alawa Ethnobotany. Aboriginal Plant Use from Minyerri, Northern Australia'. Northern Territory Botanical Bulletin No. 11. Conservation Commission of the Northern Territory, Darwin.

Wightman GM, Dixon D, Williams LLV, Injimadi Dalywaters (1992a) 'Mudburra Ethnobotany. Aboriginal Plant Use from Kulumindini (Elliot), Northern Australia'. Northern Territory Botanical Bulletin No. 14. Conservation Commission of the Northern Territory, Darwin.

Wightman GM, Roberts JG, Williams LLV (1992b) 'Mangarrayi Ethnobotany. Aboriginal Plant Use from the Elsey Area, Northern Australia'. Northern Territory Botanical Bulletin No. 15. Conservation Commission of the Northern Territory, Darwin.

Wightman GM, Gurindji elders, Frith RND, Holt S (1994) 'Gurindji Ethnobotany: Aboriginal Plant Use from Daguragu, Northern Australia'. Northern Territory Botanical Bulletin No. 18. Conservation Commission of the Northern Territory, Darwin.

Wignell E (1983) The mystery of the bunyip. *This Australia* **2**(3), 60–62.

Wilhelmi C (1857) Australian plants. *The Argus*, Melbourne. 22 April 1857, p. 6.

Wilhelmi C (1861) Manners and customs of the Australian natives, in particular of the Port Lincoln District. *Transactions of the Royal Society of Victoria* **5**, 164–203.

Wilkinson GB (1848) *South Australia: Its Advantages and its Resources. Being a Description of that Colony, and a Manual of Information for Emigrants.* John Murray, London.

Williams A, Sides T (2008) *Wiradjuri Plant Use in the Murrumbidgee Catchment.* Murrumbidgee Catchment Management Authority, Wagga Wagga, News South Wales.

Williams AN, Mooney SD, Sisson SA, Marlon J (2015) Exploring the relationship between Aboriginal population indices and fire in Australia over the last 20,000 years. *Palaeogeography, Palaeoclimatology, Palaeoecology* **432**, 49–57. doi:10.1016/j.palaeo.2015.04.030

Williams E (1988) *Complex Hunter-gatherers: A Late Holocene Example from Temperate Australia.* B.A.R, Oxford.

Williams NM, Hunn ES (Eds) (1982) *Resource Managers: North American and Australian Hunter-gatherers.* Australian Institute of Aboriginal Studies, Canberra.

Wilson M (1937) Emus and the brolgas. An Aboriginal legend. *Chronicle*, Adelaide. 3 June 1937, p. 49.

Wilson M (no date) *The Moolgewauk.* Fry Papers, South Australian Museum Archives. Published in the *Journal of the Anthropological Society of South Australia* **23**(1), 11–16.

Withers WB (1887) *The History of Ballarat.* Niven, Ballarat, Victoria.

Wiynjorrotj P, Flor S, Brown ND, Jatbula P, Galmur J, Katherine M, Merlan F, Wightman G (2005) 'Jawoyn Plants and Animals. Aboriginal Flora and Fauna Knowledge from Nitmiluk National Park and the Katherine Area, Northern Australia'. Northern Territory Botanical Bulletin No. 29. Ethnobiology Project (NRETA) and Jawoyn Association, Darwin.

Woiwod M (2012) *Coranderrk Database.* Kangaroo Ground, Australia Tarcoola Press, Victoria.

Wood JG (1870) *The Uncivilized Races, or Natural History of Man*, 2 Vols. American Publishing Company, Hartford, Connecticut.

Woodbury W (c.1854) *Letters.* Mfm M 1952. National Library of Australia, Canberra.

Woodriff D (1986) *Journal of H.M.S. Calcutta at Port Phillip.* Queensberry Hill Press, Melbourne.

Woods JD (1879) *The Native Tribes of South Australia.* ES Wigg & Son, Adelaide.

Woolner T (c.1855) *Diary.* MS 2939. National Library of Australia, Canberra.

Worsnop T (1897) *The Prehistoric Arts, Manufactures, Works, Weapons, etc., of the Aborigines of Australia.* South Australian Government Printer, Adelaide.

Wright R (1979) A modicum of taste: Aboriginal cloaks and rugs. *Newsletter - Australian Institute of Aboriginal Studies* **11**, 51–68.

Wuisang C, Jones DS (2014) Minahasan perspective on landscape custodianship: Sulawesi Indigenous landscape management and planning issues and challenges. In *IFLA 2014: Proceedings for the International Federation of Landscape Architects World Congress, International Federation of landscape Architects (IFLA)*. pp. 1–29. Kuching, Malaysia.

Yaniv Z, Bachrach U (2005) *Handbook of Medicinal Plants.* Food Products Press, Binghampton, New York.

Yen AL (2010) Edible insects and other invertebrates in Australia: future prospects. In *Forest Insects as Food: Humans Bite Back, Proceedings of a Workshop on Asia-Pacific Resources and their Potential for Development.* (Eds PB Durst, DV Johnson, RL Leslie and K Shono) pp. 65–84. FAO Regional Office for Asia and the Pacific, Bangkok.

Yen AL (2015) Insects as food and feed in the Asia Pacific region: current perspectives and future directions. *Journal of Insects as Food and Feed* **1**(1), 33–55. doi:10.3920/JIFF2014.0017

Young E, Ross H, Johnson J, Kesteven J (1991) *Caring for Country: Aborigines and Land Management.* Australian National Parks and Wildlife Service, Canberra.

Yunupingu B, Yunupingu-Marika L, Marika D, Marika B, Marika B, Marika R, Wightman G (1995) 'Rirratjinu Ethnobotany. Aboriginal Plant Use from Yirrkala, Arnhem Land, Australia'. Northern Territory Botanical Bulletin No. 21. Conservation Commission of the Northern Territory, Darwin.

Zola N, Gott B (1992) *Koorie Plants. Koorie People. Traditional Aboriginal Food, Fibre and Healing Plants of Victoria.* Koorie Heritage Trust, Melbourne.

Index